IODINE
and the *Brain*

IODINE
and the Brain

Edited by

G. Robert DeLong
Duke University Medical Center
Durham, North Carolina

Jacob Robbins
National Institutes of Health
Bethesda, Maryland

and

Peter G. Condliffe
National Institutes of Health
Bethesda, Maryland

PLENUM PRESS • NEW YORK AND LONDON

Library of Congress Cataloging in Publication Data

International Conference on Iodine and the Brain (1988: Bethesda, Md.)
    Iodine and the brain / edited by G. Robert DeLong, Jacob Robbins, and Peter G.
Condliffe.
        p.   cm.
    "Proceedings of an International Conference on Iodine and the Brain, held March 21–
23, 1988, in Bethesda, Maryland" — T.p. verso.
    Includes bibliographies and index.
    ISBN-13: 978-1-4612-8071-2   e-ISBN-13: 978-1-4613-0765-5
    DOI: 10.1007/978-1-4613-0765-5

    1. Iodine deficiency diseases — Congresses. 2. Cretinism — Nutritional aspects —
Congresses. 3. Hypothyroidism — Nutritional aspects — Congresses. 4. Brain — Diseases —
Nutritional aspects — Congresses. 5. Brain — Physiology — Congresses. I. DeLong, G.
Robert. II. Robbins, Jacob, 1922–   . III. Condliffe, Peter G., 1922–   . IV. Title.
    [DNLM: 1. Brain — physiology — congresses. 2. Iodine — deficiency — congresses. 3.
Iodine — physiology — congresses. QV 283 I61i]
RC627.I63I67   1988
612'.8042 — dc19
DNLM/DLC   88-39315
for Library of Congress   CIP

Proceedings of an International Conference on Iodine and the Brain,
held March 21–23, 1988, in Bethesda, Maryland

© 1989 Plenum Press, New York

A Division of Plenum Publishing Corporation
233 Spring Street, New York, N.Y. 10013

PREFACE

This volume contains the proceedings of a conference held at the National Institutes of Health in Bethesda on March 21-23, 1988, jointly sponsored by the International Council for Control of Iodine Deficiency Disorders (ICCIDD) and the Fogarty International Center of the National Institutes of Health. Several themes converged to make this meeting timely.

The first is an increasing awareness of iodine deficiency disorders as a world-wide problem of public health and a preventable cause of mental deficiency, and as a subject of scientific effort. Increased interest in these problems owes a great deal to accessibility to remote and under-developed areas of the world where iodine deficiency persists. As with any subject, greater scrutiny yields unexpected complexity and interest. It is true that provision of iodine, typically as iodized salt, is the necessary and sufficient preventative for iodine deficiency disorders, without including endemic cretinism. This provision is a governmental, economic and social problem. Apart from this, however, the scientific and medical problem of iodine deficiency and its effect on brain development and function is one of great interest and importance for developmental neurology and psychology. Even though the specific preventative agent is known, we do not totally understand the neurobiological questions raised. Accessibility to endemic areas makes it possible to bring modern methods of clinical evaluation and sophisticated laboratory methods to the field, resulting in a more detailed picture of the clinical and biochemical features of endemic cretinism. As always, new knowledge has challenged old and comfortable ideas and encouraged new approaches.

An example is the development of experimental animal models of iodine deficiency in an attempt to elucidate features of the disorder. Details of the in utero pathogenesis of the neurological defect in neurological cretins have long been obscure, as has the basis for the difference between neurological and myxedematous forms of endemic cretinism. These questions have been studied in experimental animals. The most striking result has been a revision of the concept that maternal thyroid hormones are not transported to the fetus. This affords one example - among several in these pages - of the ways in which studies of a specific problem can yield results of general scientific interest.

A second theme has been a focus on the neurological features. Endemic cretinism is a distinct neurological clinical entity virtually unknown to neurologists in the Western world. It is different from sporadic cretinism. Its pathogenesis, neuropathology and neurophysiology have been little studied and less understood.

A related development is the tremendous expansion of neuroscience. Thyroid hormones occupy a central place in brain development. It seems pertinent to bring the neurobiological effects of thyroid hormone disorder to the attention of the neuroscience community, especially as hypothyroidism affects nerve cell growth and connectivity, neurotransmitter levels, membrane functions, and other areas of neurobiology.

A third theme prompting this meeting was the opportunity to bring the unique, and until recently largely unrecognized, Chinese experience with iodine deficiency to the attention of the Western scientific world. Iodine deficiency and cretinism have been widespread in China and have been the subjects of major research and treatment efforts. Their studies of human material have not been equaled elsewhere and yield invaluable insights into the pathophysiology, neuropathology, and epidemiology of endemic cretinism.

A fourth theme is the realization that iodine deficiency may affect brain development and function in people in endemic areas who are not cretinous. In any deficiency disorder, one expects to find varying degrees of disability shading off to normal. Can individuals who are not frankly cretinous nevertheless have defects in psychomotor function because of iodine deficiency? This question is of obvious importance, but is very difficult to answer and presents a great challenge to the skills of psychologists and epidemiologists. Iodine-deficient populations exist in impoverished rural areas where culture and education are far different from metropolitan experience. Standard tests may have limited usefulness. It is difficult to find control populations comparable in all important respects except for iodine deficiency. The problem has been addressed in different ways in different parts of the world. Results of these studies are presented here. They underscore the importance of iodine deficiency in limiting human potential. They also illustrate the subtle and complex interaction of human brain development and intellectual and cultural capabilities.

The joining of these themes resulted in a meeting in which topics ranged from the molecular biology of the thyroid receptor to the educational and social consequences of iodine deficiency. This conference is part of the FIC program in advanced studies, a major theme of which is the study of preventable disease and the extra-scientific problems involved in the application of scientific knowledge to disease prevention. The meeting was conceived by Drs. V. Ramalingaswami, John Stanbury, and Basil Hetzel, while Dr. Ramalingaswami was a Scholar-in-Residence at the Fogarty International Center. The editors are grateful to these individuals for their vision and leadership, and to the authors for their contributions.

The scourge of endemic cretinism is still very much a reality. We hope that these efforts to understand it will contribute to the elimination of this anachronistic and preventable disease.

Robert DeLong
Jacob Robbins
Peter Condliffe

CONTENTS

IODINE AND THE BRAIN

Basil S. Hetzel

Executive Director, International Council for
Control of Iodine Deficiency Disorders,
Adelaide, Australia

It is an honour for me on behalf of the International Council for
Control of Iodine Deficiency Disorders to introduce this Conference on
Iodine and Brain which has been jointly sponsored by NIH through the
Fogarty International Center and the ICCIDD.

We are pleased to have this opportunity of drawing attention to a very
important area that has, until recently, been relatively neglected in
studies of brain development.

The ICCIDD was formed only two years ago as a global multidisciplinary
group of 400 scientists, planners, economists, technologists and others
concerned with the enormous gap that exists between our knowledge of the
effects of iodine deficiency and its correction, and the use of this
knowledge in public health programs that would prevent the effects of
iodine deficiency.

There is a widespread perception that iodine deficiency has already
been controlled in a previous generation. How different the reality is!!

Some 800 million are conservatively estimated to be at risk from
iodine deficiency disorders through living in an iodine deficient
environment. These populations are to found particularly in Asia, Africa
and Latin America. There are estimated to be no less than one third of the
population of China at risk (in excess of 300 million); some 200 million in
India, another 150 million in Africa with smaller, but very significant
populations at risk in Latin America and other parts of Asia.

The reason for persistence of iodine deficiency is that most of these
populations live within systems of subsistence agriculture which means that
they are locked into a vicious cycle of iodine deficiency as long as they
are dependent on food grown in the local iodine deficient environment.

Release from this vicious cycle can only be achieved by iodine
provided as a supplement, or through dietary diversification, as has
occurred in previous generations in Central and Southern Europe.

We now recognise that the effect of iodine deficiency is primarily due to a massive prevalence of hypothyroidism with its special effects on the brain. Recent elegant experimental studies have clearly demonstrated that hypothyroidism has a selective effect on the brain, in contrast to other organs and this has major significance for public health. It has become apparent that possibly a third of those living in a severely iodine deficient area, may have some degree of hypothyroidism affecting brain function. This indicates the importance and appropriateness of the title of this symposium on Iodine and the Brain.

In order to achieve its objective of bridging the gap between our knowledge of iodine deficiency and its application to the many millions that would benefit from it, the ICCIDD has established an organisation with a Board of some 32 members, two thirds of whom come from Third World countries, and an Executive Committee of seven with six Regional Coordinators who are responsible for each of the major World Health Organization Regions. The list of Office Bearers is shown in the accompanying appendix.

The ICCIDD has already published a monograph on the Prevention and Control of Iodine Deficiency Disorders, which is a State-of-the-Art review carried out on the occasion of the inaugural meeting held in Kathmandu, Nepal in March 1986. It has established a quarterly Newsletter which has a global circulation and maintains a network in excess of 400 members as well as several thousand others who are on the mailing list. There are a series of working groups concerned with major issues such as salt iodination, production of iodised oil, survey methods and economic aspects of IDD control programs.

In helping WHO and UNICEF to attack this problem, the ICCIDD has established in collaboration with these bodies, at the global level, an IDD Working Group responsible to the subcommittee of nutrition of the United Nations Agencies Administrative Coordinative Committee (ACC/SCN). This group has representation from WHO, UNICEF, The World Bank, and number of bilateral agencies concerned with the IDD problem. It reports annually on progress and is dedicated to developing National IDD Control Programs.

At the regional level in Africa, the ICCIDD has established a Task Force that has begun to meet the great challenge in this vast Continent where it is estimated that 100-150 million are at risk but where public health programs have barely begun!

At the country level, the ICCIDD has been involved in consultation with such countries as India, Indonesia, and Nepal in Asia, and Bolivia, Peru and Ecuador in Latin America. Regular close contact is maintained with the WHO and UNICEF in all these activities.

This new development of the International Council for Control of Iodine Deficiency Disorders is part of a new interest in the problem of iodine deficiency. This is indicated by the recent passage of a Resolution by the World Health Assembly in May 1986, calling for the Prevention and Control of Iodine Deficiency Disorders, pointing to its feasibility within a 5-10 year period. This resolution was sponsored by Australia and co-sponsored by 22 other countries and was carried unanimously.

Another important development has been the adoption of a global strategy for the prevention and control of iodine deficiency disorders by a combined United Nations Agencies group, SubCommittee of Nutrition in March 1987. As already mentioned, the ICCIDD is now a partner with the other agencies in an IDD Working Group set up by the SubCommittee of Nutrition.

This new interest, therefore, includes a scientific development in the ICCIDD, a political development in the World Health Assembly, and an organisational development with the adoption of a global strategy.

All this activity has stemmed from a new perception of the effects of iodine deficiency. The concept of the Iodine Deficiency Disorders refers to a broad spectrum of conditions which affect the fetus, the neonate, the child and the adult. Particularly important are the fetal effects, including miscarriage, stillbirth, as well as neurological cretinism which is of special interest to this symposium, as it is the reference point for studies on iodine and the brain. This condition, well known in the ancient and mediaeval worlds, was described in Diderot's Encyclopedie in 1754 when a cretin was defined as an "imbecile with a goitre down to the waist".

Up until the 1970's there was dispute as to whether cretinism was truly related to iodine deficiency. The spontaneous decline during the 19th century, without formal iodisation programs, raised this possibility. It is now recognised that this spontaneous decline was probably due to increase in iodine intake associated with social and economic development.

However, it was in the 1970's that the matter was settled: A controlled trial with iodised oil in the Western Highlands of Papua New Guinea took advantage of the demonstrable effect of and injection of iodised oil in correcting severe iodine deficiency for a period of up to 5 years. Alternate families were injected with iodised oil or saline in order to see whether indeed the correction of iodine deficiency would prevent cretinism.

It was possible to show by 1970 that cretins had disappeared from the progeny of treated mothers, but continued to appear from the untreated mothers. Although cretins had been born to seven treated mothers, six of seven of these mothers were already known to be pregnant when injected. It was therefore concluded that cretinism could be prevented if the iodine deficiency could be corrected before pregnancy.

This finding has been supported by evidence from Ecuador, Zaire and China and there is now no reasonable doubt about the relation between iodine deficiency and cretinism. It is this relationship on which more fundamental research can now be based in order to explore the mechanisms involved. It is likely that this new knowledge will be applicable over a wide area of brain physiology and brain function.

I believe that we can be confident this symposium will be a very significant landmark in the field and the development of our knowledge on iodine and the brain.

We appreciate very much the interest and help of the Fogarty International Center, N.I.H. in this joint symposium.

REGULATION OF THYROID HORMONE METABOLISM

IN THE BRAIN

P. Reed Larsen

Howard Hughes Medical Institute Laboratory and
Dept. of Medicine, Brigham and Women's Hospital
Harvard Medical School, Boston, Massachusetts

In recent years, it has become apparent that most of the 3,5,3'-triiodothyronine (T3) present in the cerebral cortical nuclei is derived from local thyroxine (T4) to T3 conversion. In this discussion, I will review our knowledge of the processes by which this situation arises and how it might serve to protect the brain from thyroid hormone deficiency in circumstances where serum T4 is reduced, such as occurs in hypothyroidism or in iodine deficiency. The review will include the enzymology of the brain deiodinases and how they respond to alterations in thyroid status, as well as our knowledge of their anatomical and cellular localization. In addition, the various levels of adaptation of thyroid hormone economy in the hypothyroid neonatal rat brain will be examined. Considerable knowledge has accumulated in this area over the last eight years. The interested reader should supplement the information in the discussion with the bibliographic citations and especially a recent comprehensive volume reviewing the subject of thyroid hormone metabolism (1).

THYROID HORMONE ACTIVATION IN THE BRAIN

In early studies of the thyroid hormone regulation of TSH release in the rat, we identified a 6-n-propylthiouracil (PTU) insensitive iodothyronine 5' monodeiodinase which efficiently converted T4 to T3 in the anterior pituitary and accounted for a significant fraction of pituitary nuclear T3 (2,3). Subsequently, we identified a similar deiodination process in the brain and found that conversion of T4 to T3 in vivo was blocked by iopanoic acid (4). Equilibrium studies performed with 125-I T4 and 131-I T3 demonstrated that in the cerebral cortex between 70 and 80% of the T3 specifically bound to the nucleus was derived from local production of T3 within the central nervous system (Figure 1). This contrasts with the circumstances in the liver and kidney where virtually all of the specifically bound nuclear T3 derives from plasma (3,5). Thus the brain is similar in this regard to the anterior pituitary and the brown adipose tissue (6). For the sake of brevity, the T3 derived from local T4 to T3 conversion in a given tissue will be termed T3(T4) and that derived from the plasma as T3(T3) throughout this review. As shown in Figure 1, the presence of a local 5' deiodinase pathway for T4 to T3 conversion resulted in nearly complete saturation of the nuclear T3 binding proteins in cerebral cortex based on an in vivo analysis. These results indicated that in the rat, T4 would serve an important role as the precursor to the active thyroid hormone in the brain.

Kaplan and Yaskowski first studied the enzymology of iodothyronine metabolism in the rat cerebral cortex, cerebellum and hypothalamus (7). They observed a striking increase in the activity of T4 5'deiodination in hypothyroidism and, in addition, a reduction in this process when animals were made hyperthyroid. They also observed enzymatic deiodination of T3 in the inner or tyrosyl ring (later termed Type III deiodination). This process was also responsive to thyroid status but the changes were opposite in direction to that of 5' iodothyronine deiodination (Table 1). This suggested the possibility that a compensatory change was occurring within the central nervous system which would, by increasing fractional T4 to T3 conversion and decreasing the rate of T3 degradation, act to sustain T3 concentrations as long as possible during T4 deficiency. Both deiodinase activities were found thoughout the central nervous system but 5' deiodination was more prominent within the cerebellum and cerebral cortex while 5 deiodination of T3 was especially high in the hypothalamus.

More recent studies of the hypothalamus have shown that the situation with respect to deiodination is likely to be more complex than originally

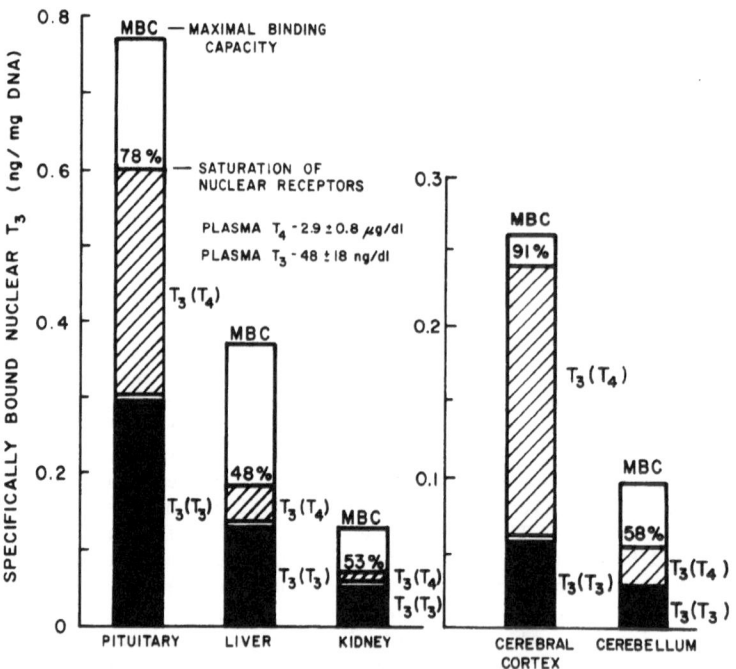

FIGURE 1.   Sources of nuclear T3 in anterior pituitary, liver, kidney, cerebral cortex and cerebellum of euthyroid rats based on results of tracer distribution studies.   The maximal T3 binding capacity of nuclear receptors for each tissue as assessed by in vivo saturation analysis is indicated by the height of the bar.   The component of nuclear T3 deriving from either plasma T3[T3(T3)] or from intracellular T4 5' monodeiodination [T3(T4)] in each tissue is indicated by the coded areas within each bar.

TABLE 1

IODOTHYRONINE DEIODINASES IN THE RAT CENTRAL NERVOUS SYSTEM

	TYPE I (5')	TYPE II (5')	TYPE III (5)
SUBSTRATE PREFERENCE	RT3>>T4>T3	T4>RT3	T3>T4
K_M FOR T4	1×10^{-6} M	1×10^{-9} M	6×10^{-9} M (T3), 37×10^{-9} M (T4)
DEIODINATION SITE	OUTER AND INNER RING	OUTER RING	INNER RING
KINETIC MECHANISM	PING-PONG	SEQUENTIAL	SEQUENTIAL
DITHIOTHREITOL (DTT)	STIMULATES	STIMULATES	STIMULATES
APPARENT K_I FOR PTU (AT 5 MM DTT)	5×10^{-7} M	4×10^{-3} M	? ($>10^{-3}$ M)
IOPANOIC ACID	INHIBIT	INHIBIT	INHIBIT
HYPOTHYROIDISM	DECREASE	INCREASE	DECREASE
HYPERTHYROIDISM	NO CHANGE	DECREASE	INCREASE

appreciated. By examination of 5' deiodinase activity in 1.2 mm punches of 1 mm slices of hypothyroid rat brain, Riskind et al observed a marked localization of 5' deiodinase activity in the arcuate nucleus and median eminence areas (8). The deiodinase activity was about 10 fold higher in these regions than in any other area of the hypothalamus and was 4 to 5 times higher than that in the frontal cortex. Of particular interest was the lack of increased deiodinase activity in the paraventricular nuclei, since these contain the cell bodies of the neurons which generate the TRH regulating TSH release. Deiodinase activity was also highest in this area of euthyroid rat brain. The arcuate nucleus/median eminence area is the site where the tuberoinfundibular neurons terminate in the neurohemeral structures and in which dopamine and growth hormone releasing hormone are synthesized. There is a high potential for modulation of the access of neuropeptides having a critical role in TSH and growth hormone release into the portal circulation in this anatomical location, though there is no direct information as to how how this might occur. Nonetheless, the marked sequestration of the 5' deiodinase in this region suggests an important physiological role. No similar data are available for the inner ring deiodinase.

Kaplan and Yaskoski found a characteristic pattern of deiodinase changes occurring during development in the rat (9). There was a peak of T4 5' deiodinase activity between two and six weeks in cerebellum, cerebrum and hypothalamus, whereas inner ring deiodination of T3 was highest at birth and fell precipitously over the first few weeks of life. The physiological importance of these patterns is not understood at present. However, the results were the first to suggest that unlike the situation in the liver, inner and outer ring deiodination would be catalyzed by different proteins, rather than being a directed function of a single enzyme.

While early studies indicated that the enzyme converting T4 to T3 would accept either T4 or reverse T3 (rT3) as a substrate for 5' deiodination, there were several difficulties with the concept that rT3 and T4 5' deiodination in the central nervous system were catalyzed by the same enzyme. First, PTU (10^{-3} M) partially inhibited the 5' deiodination of rT3, but not of T4 in euthyroid microsomes (10,11). This was observed with partially purified enzyme, as well as after in vivo PTU administration (12). Subsequent studies comparing the behavior of these two substrates under varying substrate concentrations and thyroid states allowed the discrimination of two pathways for outer ring deiodination in the cerebral cortex. 5' Deiodination of one substrate was significantly inhibited by 1 mM PTU while that of T4 was relatively insensitive (Table 1). This is demonstrated in Figure 2 showing results of in vivo studies where deiodination of T4 and rT3 are compared in euthyroid rats given either PTU or iopanoic acid (12). In Figure 2A it is apparent that PTU has no effect on T4 to T3 deiodination in the cerebral cortex, whereas Figure 2B shows approximately 80% inhibition of rT3 deiodination by the same PTU concentrations. Both pathways are inhibited by iopanoic acid. Many other experiments have demonstrated the contrast between these two 5' deiodinase activities which are summarized in Table 1. One discriminating difference is in the sensitivity of the two activities (Type I, PTU sensitive and Type II, PTU resistant) to inhibition by carboxymethylation (13). Iodoacetate

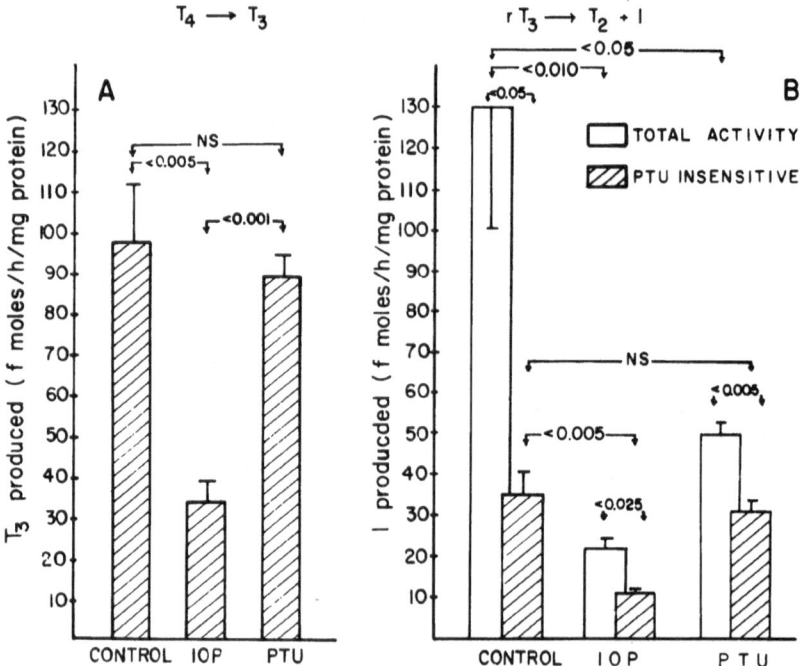

FIGURE 2. Iodothyronine 5' deiodinase activity in microsomes from rat cerebral cortex. Groups of euthyroid rats were treated with PTU or iopanoic acid. T4 to T3 conversion was measured by the production of 125-I T3 from 5 nM 125-I T4 in the presence of 15 mM DTT and 1 μM T3 (panel A). Reverse T3 deiodination was measured by the release of 125-I- from 2 nM 125-I rT3 in the presence of 15 mM DTT. PTU-insensitive deiodination refers to reactions performed in the presence of 1 mM PTU (panel B). Reprinted with permission from Silva, Leonard, Crantz and Larsen, J. Clin. Invest. 69:1176-1184, 1982.

(10^{-6} M) causes complete inhibition of PTU sensitive rT3 5' deiodination in the euthyroid cerebral cortex whereas 1000 fold higher concentrations are required to inhibit T_4 to T3 conversion in the same preparations. Thus the Type I 5' deiodinase activity in cerebral cortex, like that in the liver, requires an active sulfhydryl group, the carboxymethylation of which causes enzyme inactivation. In hypothyroid animals, most of the rT3 is deiodinated by the Type II pathway, since the Type I activity is reduced, and Type II activity increased several fold. Further studies with brain and brown adipose tissue microsomes have shown that the sensitivity of Type II activity to PTU is inversely related to the DTT concentration used during the assay, so that it is important to keep this factor in mind when assessing the sensitivity of a particular enzymatic activity to inhibition by this agent (15,16). Interested readers are referred to these references for a more thorough discussion of this complex area. While the two 5' deiodinase activities are quite distinct enzymatically, until such time as the protein sequences are determined, a definitive answer as to their structural similarities cannot be given.

Having defined critical conditions for the examination of outer ring deiodination, similar techniques were applied to the Type III or inner ring deiodinase activity (17-19). The results were anticipated by the initial studies of the enzyme homogenates. The Type III enzyme prefers T3 as a substrate, follows sequential, rather than ping-pong, kinetics and is insensitive to PTU at 10^{-3} M (Table 1). Despite the fact that T3 is the preferred substrate for the Type III activity, it is this enzyme which produces rT3 from T4 in the central nervous system.

We and others have performed studies attempting to assign the various enzyme activities to specific cell types in the central nervous system. Using primary mixed cultures of dispersed fetal rat brain cells, we demonstrated the presence of all three enzyme activities (20). The predominant pathway for iodothyronine metabolism in such preparations was Type III, although when corrections were made for the degradation of newly formed product, Type I and Type II activities were also demonstrable. Types II and III activities were influenced by the medium thyroid hormone concentration in the expected manner, but in the early studies we were unable to develop pure populations of glia or neurons. Studies by others have demonstrated Type II activity in the mouse neuroblastoma cell line NB41A3 (21) and in glial cells in primary culture (22). Astrocytes have been found to have a predominant Type III activity (23), although if mixed cultures containing both oligodendroglial cells and astrocytes are exposed to low substrate concentration (50 pM T4), the predominant deiodination pathway is phenolic ring deiodination (22). Another mouse neuroblastoma line (S-20Y) has been found to have Type I activity but has not been thoroughly evaluated for Types II or III (24). Thus it seems likely that both neurons and glial cells contain Types I and II activity and that Type III activity is present in glial cells though it may not be expressed in neurons. The flux of an iodothyronine between the two pathways, 5' versus 5 deiodination, will thus be influenced not only by thyroid status but also by the cell type in which it resides. Despite the dramatic effects of thyroid hormones, glucocorticoids have little influence on the central nervous system deiodination activities as measured in homogenates or microsomes (25,26). This contrasts with the situation in the liver where Type I activity seems to be inhibited by these perturbations.

One cannot assume, however, that thyroid hormone economy in brain can be entirely understood by the study of the deiodinases, even though some general principles can be derived from such experiments. The

TABLE 2

DIFFERENCES IN T3 ECONOMY BETWEEN LIVER AND CEREBRAL CORTEX*

CAPILLARY TRANSIT	PROLONGED	RAPID
RESULT	RAPID UPTAKE OF PLASMA T4 AND T3	SLOW CLEARANCE OF PLASMA T4 AND T3
T3 GENERATION	5'D-I ONLY; LOW EFFICIENCY BUT LARGE QUANTITY AND T4 POOL	5'D-II, SMALL QUANTITY OF ENZYME BUT HIGH EFFICIENCY; 5'D-I PRESENT BUT INSIGNIFICANT
RESULT	HIGH T3 PRODUCTION	COMPARABLE T3 PRODUCTION ON A WEIGHT BASIS
T3 DISPOSAL PATHWAYS	CONJUGATION-EXCRETION, CONJUGATION-DEIODINATION, EXIT TO PLASMA	5-DEIODINATION, EXIT TO PLASMA (?)
RATE	RAPID	SLOW
DIRECT CONSEQUENCE	SHORT RESIDENCE TIME OF LOCALLY PRODUCED T3	PROLONGED RESIDENCE TIME OF LOCALLY PRODUCED T3
HOMEOSTATIC REGULATION PRIORITY	SYSTEMIC	LOCAL
GLOBAL RESULTS OF THE ABOVE DIFFERENCES		
SOURCE OF NUCLEAR T3	80% PLASMA, 20% LOCAL PRODUCTION	20% PLASMA, 80% LOCAL PRODUCTION
RESPONSE TO HYPOTHYROIDISM	DECREASE T4 CONSUMPTION	INCREASE IN T4 TO T3 CONVERSION AND DECREASE IN T3 DISPOSAL
RESPONSE TO HYPERTHYROIDISM	INCREASE T4 CONSUMPTION (? BENEFICIAL)	DECREASE IN T4 TO T3 CONVERSION AND INCREASE T3 DISPOSAL

*MODIFIED FROM SILVA AND LARSEN 1986 (27).

TABLE 3

COMPARISON OF IN VIVO AND IN VITRO ESTIMATES OF SPECIFIC HIGH AFFINITY NUCLEAR T3 BINDING SITES IN RAT CEREBRAL CORTEX AND LIVER

	In Vivo	In vitro
Cerebral Cortex		
MBC (ng T3/mg DNA)	0.27	0.97
K_D (M)	1.5×10^{-11}*	5.6×10^{-10}
Liver		
MBC (ng T3/mg DNA)	0.76	0.77
K_D (M)	6.5×10^{-12}*	4.6×10^{-10}

*Assumes a free hormone fraction of 0.42%. Data are from Kolodny et al, 1985 (31).

reason for this is that the entry of thyroid hormone, particularly T3, into the brain is markedly different from its entry into a tissue such as the liver. The differences in this uptake mechanism influence the source of T3 in the brain and are in part the explanation for the predominance of T3(T4) in this tissue. Principal factors to be considered in comparing the physiology of the thyroid hormone activation process in liver and brain are presented in Table 2 (27). A most striking difference is the fact that thyroid hormones are rapidly taken up by liver, whereas plasma clearance of T3 and T4 is very low in the brain, in part due to the rapid capillary transit time in this tissue. A consequence of this is that plasma T3 enters the brain with considerable difficulty, resulting in a much lower tissue to plasma T3 ratio for brain than for the liver. When combined with the presence of the Type III deiodinase, it becomes quite difficult for plasma T3 to enter the nuclear compartment. In fact, preliminary studies in our laboratory by Dr. Peter Rudas have suggested that the residence time of T3(T4) in the brain is much longer than that of T3(T3), as if the former were produced in a compartment which is less accessible to deiodination and/or excretory pathways than is the T3 entering directly from the plasma. The T3 produced from T4 in liver seems to leave the tissue promptly. The anatomical correlations of these kinetic studies have not been made at the present time, but presumably will be derived from comparative studies of the in situ distribution of the deiodinases. A further important difference between the two tissues is that in hypothyroidism, 5'D-I activity is reduced in liver whereas the opposite occurs in brain. In hyperthyroidism, the changes are opposite in both tissues. Thus the liver responds in a systemic fashion to alterations in thyroid status while in the brain the maintenance of the tissue concentrations of T3 within a narrow range is the highest priority (28-30).

One result of these unique features of T3 metabolism in brain is shown in Table 3. If in vivo saturation techniques are used to determine the binding capacity of the T3 receptors in cerebral cortex, a much lower maximal binding capacity is calculated than is the case in vitro (5,31). This is not the case in liver where the results of binding capacity determinations are identical in vivo and in vitro. Our results showed that the in vitro binding capacity for brain nuclei was about 1 ng/mg DNA, a level which is comparable to that in the liver, whereas in vivo binding studies suggested that it was about 0.3 ng/mg, about 50% that of hepatic nuclei. This is of some interest given the high level of expression of c-erb-A mRNA for a T3 binding protein in brain (32). However, even given the higher numbers of binding by in vitro studies, the erb-A expression is far higher than the T3 binding capacity in these tissues for this cDNA. It is of interest that high affinity T3 binding activity is 8 to 10 fold higher in neuronal than in glial nuclei (31,33,34).

THE RESPONSE OF THE BRAIN TO HYPOTHYROIDISM

Within 24 hours of thyroidectomy there is a 4 to 6 fold increase in Type II activity in cerebral cortex, whereas no change is found in liver Type I or central nervous system Type III activity (35). Figure 3 demonstrates that there is a rapid recognition and response of the

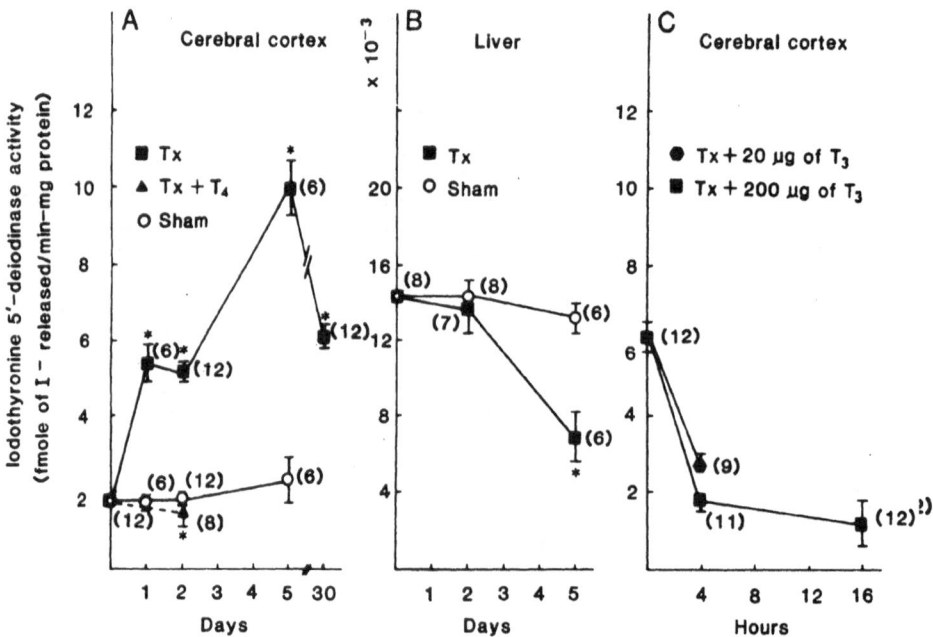

FIGURE 3. Effects of thyroidectomy or thyroid hormone treatment on cerebral cortex and hepatic iodothyronine 5' deiodination. Panel A: Time course of changes in cerebral cortex 5' deiodination after thyroidectomy. Assays were performed using 125-I rT3 as substrate in the presence of 15 mM DTT. Panel B: Assays were performed using 125-I rT3 as substrate in the presence of 1 mM DTT. Note the reaction rates in liver are approximately 10^3 times those in the cerebral cortex. Panel C: Effects of T3 administration on cerebral cortex 5' deidoination in chronically hypothyroid rats. Enzyme assays were performed as in panel A. In all experiments, the number of animals per group is indicated in parentheses. * indicates p<0.001. Tx, thyroidectomized rats. Reprinted from Leonard, Kaplan, Visser, Silva and Larsen, Science 214:571,1981, by permission of the American Association for the Advancement of Science.

TABLE 4

EFFECT OF T4 INJECTIONS
ON SERUM AND CEREBROCORTICAL IODOTHYRONINE CONCENTRATIONS
AND CEREBROCORTICAL 5'D-II ACTIVITIES
IN HYPOTHYROID RATS*

	Dose of T4 (μg/100g BW)* (Given i.v. 4 hrs earlier)		
	0.2	0.6	1.6
Serum T4 (ng/ml)	8.6	35	58
Cerebrocortical Iodothyronines			
T4 (ng/g)	0.27	1.1	3.9
rT3 (pg/g)	3.4	18	96
Estimated 5'D-II Suppression (%)			
Total	50	70	90
From Cortical rT3 Alone	0	20	50

*Obregon et al 1985 (39), Silva and Leonard 1985 (36).

cerebral cortex to a reduction in plasma thyroid hormone. It is also possible to reduce the elevated 5'D-II level in the hypothyroid brain acutely by administration of intravenous iodothyronine. In studies comparing T4, T3 and rT3 it was found that the dose causing 50% reduction in Type II activity in hypothyroid cerebral cortex was 0.2 ug T4/100 g BW, whereas that for T3 was 5 ug/100 g BW (36). In fact, T3 was even less effective than rT3 which had an ED_{50} of 2 ug/100 g BW. This suggested that regulation of 5'D-II activity in brain (and pituitary) was not related to transcriptionally mediated processes, a conclusion that was further substantiated by the fact that cycloheximide did not inhibit the response to T3. Kinetic studies of Type II enzyme in the cerebral cortex of rats given T3 indicated that the turnover rate of the enzyme was markedly reduced in hypothyroidism and we found that substrates such as T4 or rT3 could accelerate enzyme inactivation by as yet unexplained mechanisms (37).

While rT3 was less potent than was T4 in suppressing enzyme activity in the cerebral cortex of hypothyroid rats, these studies did not attempt to deal with the rapid plasma clearance of this iodothyronine. The clearance of rT3 is at least 30 fold faster than that of T4 and it was conceivable that locally produced rT3 could be important in the physiological regulation of the deiodinase activity in the intact cerebral cortex especially given the high Type III activity in this tissue. Accordingly, Obregon et al correlated tissue rT3 concentrations with Type II activity after administration of intravenous rT3 to hypothyroid animals (38). Reverse T3 caused an immediate inhibition of 5'D-II activity. This suppression of 5'D-II activity required the persistence of rT3 in the cerebral cortex and there was a log/linear relationship between the cerebrocortical rT3 concentration and the suppression of deiodinase activity up to the level of 1 ng/g. The EC_{50} for suppression of cortical 5'D-II activity was approximately 50 pg rT3/g cerebral cortex. To determine whether this was a physiologically meaningful concentration, Obregon et al also assessed the sources of rT3 in cerebral cortex using tracer T4 and rT3 infusions employing techniques that were used earlier for studies of local T3 production in

TABLE 5

PARAMETERS OF TISSUE-PLASMA EXCHANGE OF T3
IN VARIOUS TISSUES OF 2-WEEK-OLD RATS[*]

		Cerebral Cortex	Liver	Kidney
Fractional removal	Eu	0.26	0.98	0.97
rates (hr^{-1})	Hypo	0.07	0.67	0.46
Equilibrium T/P	Eu	1.5	5.1	5.8
(ml/g)	Hypo	1.6	3.8	7.2
Uptake from plasma	Eu	0.38	5.0	5.6
[(ml/g)xhr^{-1}]	Hypo	0.10	2.5	3.3

[*]Silva and Matthews 1984 (40).

this tissue (39). Especially critical to the studies was the affinity chromatographic technique for specific isolation of rT3, since direct chromatographic methods did not allow easy resolution of the small quantities of locally generated rT3 from the large amounts of the labelled T4 precursor. Results of these studies showed that cortical T4 concentrations were about 1000 pg/g in euthyroid rats and that virtually all of the rT3 present, 30 pg/g, was derived from local inner ring deiodination of T4. Less than 1% of cerebral cortical rT3 could be accounted for by plasma rT3. Obregon et al also evaluated the rT3 concentrations in the cerebral cortex in hypothyroid rats given increasing doses of T4 to determine whether the rT3 produced locally could explain the suppression of enzyme activity in the cerebral cortex (Table 4). At the ED_{50} for T4 (0.2 ug/100 g BW) for Type II enzyme suppression there was only 3.4 pg rT3/g cerebral cortex, an insufficient concentration to affect deiodinase activities. With a three times larger T4 dose, rT3 levels rose to 18 pg/g, sufficient to cause about 20% inhibition of deiodinase activity. The deiodinase suppression was 70% at this T4 concentration. This is an especially relevant T4 dose since the serum T4 concentration four hours after this dose is 35 ng/ml, a normal value for euthyroid rats. These results allowed the conclusion that virtually all of the Type II deiodinase suppressive effect of T4 in the euthyroid cerebral cortex derives from T4 per se with a maximum of 20% owing to the rT3 locally derived from it. Thus, we are unable to assign a significant physiological role for the rT3 in the cerebral cortex as a regulator of 5' deiodinase activity.

As indicated above, in vitro studies using glial or mixed neuronal glial cell preparations have indicated that it is possible to regulate Type III activity in vitro by addition of thyroid hormones (20,23). The mechanism for this increase is unknown, although it appears to be a consequence of an increase in Vmax as assessed by kinetic studies. This increased activity could be new enzyme synthesis or activation of proenzyme. Changes in Type III activity require more time than do comparable but inverse changes in 5'D-II but nothing further can be concluded about the intermediate processes at the present time.

In order to understand the total responsiveness of the brain to hypothyroxinemia one must integrate the consequences of all of the changes which occur under these circumstances. For example, in Table 5 are shown the changes in T3 uptake in two week old hypothyroid rats as demonstrated in the studies of Silva and Matthews (40). The fractional

TABLE 6

EFFECTS OF TWO WEEKS OF HYPOTHYROIDISM
ON CEREBROCORTICAL T4 AND T3 METABOLISM IN NEONATAL RATS*

	Euthyroid	Hypothyroid	Hy/Eu
Integrated $[^{125}I]$T4 [(%dose/g)xhr)]	16.1	2.0	0.12
Fractional T3 disposal rate (hr^{-1})	0.26	0.07	0.26
Fractional conversion rate (hr^{-1})	0.39	2.40	6.13
Integrated local $[^{125}I]$T3(T4) [(%dose/g)xhr]	25.5	69.0	2.74

*Silva and Matthews 1984 (40).

removal of T3 from plasma in hypothyroidism is reduced in cerebral cortex, liver and kidney, leading to an overall reduction in tissue T3 uptake. A similar qualitative change occurs in the fractional removal of T4 from serum. This may well be due to changes in blood flow as a consequence of the hypothyroidism or other nonspecific effects, but the changes are opposite in direction to what would be desirable from a teleological point of view.

However, in Table 6 are shown the consequences of the compensatory changes in inner and outer ring deiodination rates on the integrated T4 and T3 economy in the cerebral cortex in the hypothyroid neonates (40). There is a 90% reduction in T4 uptake by the cerebral cortex. The rate of T3 disposal from the tissue is also reduced to about 26% of its euthyroid level. The tissue half life of T3 is prolonged in part, at least, due to a decrease in Type III enzyme activity. The fractional conversion rate of T4 to T3 is increased 6 fold in these animals and the combined effect of the decrease in T3 disposal rate and the increase in fractional T4 to T3 conversion is a 2.7 fold increase in integrated local T3(T4) concentration. It should be mentioned that Dratman et al (41) did not observe significant changes in T3 turnover in hyper- and hypothyroidism in rat brain, an observation at variance with those generally reported (29,30) and with the above mentioned changes in Type III activity (7,19).

These changes show the complexity of the response of this organ to the stress of hypothyroidism. A similar pattern would presumably also apply to the events occurring during iodine deficiency since it appears to be a reduction in T4 per se which is the primary signal at least for induction of the changes in the Type II 5' deiodination. In fact, some of the changes in blood flow and plasma hormone uptake in iodine deficiency could be much less since these are circumstances in which a low T4 is not accompanied by a low serum T3 which could lead to myocardial hypothyroidism and reduced cardiac output. In other studies, we have found that thyroid hormone dependent enzymes in the central nervous system remain normal in hypothyroid neonatal rats until serum T4 reaches very low levels (42). Furthermore, single injections of only 60 ng of T4 to hypothyroid neonatal rats cause a 3 fold greater increase in cerebrocortical T3 than does a similar injection in two week old euthyroid pups. It is also possible to increase the thyroid hormone dependent enzyme, aspartic transaminase, in the hypothyroid central nervous system by giving T4 without affecting hepatic mitochondrial alpha-glycerophosphate dehydrogenase (42). All of these changes would appear to make excellent teleological sense and demonstrate the advantages of a system for local T3 production which is responsive to a reduction in hormone supply.

If we may extrapolate from these results to the clinical situation, we conclude that maintenance of optimal intracellular T3 concentrations in the central nervous system requires T4 as a substrate and a series of adaptations in the metabolism of this prohormone and T3, which then compensates for the plasma hypothyroxinemia. To the extent that the compensatory changes described in the cerebral cortex of the hypothyroid rat do not occur in the human, a reduction of serum T4 in iodine deficient persons could present a significant threat to the thyroid status of the cerebral cortex, even if serum T3 remained at normal concentrations. There are no data with respect to the presence of a local T4 to T3 conversion system in the human central nervous system. However, the similarity of the responses of the pituitary-thyroid axis to hypothyroidism and iodine deficiency in the rat and man suggests that the Type II deiodinase is common to both species (28,43). Presumably then, the concepts regarding intracerebral thyroid hormone metabolism derived from those experiments in the rat are also relevant to man.

REFERENCES

1. G. Hennemann, "Thyroid hormone metabolism," Marcel Dekker, Inc., New York, 1986.
2. R. G. Cheron, M. M. Kaplan, and P. R. Larsen, Physiological and pharmacological influences on thyroxine to 3,5,3'-triiodothyronine conversion and nuclear 3,5,3'-triiodothyronine binding in rat anterior pituitary, J. Clin. Invest., 64:1402 (1979).
3. J. E. Silva, T. E. Dick, and P. R. Larsen, The contribution of local tissue thyroxine monodeiodination to the nuclear 3,5,3'-triiodothyronine in pituitary, liver, and kidney of euthyroid rats, Endocrinology, 103:1196 (1978).
4. F. R. Crantz and P. R. Larsen, Rapid thyroxine to 3,5,3'-triiodothyronine conversion and nuclear 3,5,3'-triiodothyronine binding in rat cerebral cortex and cerebellum, J. Clin. Invest., 65:0000 (1980).
5. F. R. Crantz, J. E. Silva, and P. R. Larsen, An analysis of the sources and quantity of 3,5,3'-triiodothyronine specifically bound to nuclear receptors in rat cerebral cortex and cerebellum, Endocrinology, 110:367 (1982).
6. A. C. Bianco and J. E. Silva, Nuclear 3,5,3'-triiodothyronine (T3) in brown adipose tissue: receptor occupancy and sources of T3 as determined by in vivo techniques, Endocrinology, 120:55 (1987).
7. M. M. Kaplan and K. A. Yaskoski, Phenolic and tyrosyl ring deiodination of iodothyronines in rat brain homogenates, J. Clin. Invest., 66:551 (1980).
8. P. N. Riskind, J. M. Kolodny, and P. R. Larsen, The regional hypothalamic distribution of type II 5'-monodeiodinase in euthyroid and hypothyroid rats, Brain Res., 420:194 (1987).
9. M. M. Kaplan and K. A. Yaskoski, Maturational patterns of iodothyronine phenolic and tyrosyl ring deiodinase activities in rat cerebrum, cerebellum, and hypothalamus, J. Clin. Invest., 67:1208 (1981).

10. T. J. Visser, J. L. Leonard, M. M. Kaplan, and P. R. Larsen, Different pathways of iodothyronine 5'-deiodination in rat cerebral cortex, Biochem. Biophys. Res. Comm., 101:1297 (1981).

11. T. J. Visser, J. L. Leonard, M. M. Kaplan, and P. R. Larsen, Kinetic evidence suggesting two mechanisms for iodothyronine 5'-deiodination in rat cerebral cortex, Proc. Natl. Acad. Sci., 79:5080 (1982).

12. J. E. Silva, J. L. Leonard, F. R. Crantz, and P. R. Larsen, Evidence for two tissue-specific pathways for in vivo thyroxine 5'-deiodination in the rat, J. Clin. Invest., 69:1176 (1982).

13. T. J. Visser, S. Frank, and J. L. Leonard, Differential sensitivity of brain iodothyronine 5'-deiodinases to sulfhydryl-blocking reagents, Mol. Cell. Endocr., 33:321 (1983).

14. J. E. Silva, M. B. Gordon, F. R. Crantz, J. L. Leonard, and P. R. Larsen, Qualitative and quantitative differences in the pathways of extrathyroidal triiodothyronine generation between euthyroid and hypothyroid rats, J. Clin. Invest., 73:898 (1984).

15. J. E. Silva, S. Mellen, and P. R. Larsen, Comparison of kidney and brown adipose tissue iodothyronine 5'-deiodinases, Endocrinology, 121:650 (1987).

16. A. Goswami, and I. N. Rosenberg, Iodothyronine 5'-deiodinase in brown adipose tissue: thiol activation and propylthiouracil inhibition, Endocrinology, 119:916 (1986).

17. M. M. Kaplan, T. J. Visser, K. A. Yaskoski, and J. L. Leonard, Characteristics of iodothyronine tyrosyl ring deiodination by rat cerebral cortical microsomes, Endocrinology, 112:35 (1983).

18. K. Tanaka, M. Inada, H. Ishii, K. Naito, M. Nishikawa, Y. Mashio, and H. Imura, Inner ring monodeiodination of thyroxine and 3,5,3'-L-triiodothyronine in rat brain, Endocrinology, 109:1619 (1981).

19. M. M. Kaplan, U. D. McCann, K. A. Yaskoski, P. R. Larsen, and Jack L. Leonard, Anatomical distribution of phenolic and tyrosyl ring iodothyronine deiodinases in the nervous system of normal and hypothyroid rats, Endocrinology, 109:397 (1981).

20. J. L. Leonard and P. R. Larsen, Thyroid hormone metabolism in primary cultures of fetal rat brain cells, Brain Res., 327:1 (1985).

21. D. L. St. Germain, Hormonal control of a low Km (Type II) iodothyronine 5'-deiodinase in cultured NB41A3 mouse neuroblastoma cells, Endocrinology, 119:840 (1986).

22. F. Courtin, F. Chantoux, and J. Francon, Thyroid hormone metabolism by glial cells in primary culture, Mol. Cell. Endocr., 48:167 (1986).

23. R. R. Cavalieri, L. A. Gavin, R. Cole, and J. DeVellis, Thyroid hormone deiodinases in purified primary glial cell cultures, Brain Res., 364:382, (1986).

24. M. Chacon, S. R. Max, J. A. Kirshner, and J. T. Tildon, Thyroid hormone actions on a cholinergic neuroblastoma cell line (S-20Y), J. Neurochem., 47:1604 (1985).

25. M. M. Kaplan and K. A. Yaskoski, Effects of congenital hypothyroidism and partial and complete food deprivation on phenolic and tyrosyl ring iodothyronine deiodination in rat brain, Endocrinology, 110:761 (1982).

26. U. D. McCann, E. A. Shaw, and M. M. Kaplan, Iodothyronine deiodination reaction types in several rat tissues: effects of age, thyroid status, and glucocorticoid treatment, Endocrinology, 114:1513 (1984).

27. J. E. Silva and P. R. Larsen, Regulation of thyroid hormone expression at the prereceptor and receptor levels, in Thyroid hormone metabolism, G. Hennemann, ed., Marcel Dekker, Inc., New York, 1986.

28. P. R. Larsen, J. E. Silva, and M. M. Kaplan, Relationships between circulating and intracellular thyroid hormones: physiological and clinical implications, Endocrine Rev., 2:87 (1981).

29. J. van Doorn, D. van der Heide, and F. Roelfsema, The contribution of local thyroxine monodeiodination to intracellular 3,5,3'-triiodothyronine in several tissues of hyperthyroid rats at isotopic equilibrium, Endocrinology, 115:174 (1984).

30. J. van Doorn, F. Roelfsema and D. van der Heide, Conversion of thyroxine to 3,5,3'-triiodothyronine in several rat tissues in vivo: the effect of hypothyroidism, Acta Endocrinol., 113:59 (1985).

31. J. M. Kolodny, P. R. Larsen, and J. E. Silva, In vitro 3,5,3'-triiodothyronine binding to rat cerebrocortical neuronal and glial nuclei suggests the presence of binding sites unavailable in vivo, Endocrinology, 116:2019 (1985).

32. C. C. Thompson, C. Weinberger, R. Lebo, and R. M. Evans, Identification of a novel thyroid hormone receptor expressed in the mammalian central nervous system, Science, 237:1610 (1987).

33. J. M. Kolodny, J. L. Leonard, P. R. Larsen, and J. E. Silva, Studies of nuclear 3,5,3'-triiodothyronine binding in primary cultures of rat brain, Endocrinology, 117:1848 (1985).

34. D. Gullo, A. K. Sinha, R. Woods, K. Pervin, and R. P. Ekins, Triiodothyronine binding in adult rat brain: compartmentation of receptor populations in purified neuronal and glial nuclei, Endocrinology, 120:325 (1987).

35. J. L. Leonard, M. M. Kaplan, T. J. Visser, J. E. Silva, and P. R. Larsen, Cerebral cortex responds rapidly to thyroid hormones, Science, 214:571 (1981).

36. J. E. Silva, and J. L. Leonard, Regulation of rat cerebrocortical and adenohypophyseal Type II 5'-deiodinase by thyroxine, triiodothyronine, and reverse triiodothyronine, Endocrinology, 116:1627 (1985).

37. J. L. Leonard, J. E. Silva, M. M. Kaplan, S. A. Mellen, T. J. Visser, and P. R. Larsen, Acute posttranscriptional regulation of cerebrocortical and pituitary iodothyronine 5'-deiodinases by thyroid hormone, Endocrinology, 114:998 (1984).

38. M.-J. Obregon, P. R. Larsen, and J. E. Silva, The role of 3,3'5'-triiodothyronine in the regulation of Type II iodothyronine 5'-deiodinase in the rat cerebral cortex, Endocrinology, 119:2186 (1986).

39. M.-J. Obregon, P. R. Larsen, and J. E. Silva, Plasma kinetics, tissue distribution, and cerebrocortical sources of reverse triiodothyronine in the rat, Endocrinology, 116:2192 (1985).

40. J. E. Silva, and P. S. Matthews, Production rates and turnover of triiodothyronine in rat-developing cerebral cortex and cerebellum, J. Clin. Invest., 74:1035 (1984).

41. M. B. Dratman, F. L. Crutchfield, J. T. Gordon, and A. S. Jennings, Iodothyronine homeostasis in rat brain during hypo- and hyperthyroidism, Am. J. Physiol., 245:E185 (1983).

42. J. E. Silva, and P. R. Larsen, Comparison of iodothyronine 5'-deiodinase and other thyroid-hormone-dependent enzyme activities in the cerebral cortex of hypothyroid neonatal rat, J. Clin. Invest., 70:1110 (1982).

43. P. R. Larsen, Thyroid-pituitary interaction: feedback regulation of thyrotropin secretion by thyroid hormones, New Engl. J. Med., 306:23 (1982).

THYROID HORMONE RECEPTOR RELATED mRNAs

Vera M. Nikodem and Tomoaki Mitsuhashi

National Institute of Diabetes, Digestive & Kidney Diseases
National Institutes of Health

Although thyroid hormone affects growth, differentiation and develop-ment in multicellular organisms, the molecular basis of its action remains unclear. A striking effect of thyroid hormone on differentiation and matu-ration of the developing brain has been documented, and will be further discussed during this symposium. The diverse effects of thyroid hormone can be mediated by multiple mechanisms at different cellular levels. Sev-eral biochemical parameters: O_2 consumption, high affinity-low capacity binding sites for 3,3',5-triiodo-L-thyronine (T_3) in the nucleus, altered enzymatic activities of mitochondrial α glycerophosphate dehydrogenase and cytosolic malic enzyme have been used to characterize tissue responsiveness to T_3. Some of the published results are summarized in Table 1.

Table 1. <u>Tissue responsiveness to thyroid hormone.</u> Effect of thyroid hormone on oxygen consumption[1]; and malic enzyme mRNA levels[3]; compared to nuclear T_3-binding capacity[2]. Data for selected tissues in adult rats were adapted from the cited refer-ences. Oxygen consumption (cu.mm./mg wet weight/h) is the ratio in euthyroid versus thyroidectomized (Tx) rats. The fold of induction of malic enzyme mRNAs was calculated from values obtained from tissues of euthyroid rats and rats treated for 10 days with 15µg T_3/100 g b.w. The binding capacity was measured in normal rats.

TISSUE	O_2 CONSUMPTION Eu/Tx	BINDING CAPACITY ng T_3/mg DNA	ME mRNA T_3/Eu
Liver	1.29	.61	≈ 12
Heart	1.62	.40	4
Kidney	1.16	.53	3
Brain	1.08	.27	1
Spleen	1.01	.018	1
Testis	0.97	.0023	1

O$_2$ consumption is increased by thyroid hormone in liver, heart, and kidney, but not in brain, spleen or testis. The former tissues show a high nuclear binding capacity for T$_3$, when examined as whole tissues, while brain contains substantially fewer binding sites and spleen and testis very low numbers of binding sites for the hormone. We have chosen rat malic enzyme (ME) as a model system to gain insight into the molecular mechanism by which T$_3$ regulates protein synthesis. The last column in Table 1 shows that the increase in ME mRNA level was detected only in those tissues in which O$_2$ consumption was altered by thyroid hormone and those containing a high T$_3$ binding capacity[4]. Further studies in our laboratory revealed that part of the approximately 12-fold increase in the ME mRNA in liver is due to ≈ 4-fold stimulation of transcriptional activity of the gene[4]. A greater effect is manifested subsequent to transcription and is caused by stabilization of the ME primary transcript (unpublished data). Thus, it appears that in liver, T$_3$ regulates ME mRNA levels by at least two mechanisms. One mechanism is the regulation of gene transcription, which we assume is due in some way to the binding of the thyroid hormone-receptor complex to a regulatory element of the ME-sensitive gene. To elucidate the molecular basis by which transcriptional activation of the ME gene occurs, we planned to clone the liver thyroid hormone receptor mRNA. However, elegant studies of Evans'[5] and Vennstrom's[6] laboratories demonstrated that the "in vitro" translation products of an erb-A proto-oncogene have binding characteristics of the T$_3$ receptor. T$_3$ receptor cDNAs have been isolated from chicken embryo[6], human placenta[5], and rat brain libraries[7]. Sequence comparison among c-erb-A cDNAs and mutation analyses of steroid receptors permitted the identification of the DNA- and hormone-binding domains and revealed that these trans-acting factors are members of a superfamily related to the viral oncogene erb-A[8].

We used the Pst fragment of v-erb-A to screen 4 independent rat liver cDNA libraries of about 5x10^6 phage recombinants. No positive signal was obtained, although rat liver thyroid hormone receptor cDNA was recently isolated in Dr. Towle's laboratory (personal communication). Then, we screened a rat brain cDNA library with the same ^{32}P-labeled fragment used to screen the rat liver libraries. From ≈ 4x10^5 recombinants, 12 positive clones were obtained and were further characterized by restriction endonuclease mapping[9]. Fig. 1 shows restriction patterns of the Bgl I restricted putative T$_3$ receptor cDNAs. Two different patterns were obtained. This suggests that two sets of clones were isolated, characterized by the presence or absence of an internal Bgl I site (two bands or one band) and designated as variants I and II of rTRα (rat brain T$_3$ receptor recently cloned by Thompson et al.[7]). A representative from each set was sequenced by the dideoxy-chain termination method[10] and nucleotide and predicted amino acid sequences for the vI and vII clones revealed that the 1910 nucleotide sequence of the longer cDNA (vI) contains an open reading frame encoding a protein of 454 amino acids with the translation initiation codon at nt 78. The shorter cDNA (vII) starts at nt 413 of vI and then lacks 117 nt from 1074 to 1190. Both variant forms share an identical 3' noncoding sequence (data not shown). Comparison of amino acid sequences of v-erb-A, rTRα, cTRα, hTRβ and the two variant forms we have isolated (Mitsuhashi et al.[11]) is shown in Fig. 2. Comparative analyses of rTRα and vI, vII revealed striking homology in the presumed DNA binding domain (100%) and in the first 180 amino acids (vertical dashed line) of the T$_3$ binding domain (99%). Extensive amino acid sequence similarities were found with cTRα, hTRβ and v-erb-A up to the divergent point, amino acid residue 368, 419, 354, respectively, located in the hormone binding regions, as indicated by the vertical dashed line in Fig. 2. In the unique region of rTRα vI (amino acid residue 332-454), no sequence similarity was found with any amino acid sequence available in the protein databank, except a stretch of 5 amino acids which is conserved, with an additional 4 amino acids, in all protein

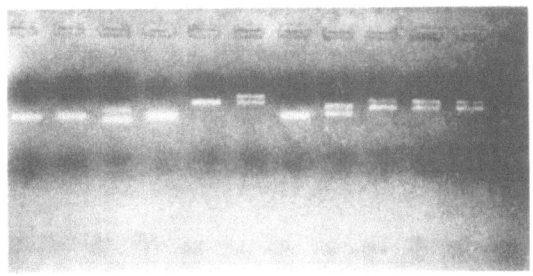

Fig. 1. Twelve clones isolated from a rat brain cDNA
library with ^{32}P-labeled Pst/Pst fragment of v-erb-A
were digested with Bgl I endonuclease and electro-
phoresed on an agarose gel.

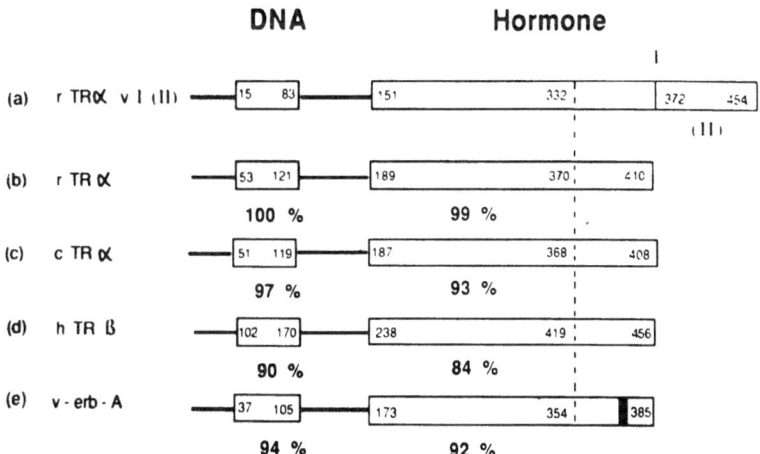

Comparison of Amino Acid Sequences of erb-A Proteins

Fig. 2. Alignment of the amino acid sequences of v-erb-A,
human placenta (hTRβ), chicken embryo (cTRα), and rat
brain (rTRα) thyroid hormone receptors with rTRα vI and
vII. The schematic structure of rTRα vI and vII is
shown (a), with numbers designating amino acid residues.
The sequence similarity of erb-A proteins within open
boxes up to the vertical dashed line, with the corresponding region of rTRα vI and vII is indicated by
percentages.

products of other c-erb-A's but is absent from the v-erb-A protein, indicated by a solid bar in v-erb-A.

The diagram in Fig. 3 depicts in detail the differences at the 5' and 3' coding regions among rTRα and vI, vII and a recently obtained cDNA, designated as vI'. Rat TRα is encoded by 410 amino acids. The translation

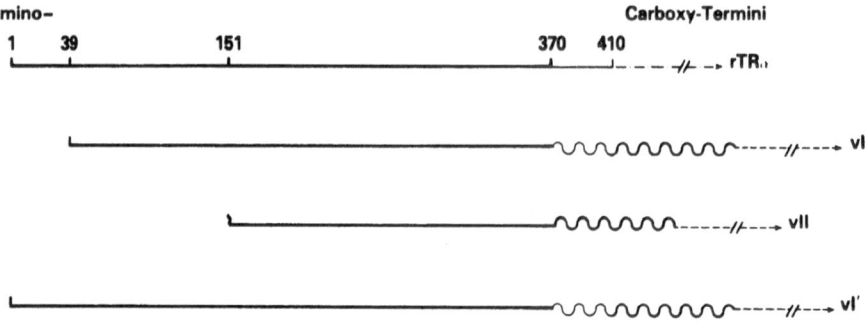

Schematic Structures of the Coding Region of rTRα and Variants

Fig. 3. Amino acid sequences were aligned with numbers referring to the amino acid residues of rTRα. The solid or wavy lines indicate the common sequences; 3' untranslated regions are designated by dot-dashed or dashed lines.

initiation codon of the vI is located at amino acid 39 of rTRα. The 5' end of vII starts at amino acid 151 of rTRα, corresponding to the amino acid isoleucine and thus probably does not represent an authentic 5' end. The amino terminus of vI' coincides with that of rTRα. At amino acid 370 of rTRα, the homology abruptly disappears. The last 40 amino acids of rTRα are substituted by 122 amino acids (wavy line) in vI and vI', whereas vII contains only the last 83 amino acids of vI and vI' (heavy wavy line). Interestingly, there is a consensus splicing junction AG:G at each breakpoint, suggesting that mRNAs of variant forms might be generated by alternative splicing of transcripts from the receptor gene. Therefore a rat genomic library was screened with a probe common to the 3' ends of both rTRα and variants. Nucleotide sequence analysis revealed that the same genomic clone contains common and unique sequences not only to the 3' end of rTRα but to the variants as well. The formation of the 3' ends of receptor and vI,II mRNAs is shown in Fig. 4. To generate rTRα vI and vII mRNAs, the internal donor site at the a-b junction (the dashed line in Fig. 2) is utilized. The exclusion of the b region as part of the intron results in the formation of rTRα vI. In the case of rTRα vII, the internal acceptor site at the c-d junction is utilized and the c region is excluded as well. In either case, splice junction sequences conform to the GT/AG rule, i.e., spliced out sequences begin at the 5' end with the dinucleotide GT and terminate with the dinucleotide AG. It appears that the putative polyadenylation signal at the 3' end of the b region generates rTRα.

To determine whether the variant forms of rTRα encode a protein that specifically binds thyroid hormones, rTRα vI cDNA was cloned into the pGEM

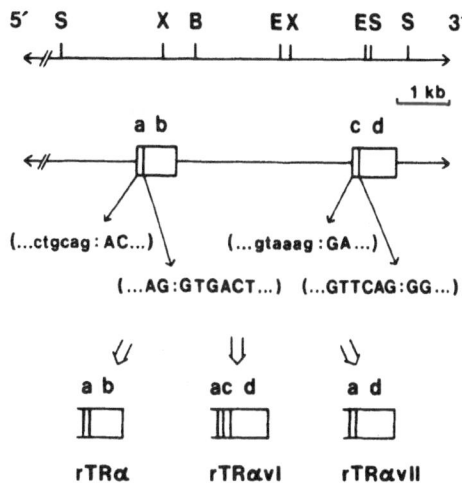

Generation of Various mRNAs from the rTR₁ Gene

Fig. 4. Restriction map of the genomic clone encompassing the 3' ends of coding, noncoding and intronic
sequence of rTRα and its variant forms is shown at
the top. The boxed regions and upper case letters
in parentheses designate cDNA sequences, whereas lower
case letters in parentheses indicate sequence of introns.
The a-b junction corresponds to the amino acid residue
370 and 332 of rTRα and rTRα vI, vII, respectively.

3 expression vector to allow "in vitro" transcription with the T7 RNA polymerase followed by translation in a reticulocyte lysate. "In vitro" translation products of rTRα vI correspond to molecular masses 55, 44, and 40 kD
as determined by NaDodSO$_4$-PAGE (Fig. 5, lane 2). The translation efficiency was comparable to that of hTRβ cDNA (lane 1). The products of "in
vitro" translation of vI and vI' were then analysed for binding to thyroid
hormones. Five µl of the "in vitro" translation mixture programmed with
the "in vitro" transcript of rTRα (Fig. 6 left columns), rTRα vI' (middle
columns), and reticulocyte lysate alone (right columns), were incubated
with increasing concentrations of labeled T$_3$ in the absence or presence of
1µM unlabeled T$_3$. As can be seen, definite specific binding was detected
only for rTRα (open boxes).

Neither T$_3$, L-thyroxine, nor 3,3',5'-triiodo-L-thyronine bound specifically to "in vitro" translation products derived from either variant of
rTRα. The binding was indistinguishable from that obtained with reticulocyte lysate alone. Hence, the variant forms of rTRα do not bind thyroid
hormone despite their extensive sequence homology with the receptor. They
will however, be detected with probes which do not discriminate between
the variant and the receptor messages. Since Thompson et al.[7] reported
that T$_3$ receptor mRNA is abundantly expressed in brain, contrary to the low
receptor level determined by ligand binding studies we have examined the
levels of rTRα vI and vII mRNAs employing various restriction fragments and
synthetic oligonucleotide probes which discriminate between variant and
receptor messages.

The following figures show results of Northern analyses of poly(A)$^+$
RNA prepared from various rat tissues and hybridized with different probes.
In Fig. 7, the structure coding sequence of vI is illustrated by the bar at
the top with the filled portion designating sequences in common with the
receptor to the breakpoint shown by a triangle; therefore, probe A used in

23

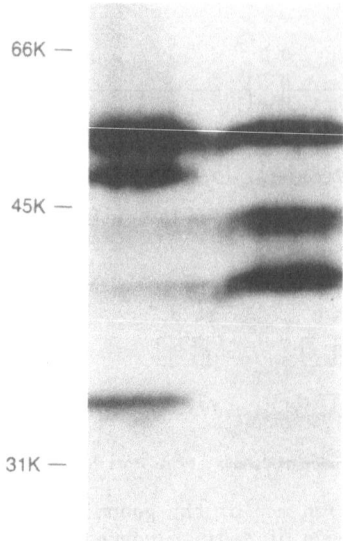

1   2

66K —

45K —

31K —

Fig. 5. NaDodSO$_4$-PAGE of in vitro translation
products of hTRβ (lane 1) and rTRα vI (lane 2).
The pGEM3 expression vector, T7 RNA polymerase
and reticulocyte lysate was used as suggested
by the supplier (Promega). Molecular masses
of protein standards are indicated on the left.

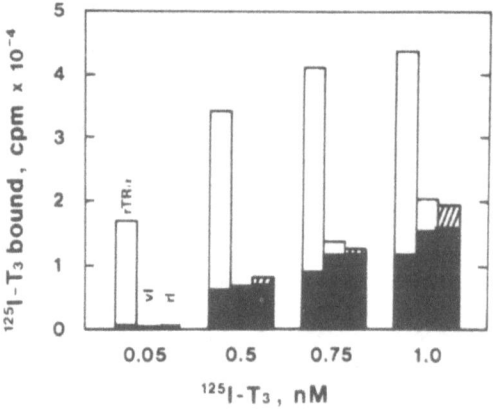

Thyroid Hormone Binding Characteristics of the in *Vitro*
Translation Products of rTRα, rTRα vI and Reticulocyte lysate

Fig. 6. Five μl of in vitro translation mixture programmed with in vitro transcripts of rTRα (left columns),
rTRα vI (middle columns), or reticulocyte lysate alone
(right columns) were incubated with increasing concentrations of labeled T$_3$ (open & shaded bar). One μM unlabeled
T$_3$ was added to determine nonspecific binding (filled
bars).

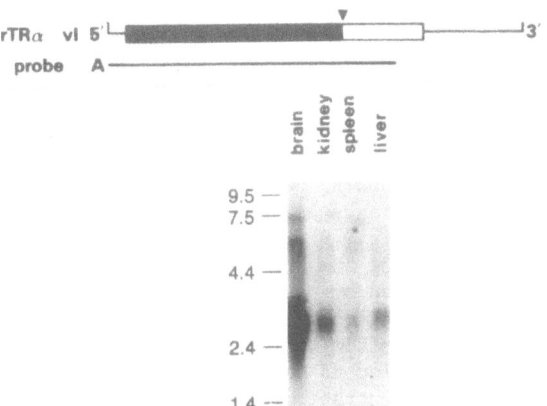

rTRα vI 5'

probe A

9.5 —
7.5 —

4.4 —

2.4 —

1.4 —

Northern Analysis of 10μg Poly(A)⁺ RNA from Various
Rat Tissues, cDNA Probe A is Common to rTRα and vI,II

Fig. 7. Schematic representation of the structure of rTRα
vI is shown at the top. The bar shows the coding sequence
with the filled in portion common to rTRα. The probe A
sequence is marked with a solid line. Total cellular RNA
was prepared using guanidinium thiocyanate[12] and then
chromatographed on oligo (dT)-cellulose[9].

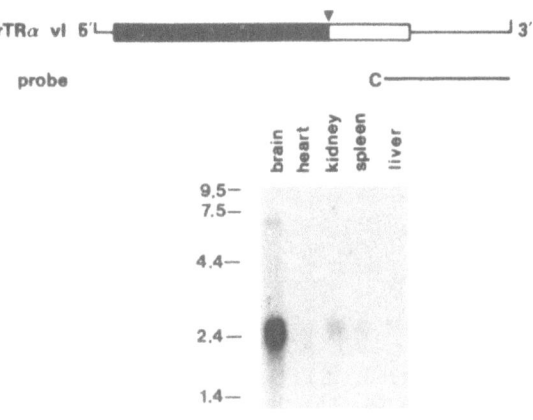

rTRα vI 5'

probe   C

9.5—
7.5—

4.4—

2.4—

1.4—

Northern Analysis of 5μg Poly(A)⁺ from Various
Rat Tissues, cDNA Probe C is Unique to the Variants.

Fig. 8. Schematic representation of the structure of rTRα
vI is shown at the top and is described in the legend to
Fig 7. The probe C sequence is designated by a solid line.
The preparations of RNA are as in Fig 7.

Northern analysis, shown above, hybridizes to both the variants and the
receptor mRNAs. Probe A hybridized with rat brain poly(A)⁺ RNAs ≈ 2.6,
5.4, and 6.8 kb in size, the 2.6 kb message being most abundant. These
messages are also present in kidney, spleen and liver. Hybridization of
poly(A)⁺ RNA prepared from brain, heart, kidney, spleen, and liver with

probe C (unique to the variants) revealed the presence of rTRα vI, and II mRNAs in all tested tissues, but much more abundantly in brain and least in liver (Fig. 8). This clearly demonstrates that the detection of mRNAs using probes which cannot discriminate between the receptor and its variant forms do not reflect T_3 receptor message levels.

Lastly, we used cDNA probes B, common to the receptor and its variants, and C, unique to variants, to estimate their relative abundance in brain. As can be seen in Fig. 9, signals of similar intensity were detected, strongly suggesting the predominance of variant messages in rat brain. This was then supported by results obtained with synthetic oligonucleotide probes D, hybridizing only to the variants and E, specific to the receptor only (sequence taken from ref. 8). A strong signal was seen with probe D, while probe E failed to detect any mRNA under the tested conditions, implying that the concentration of rat brain receptor is very low and a probe of much higher specific activity and more than 5μg of poly(A)$^+$ RNA are necessary in order to detect the receptor sequences. Thus, these results show that only the variant forms of c-erb-A mRNA are expressed at high levels in brain and resolve the contradiction between previously reported findings that the T_3 receptor is present in a low amount in brain as determined by Scatchard analysis[13] and those recently reported by Thompson et al.[7] using an rTRα cDNA probe.

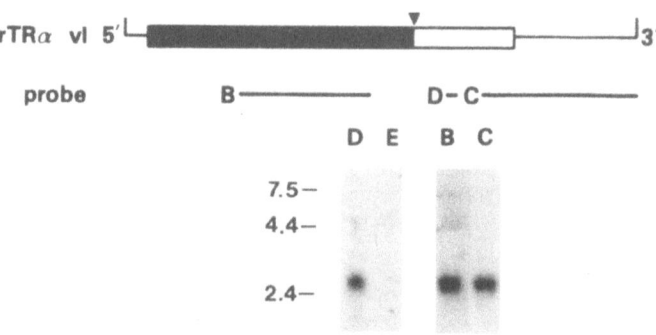

**Northern Analysis of 5 μg Rat Brain Poly(A)⁺ RNA
hybridized with cDNA (B,C) or Oligonucleotide (D,E) Probes**

Fig. 9. See the legend to Fig. 7. The probes C and D are unique to the variants, the probe E is unique to the receptor whereas the probe B is common for both sequences studied. The cDNA probes B and C were labeled to specific activities ≈ 7-10x10^8 cpm/μg. The oligonucleotide probes E and D (42-mers) were labeled to ≈ 1x10^8 cpm/μg.

A question arises as to whether the variant forms play any physiological role. They may bind ligands other than thyroid hormones. Alternatively, the presence of an intact DNA-binding domain may allow the variants to act as transcriptional factors independent of T_3, or even to compete with the T_3 receptor complex for site specific binding to the thyroid hormone responsive element. If these variants of c-erb-A have some modulatory functions in T_3 action, the mechanism of thyroid hormone action might be more complicated than previously thought, and regulation at the level of alternative splicing could play an important role especially during development.

In conclusion: 1) We have isolated two cDNA clones encoding proteins with virtually identical amino-acid sequences with the rat brain T_3 receptor, except the last 40 amino acids in the carboxy termini; 2) Isolation of genomic clones of these cDNAs and Southern analysis indicates that these 3 different mRNAs are generated from a single gene by alternative splicing of the primary transcripts of the rat brain T_3 receptor gene; 3) The "in vitro" translation products of the variant forms we have isolated lack hormone binding activity; 4) Using probes which discriminate between T_3-binding and variant forms revealed that the rat brain T_3 receptor is not predominantly expressed in this tissue in accord with previous ligand binding studies; 5) The presence of an intact DNA binding domain in the variants suggests their possible involvement in T_3 action.

ACKNOWLEDGEMENTS

We are grateful to Drs. C. Weinberger and R.M. Evans for providing us with hTRβ and rTRα cDNAs. We thank Drs. J.E. Rall, J. Robbins, and K. Petty for discussions and Dr. G. Tennyson for assistance in sequencing.

REFERENCES

1. S.P. Barker, and H.M. Klitgaard, Metabolism of tissues excised from thyroxine-injected rats, Am. J. Physiol. 170:81 (1952).
2. J.H. Oppenheimer, The nuclear receptor-triiodothyronine complex: relationship to thyroid hormone distribution, metabolism, and biological action, in: "Molecular Basis of Thyroid Hormone Action", Academic Press, (1983).
3. B. Dozin, M.A. Magnuson, and V.M. Nikodem, Tissue-specific regulation of two functional malic enzyme mRNAs by triiodothyronine, Biochemistry Sept 24:5581 (1985).
4. B. Dozin, M.A. Magnuson, and V.M. Nikodem, Thyroid hormone regulation of malic enzyme synthesis, J. Biol. Chem. 261:10290 (1986).
5. C. Weinberger, C.C. Thompson, E.S. Ong, R. Lebo, D.J. Gruol, and R.M. Evans, The c-erb-A gene encodes a thyroid hormone receptor, Nature 324:641 (1986).
6. J. Sap, A. Munoz, K. Damm, Y. Goldberg, J. Ghysdael, A. Leutz, H. Beug, and B. Vennstrom, The c-erb-A protein is a high-affinity receptor for thyroid hormone, Nature 324:635 (1986).
7. C.C. Thompson, C. Weinberger, R. Lebo, R.M. Evans, Identification of a novel thyroid hormone receptor expressed in the mammalian central nervous system, Science 237:1610 (1987).
8. S. Green, and P. Chambon, A superfamily of potentially oncogenic hormone receptors, Nature 324:61 (1986).
9. T. Maniatis, E.F. Fritsch, and J. Sambrook, "Molecular Cloning: A Laboratory Manual", CSH Publishers, (1987).
10. F. Sanger, S. Nicklen, and A.R. Coulson, DNA sequencing with chain-terminating inhibitors, Proc Natl. Acad. Sci. USA 74:5463 (1977).
11. T. Mitsuhashi, G.E. Tennyson, and V.M. Nikodem, Alternative splicing generates novel messages encoding rat c-erb-A proteins which do not bind thyroid hormone, Proc. Natl. Acad. Sci. USA (in press).
12. J.M. Chirgwin, A.E. Przybyla, R.J. MacDonald, and W.J. Rutter, Isolation of biologically active ribonucleic acid from sources enriched in ribonuclease, Biochemistry 79:5294 (1979).
13. J.H. Oppenheimer, H.L Schwartz, and M.I. Surks, Tissue differences in the concentration of triiodothyronine nuclear binding sites in the rat: liver, kidney, pituitary, heart, brain, spleen and testis, Endocrinology 95:897 (1974).

A NEURAL THYROID HORMONE RECEPTOR GENE

Cary Weinberger[O], David J. Bradley[O#], Linda S. Brady[+],
Catherine C. Thompson[*] and Ronald M. Evans[*#]

[O]Laboratory of Cell Biology, [#]Howard Hughes Medical
Institute, and [+]Unit on Functional Neuroanatomy, NIMH,
Bethesda, MD 20892. [*]Gene Expression Laboratory, The Salk
Institute for Biological Studies, San Diego, CA 92138.

One of the prominent questions surrounding tissue-specific gene
activation is how a single hormone type such as thyroxine can have such
diverse physiological effects. Generally, two mechanisms contribute to the
particular variety of proteins synthesized either during development or in
response to required physiological changes. On the one hand, specific DNA
tertiary structure induced by associated nuclear proteins probably presets the
transcriptional activity of target cell gene networks (1). An additional
constraint is likely provided by hormones or growth factors mediating
changing gene expression patterns (2). Each cell produces distinct receptor
proteins which determine the effective response to hormonal stimulation.
In this manner, both the ontogenetic history of a particular cell type and the
hormone receptor field, or its distribution in specific cell types, limit the scope
of induced proteins during animal development and homeostasis.

Steroid and thyroid hormones carry out this latter transcriptional
discriminatory function by specific ligand binding to discrete nuclear receptor
proteins (3, 4). It is thought that the hormone binding triggers a receptor
structural change which unmasks its latent DNA-binding properties (5-7).
The higher affinity for DNA enhancer elements, regions specifically bound by
activated receptors, is then translated into a transcriptional modulatory mode
by a largely unknown mechanism. Using GH_1 cells, growth hormone
synthesis was shown to be increased in a synergistic fashion by both
glucocorticoids and thyroid hormones occuring primarily at the
transcriptional level (8-10).

Tissue-specific activation of gene expression is perhaps no more
apparent than in the study of thyroid hormones. They exert a panoply of
effects including modulating cardiac function, regulation of enzyme activities
such as malic enzyme and Na^+/K^+-ATPase, and the stimulation of cell
growth, thermogenesis and oxygen consumption (11). Most of these
regulated activities are tissue-specific so that oxygen uptake in peripheral
tissues is stimulated by thyroid hormones, while in the brain no change is
detected. Yet the brain, when deprived of thyroid hormone, succumbs to

dramatic developmental defects such as a marked reduction of neuronal arborization in the neonatal cerebellum (12). In addition, thyroid hormone endocrinopathies often manifest as various mood disorders suggesting they play a role in modulating normal neuronal activity (13).

Since the brain contains receptors and is a developmental and homeostatic target for thyroid hormones, a mechanism must exist which discriminates between neural and peripheral tissue-specific functions under thyroid hormone control. We describe here the properties of a thyroid hormone receptor predominantly expressed in the rat brain. We also present in situ hybridization histochemistry data outlining the brain subregions containing this receptor mRNA. Its characterization implies a multiple receptor system which may help to explain some of the tissue-specific physiological effects of thyroid hormones.

The c-Erb-A Gene Encodes a Thyroid Hormone Receptor

Biochemical characterization of thyroid hormone-binding species from various tissue sources revealed these to be proteins with high-affinity for thyroid hormones which are found predominantly in cell nuclei (14-16). Purification of the thyroid hormone receptor protein has been hampered by its low abundance in cells, but biochemical analysis and photoaffinity labelling studies indicate that two nuclear polypeptides of 57 and 47 kD specifically bind ^{125}I-T_4 (17-20).

Isolation of cDNAs encoding the glucocorticoid, estrogen, progesterone and vitamin D receptors using specific antibodies against each revealed their amino acid similarity to the v-erb-A oncogene product (21-24). This retroviral product encoded by avian erythroblastosis virus does not cause transformation per se, but facilitates transformation of chicken erythroblasts by preventing their spontaneous differentiation and decreasing the latency of leukemias (25). Its structural similarity with steroid receptors suggested that erb-A may encode a trans-acting factor. Identification of the cellular homolog of the viral erb-A oncogene product as a protein which binds thyroid hormones with high affinity indicates that it may behave as a transcriptional activator protein (26, 27). In addition, c-erb-A is localized in the nucleus and the polypeptide synthesized in vitro binds to the thyroid hormone-responsive DNA element in the rat growth hormone gene providing further support for its assignment as a functional thyroid hormone receptor (26, 28). Multiple c-erb-A genes identified in the human genome point to the possibility of different receptor species encoded by perhaps as many as four distinct genes (27, 29, 30).

Isolation of a Rat c-Erb-A cDNA

Human c-erb-A cDNA pheA4/12 (27) isolated from placenta and encoding a protein of 456 amino acids was used to screen a rat brain cDNA library (31). A unique cDNA rbeA12 with an open reading frame of 410 amino acids was identified which maintained 85% amino acid identity overall with the human gene product (Figure 1). Homologous regions between the human and rat receptors were confined to those sequences

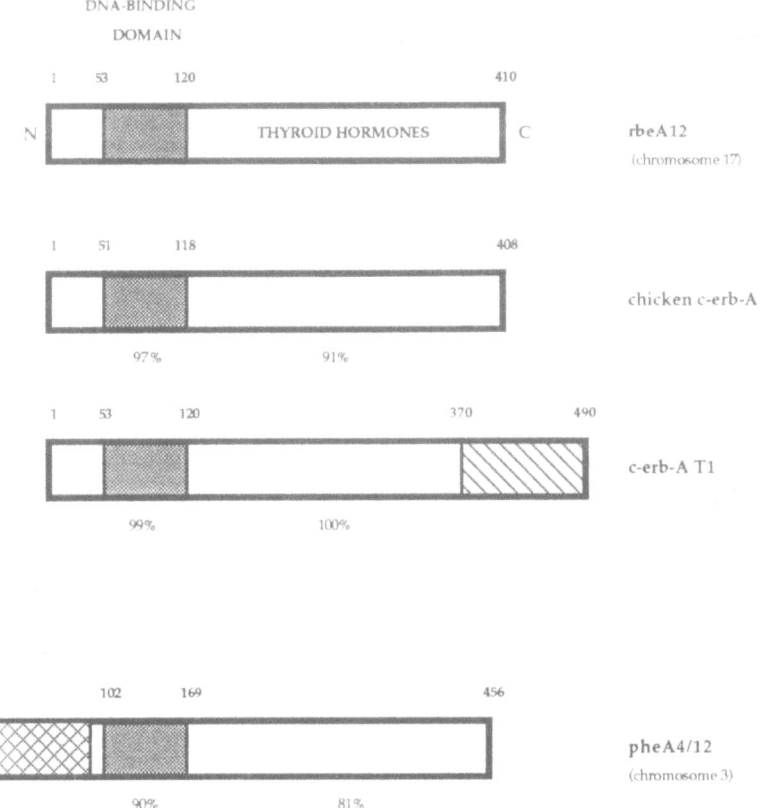

Figure 1. Schematic comparison between various thyroid hormone receptor gene products. Boxes with numbers above indicate the predicted amino acids encoded by identified c-erb-A cDNAs. The degree of amino acid identity between the vertical lined region of a given c-erb-A and the rat species is shown below the boxed regions. The three "neural-type" receptors which map to human chromosome 17 (26, 31, 32) are physically separated from the receptor identified from human placenta and mapped to chromosome 3. Shaded portions specify the region of conserved cysteine residues forming the DNA binding domain found in all steroid and thyroid hormone receptors. The hormone binding region is also indicated. The cDNA c-erb-A T1 contains a carboxyl terminus divergent from the rat species at amino acid 370 and is represented by diagonally-lined region (32). Non-homologous amino terminal residues of pheA4/12 are indicated by hatching (27). N and C refer to amino and carboxy-termini, respectively.

including a cysteine-rich stretch of amino acids comprising the DNA binding domain and downstream into the hormone binding region. However, the amino terminal residues were completely divergent upstream of the conserved cysteine-rich region. These tentative functional domains were assigned by their homology with those experimentally-determined domains of the glucocorticoid and estrogen receptors (7, 33, 34). In addition, a systematic comparison of hormone binding properties of the viral oncogene product and the chicken c-erb-A polypeptide reveal that the hormone binding domain is in the carboxyl terminus (35).

To examine whether the human placenta and rat c-erb-A cDNAs represented different genes, parallel samples of human placenta DNA digested with multiple restriction endonucleases were hybridized with the cysteine-rich regions from both erb-A species (31). Different hybridization patterns were found consistent with each species corresponding to a distinct gene. Chromosome assignment was determined by hybridization of the rat brain c-erb-A cDNA with laser-sorted human chromosomes. This rat brain c-erb-A detected a homologous human gene only on chromosome 17, not on chromosome 3 where the human placenta c-erb-A gene had been previously mapped (27, 31).

Variants of the rat neural c-erb-A cDNA rbeA12 have also been described (33). A human testis cDNA c-erb-A T1 has been isolated which is identical with rbeA12 at the amino acid level suggesting that they correspond to the same gene (Figure 1). We have also identified a homologous human gene erbA8.7 from a kidney cDNA library which is identical to erb-A T1 (unpublished observations). Both the rat and human clones hybridize with the same human genomic fragments by Southern analysis. They are different only at the carboxyl termini where their sequences diverge following amino acid residue 370. The strong homology upstream of the point of divergence suggests that the variant thyroid hormone receptor (T_3R) is derived by alternative RNA processing of a single precursor transcript.

Biochemical Characterization of the Rat Brain c-Erb-A

The polypeptide encoded by the rat c-erb-A cDNA rbeA12 was tested for its ability to bind thyroid hormones (31). Transcripts were synthesized in vitro using SP6 polymerase and translated with rabbit reticulocyte lysates. The rat brain c-erb-A polypeptides generated in vitro bound [^{125}I] 3,5,3'-triiodo-L-thyronine with a dissociation constant of 3 X 10^{-11} M, approximating the binding affinity for the thyroid hormone receptor isolated from tissue culture cells (36, 37). Competition for ^{125}I-T_3 with various thyroid hormone analogs was similar to that found for the human placenta and the chicken c-erb-A polypeptides and excess aldosterone, estrogen, progesterone, testosterone or vitamin D_3 failed to compete for labelled thyroid hormone binding (26, 27, 31).

Rat Thyroid Hormone Receptor mRNA is Abundant in the Brain

We used a DNA fragment derived from the rat brain T_3R gene spanning the cysteine-rich DNA binding region to probe corresponding RNAs from various rat tissues (31). RNA transcripts of 2.6 kb and a variably detected message of 4.4 kb were found with this probe. Presence of the two RNAs may signify two discrete genes, a precursor transcript or alternative RNA processing from a single gene. The T_3R mRNA is most enriched in rat brain, about 10 fold higher than any other tissue tested. It is also found in the heart and kidney in lower amounts, but also detectable in lung, gut, spleen and testes. It is notably absent in liver tissue which could suggest that an additional discrete hepatic receptor gene remains to be identified.

Regional Brain Localization of Thyroid Hormone Receptor mRNA

In order to identify regional structures responsible for the brain hybridization, an anti-sense 48-mer oligonucleotide probe was synthesized corresponding to the rat c-erb-A coding region between the DNA and hormone binding domains. It was 3' end-labelled with ^{35}S-dATP using terminal deoxynucleotidyltransferase (38) and hybridized with RNAs from serial sagittal sections of an adult male rat brain (Figure 2, A-F progressively corresponding to sections from medial to lateral dissection planes). The hybridization pattern demonstrates T_3R mRNA in most brain regions, particularly those areas containing heavy densities of cell nuclei by thionine dye staining methods. Specifically, the mRNA is prominently detected in the outer and inner layers of the cerebral cortex, possibly corresponding to cortical layers II and V, the olfactory cortex, the hippocampus and amygdala, and the granular cell layer of the cerebellum. Intermediate signal was present in the striatum, the pontine nuclei and the nucleus tractus solitarius, hypothalamus

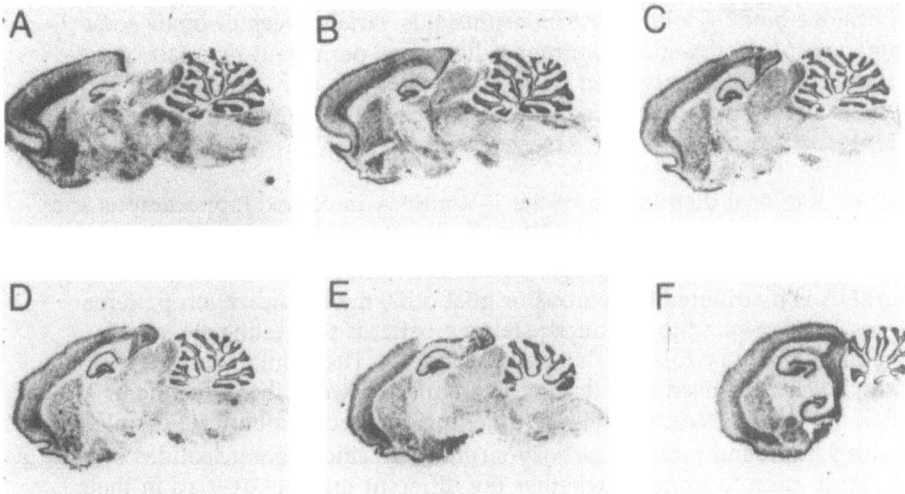

Figure 2. In situ hybridization histochemistry of rat brain sagittal sections using rat neural T_3R-specific oligonucleotide. The euthyroid brain from an adult male Sprague-Dawley rat (250 g) was fixed for in situ hybridization (38) and probed using an ^{35}S-labelled oligonucleotide complementary to bases 682-729 from rbeA12 (31). Hybridized sections were exposed to Kodak XAR5 film for 15 days at room temperature. Sections from medial (A) to lateral (F) dissection planes are presented. Greatest silver grain density corresponds to highest levels of T_3R mRNA.

and midbrain areas including colliculi. The hybridization signal was relatively low in structures such as the thalamus and the middle layers of the cortex, and undetectable in fiber tract areas such as the corpus callosum and anterior commissure.

Functional Importance of Multiple T_3 Receptors

The unexpected finding that two T_3Rs are identified in humans may shed new light on the diverse roles of thyroxine in animal development and physiology. Multiple T_3Rs may signify that discrete subsets of genes are under control of particular receptor proteins. This may be one reason why oxygen consumption in brain is not markedly affected by increased thyroid hormone levels whereas cells in peripheral tissues respond more robustly (39). Isolation of other thyroid hormone receptor genes may clarify whether each receptor species activates different gene subsets in target tissues or whether combinations of subtypes confer transcriptional specificity.

Characterization of alternatively processed neural thyroid hormone receptors at their carboxyl termini may suggest different hormone binding affinities for each which may have distinct functional consequences. Results from Nikodem et al, presented at this conference suggest that the predominant variant form (490 amino acids, corresponding to erb-A T1) of the neural T_3R does not bind hormone. Our hormone binding studies with erbA8.7 agree with this finding. We cannot rationalize these results with previous hormone binding data on erb-A T1 which binds both T_3 and T_4 (33). It is possible that a single mutation is responsible for our inability to detect hormone binding with erbA8.7. Either this variant receptor binds some unknown or unidentified hormonal ligand or perhaps it maintains some transcriptional activity even without the presence of hormone. Alternatively it may code for a receptor-like molecule with transcriptional repressor function.

Regional distribution of the T_3R mRNA in cortex, hippocampus and cerebellum is consonant with ^{125}I-T_3 autoradiographic binding evidence from adult rat brains (40). Although we have not established whether the receptor mRNA is distributed in neurons or glial cells, the hybridization patterns grossly correspond to cell nuclei staining patterns suggesting a general neuromodulatory role for thyroid hormones. The results with the oligonucleotide used here do not distinguish between the authentic or variant T_3R mRNAs. It will be more informative to monitor T_3R mRNA with variant and receptor carboxy-terminal specific oligonucleotides by in situ hybridization to examine whether the different mRNAs overlap in their expression patterns. In addition, what effects altered thyroid states might exert on T_3R mRNA levels remains an area for future investigation.

ACKNOWLEDGMENTS

Our most grateful thanks go to Drs. Eva Mezey and Miles Herkenham for their indulgence and help in describing rat neuroanatomy. In situ hybridization results would not have been possible without the oligonucleotide synthesis expertly performed by Dr. Michael J. Brownstein.

REFERENCES

1. Emerson, B. M., Lewis, C. D., and Felsenfeld, G. (1985). Interaction of specific nuclear factors with the nuclease-hypersensitive region of the chicken adult β-globin gene: nature of the binding domain. Cell 41: 21-30.
2. Yamamoto, K. R. and Alberts, B. M. (1976). Steroid receptors: elements for modulation of eukaryotic transcription. Ann. Rev. Biochem. 45: 721-746.
3. Rousseau, G. G. (1975). Interaction of steroids with hepatoma cell: molecular mechanisms of steroid hormone action. J. Steroid Biochem. 6: 75-89.
4. Oppenheimer, J. H., Schwartz, H. L., Mariash, C. N., Winlaw, W. B., Wong, N. C. W., and Freake, H. C. (1987). Advances in our understanding of thyroid hormone action at the cellular level. Endocrine Reviews 8: 288-308.
5. Hollenberg, S. M., Giguere, V., Segui, P., and Evans, R. M. (1987). Colocalization of DNA-binding and transcriptional activation functions in the human glucocortiocoid receptor. Cell 49: 39-46.
6. Godowski, P. J., Rusconi, S., Miesfeld, R., and Yamamoto, K. R. (1987). Glucocorticoid receptor mutants that are constitutive activators of transcriptional enhancement. Nature 325: 365-368.
7. Kumar, V., Green, S., Stack, G., Berry, M., Jin, J.-R., and Chambon, P. (1987). Functional domains of the human estrogen receptor. Cell 51: 941-951.
8. Shapiro, L. E., Samuels, H. H., and Yaffe, B. M. (1978). Thyroid and glucocorticoid hormones synergistically control growth hormone mRNA in cultured GH_1 cells. Proc. Natl. Acad. Sci. USA 75: 45-49.
9. Spindler, S. R., Mellon, S. H., and Baxter, J. D. (1982). Growth hormone gene transcription is regulated by thyroid and glucocorticoid hormones in cultured rat pituitary tumor cells. J. Biol. Chem. 257: 11627-11632.
10. Evans, R. M., Birnberg, N. C., and Rosenfeld, M. G. (1982). Glucocorticoid and thyroid hormones transcriptionally regulate growth hormone gene expression. Proc. Natl. Acad. Sci. USA 79: 7659-7663.
11. Oppenheimer, J. H. and Samuels, H. H., eds. (1983). "Molecular Basis of Thyroid Hormone Action," Academic Press, New York.
12. Dussault, J. H. and Ruel, J. (1987). Thyroid hormones and brain development. Ann. Rev. Physiol. 49: 321-334.
13. Prange, A. J., Wilson, K., Rubin, A. and Lipton, M. A. (1969). Enhancement of imipramine antidepressant activity by thyroid hormones. Am. J. Psychiat. 121: 457-469.
14. Schadlow, A. R., Surks, M. I., Schwartz, H. L., and Oppenheimer, J. H. (1972). Specific triiodothyronine binding sites in the anterior pituitary of the rat. Science 176: 1252-1254.
15. Oppenheimer, J. H., Koerner, D., Schwartz, H. L., and Surks, M. I. (1972). Specific nuclear triiodothyronine binding sites in rat liver and kidney. J. Clin. Endocrinol. Metab. 35: 330-333.
16. Samuels, H. H. and Tsai, J. S. (1973). Thyroid hormone action in cell culture: demonstration of nuclear receptors in intact cells and isolated nuclei. Proc. Natl. Acad. Sci. USA 70: 3488-3492.

17. De Groot, L. J., Refetoff, S., Strausser, J., and Barsano, C. (1974). Nuclear triiodothyronine-binding protein: partial characterization and binding to chromatin. Proc. Natl. Acad. Sci. USA 71: 4042-4046.

18. Latham, K. R., Ring, J. C., and Baxter, J. D. (1976). Solubilized nuclear "receptors" for thyroid hormones. J. Biol. Chem. 251: 7388-7397

19. Nikodem, V. M., Cheng, S.-Y., and Rall, J. E. (1980). Affinity labeling of rat liver thyroid hormone nuclear receptor. Proc. Natl. Acad. Sci. USA 77: 7064-7068.

20. Pascual, A., Casanova, J., and Samuels, H. H. (1982). Photoaffinity labeling of thyroid hormone nuclear receptors in intact cells. J. Biol. Chem. 257: 9640-9647.

21. Weinberger, C., Hollenberg, S. M., Rosenfeld, M. G., and Evans, R. M. (1985). Domain structure of the human glucocorticoid receptor and its relationship to the v-erb-A oncogene product. Nature 318: 670-672.

22. Green, S., Walter, P., Kumar, V., Krust, A., Bonert, J. M., Argos, P., and Chambon, P. (1986). Human estrogen receptor cDNA: sequence, expression and homology to v-erb-A. Nature 320: 134-139.

23. Jeltsch, J. M., Krozowski, Z., Quirin- Stricher, C., Gronemeyer, H., Simpson, R. J., Garnier, J. M., Krust, A. Jacob, F., and Chambon, P. (1986). Cloning of the chicken progesterone receptor. Proc. Natl. Acad. Sci. USA 83: 5424-5428.

24. Conneely, O. M., Sullivan, W. P., Toft, D. O., Birnbaumer, M., Cook, R. G., Maxwell, B. L., Zarucki-Schulz, Greene, G. L., Schrader, W. T., and O'Malley, B. W. (1986). Molecular cloning of the chicken progesterone receptor. Science 233: 767-770.

25. Frykberg, L., Palmieri, S., Beug, H., Graf, T., Hayman, M. J., and Vennstrom, B. (1983). Transforming capacities of avian erythroblastosis virus mutants deleted in the erbA or erbB oncogenes. Cell 32: 227-238.

26. Sap, J., Munoz, A., Damm, K., Goldberg, Y., Ghysdael, J., Leutz, A., Beug, H., and Vennstrom, B. (1986). The c-erb-A protein is a high-affinity receptor for thyroid hormone. Nature 324: 635-640.

27. Weinberger, C., Thompson, C. C., Ong, E. S., Lebo, R., Gruol, D. J., and Evans, R. M. (1986). The c-erb-A gene encodes a thyroid hormone receptor. Nature 324: 641-646.

28. Glass, C. K., Franco, R., Weinberger, C., Albert, V., Evans, R. M., and Rosenfeld, M. G. (1987). A c-erb-A binding site in the rat growth hormone gene mediates transactivation by thyroid hormone. Nature 329: 738-741.

29. Jansson, M. Philipson, L., and Vennstrom, B. (1983). Isolation and characterization of multiple human genes homologous to the oncogenes of avian erythroblastosis virus. EMBO J. 2: 561-565.

30. Spurr, N. K., Solomon, E., Jansson, M., Sheer, D., Goodfellow, P. N., Bodmer, W. F., and Vennstrom, B. (1984). Chromosomal localisation of the human homologues of the oncogenes erbA and B. EMBO J. 3: 159-163.

31. Thompson, C. C., Weinberger, C., Lebo, R., and Evans, R. M. (1987). Identification of a novel thyroid hormone receptor expressed in the mammalian central nervous system. Science 237: 1610-1614.

32. Benbrook, D. and Pfahl, M. (1987). A novel thyroid hormone receptor encoded by a cDNA clone from a human testis library. Science 238: 788-791.

33. Giguere, V., Hollenberg, S. M., Rosenfeld, M. G., and Evans, R. M. (1986). Functional domains of the human glucocorticoid receptor. Cell 46: 645-652.
34. Rusconi, S. and Yamamoto, K. R. (1987). Functional dissection of the hormone and DNA binding activities of the glucocorticoid receptor. EMBO J. 6: 1309-1315.
35. Munoz, A., Zenke, M., Gehring, U., Sap, J., Beug, H., and Vennstrom, B. (1988). Characterization of the hormone-binding domain of the chicken/thyroid hormone receptor protein. EMBO J. 7: 155-159.
36. Koerner, D., Surks, M. I., and Oppenheimer, J. H. (1974). In vitro demonstration of specific triiodothyronine binding sites in rat liver nuclei. J. Clin. Endocrinol. Metab. 38: 706-709.
37. Spindler, B. J., MacLeod, K. M., Ring, J., and Baxter, J. D. (1975). Thyroid hormone receptors: binding characteristics and lack of hormonal dependenct for nuclear localization. J. Biol. Chem. 250: 4113-4119.
38. Young, W. S., III, Bonner, T. I., Brann, M. R. (1986). Mesencephalic dopamine neurons regulate the expression of neuropeptide mRNAs in the rat forebrain. Proc. Natl. Acad. Sci. USA 83: 9827-9831.
39. Ismail-Beigi, F. and Edelman, I. S. (1971). The mechanism of calorigenic action of thyroid hormone: stimulation of Na^+ and K^+-activated adenosine triphosphatase activity. J. Gen. Physiol. 57: 710-722.
40. Dratman, M., Futaesaku, Y., Crutchfield, F. L., Berman, N., Payne, B., Sar, M., and Stumpf, W. E. (1982). Iodine-125-labeled triiodothyronine in rat brain: evidence for localization in discrete neural systems. Science 215: 309-312.

THYROID HORMONE TRANSPORT FROM BLOOD INTO BRAIN CELLS

Jacob Robbins, Edison Goncalves, Mark Lakshmanan,
and Daniels Foti*

*Clinical Endocrinology Branch, NDDK, NIH
Bethesda, Maryland 20892

The phenomenon of thyroid hormone transport encompasses a number of sequential steps beginning with their secretion by the thyroid follicles and ending with their distribution to multiple sites of metabolism and action within virtually all the body's organs. These steps include equilibration with multiple carrier proteins in the plasma, transcapillary passage of both the hormones and the carrier proteins into extravascular spaces, entry into cells through their external membranes, and translocation to subcellular compartments and organelles. Some that are the subject of current inquiry or controversy, include the interaction of the hormones with "minor" transport protein in plasma, the kinetics of transcapillary movement, the presence of specific receptors for the carrier proteins and the hormones on endothelial and epithelial cell membranes, and specific transport mechanisms within cell organelles.

In most organs, the capillaries or sinusoids are permeable in varying degree to large molecules, so that the hormone-carrier protein complexes can penetrate the endothelium and present themselves to the epithelial membranes. In contrast, the endothelial cells in central nervous system capillaries are completely surrounded by tight junctions, creating the blood-brain barrier that severely limits passage of macromolecules and hydrophyllic molecules[1,2]. Other elements that contribute to the barrier include scanty pinocytotic vesicles in brain endothelial cells and biochemical factors such as active clearance from CSF or brain parenchyma to blood[3]. Although a barrier to macromolecular transport has been clearly demonstrated at the endothelial surface[1,2] this does not necessarily pertain to small, lipophyllic molecules. In theory, the lipophyllic thyroid hormones, once they are dissociated from the carrier proteins, could penetrate this barrier. In vivo infusion or perfusion studies[4,5], however, have shown that the passage of T_4 and T_3 from blood to brain is saturable. It is also stereospecific in that brain uptake of L-T_3 is 3-fold greater than D-T_3[6], and it has been postulated that this selective transport is a property of the capillary endothelium. Since the brain capillaries are in close contact with glial foot processes over virtually their entire surface[1,2,7], it is also possible that the specific transport of thyroid hormones occurs at the glial plasma membrane. As we will show later, specific thyroid hormone transport into glial cells does indeed occur. Until similar studies are done with isolated brain endothelial cells, the exact site of this selective transport at the blood-brain barrier must remain open.

Inasmuch as the concentration of unbound hormone in plasma is too low to account for the amount of hormone that crosses the capillary bed, it has been postulated that a special mechanism must exist to promote release of hormone from the bound state, in particular from plasma albumin[8]. On the other hand, it can be clearly shown that the dissociation rate from each of the transport proteins is fast enough to replenish the pool of unbound hormone during capillary transit[9]. Furthermore, blood cells in the capillary lumen result in mixing of the contents[10] thus preventing laminar flow at the capillary wall and local depletion of the free hormone pool. Since the rate of dissociation from albumin is faster than from PA and, especially from TBG[9], it can be expected that much, if not most, of the T_4 and T_3 that leaves the capillary is derived from the albumin-bound pool even though albumin binds only a small fraction of the total hormone.

The surface area of the capillary bed in the central nervous system is estimated to be 5000 times greater than that of the choroid plexus[11]. Largely for this reason, capillary transport is envisioned as the main route of entry of the thyroid hormones[7]; however, the role of the choroid plexus-cerebrospinal fluid system is poorly understood. It has long been known that T_4, T_3 and their transport proteins are found at low concentrations in the CSF[12]. Interestingly, free T_4 and free T_3 concentrations in CSF are several-fold higher than in plasma[13]. A major constituent of CSF proteins is prealbumin (transthyretin)[12,13], and the ratio of PA to other proteins in CSF is at least 10 times greater than expected from the ratios in plasma[14-16]. This has now been explained by the surprising finding that the epithelial cells of the choroid plexus synthesize PA and apparently secrete it directly into the CSF in the lateral, third and fourth ventricles[17,18]. It has therefore been proposed that this constitutes a special mechanism for the distribution of thyroxine in the CNS[19]. About one-third of the CSF is generated in the ventricles and then flows into the subarachnoid space and the spinal canal[20]. It is thus possible that T_4 is carried from plasma to CSF by newly synthesized PA and thence, unidirectionally, to other regions of the CNS. Kinetic studies in rats[19] showed that intravenously injected T_4 was rapidly and intensely concentrated in the choroid plexus, and then accumulated more slowly in regions of the brain. T_3, which has a lower affinity for PA, was concentrated much less intensely in the choroid plexus, and decreased rapidly with time. The uptake of both T_4 and T_3 into choroid plexus appeared to be nonsaturable.

Kinetic studies in baboons[21] have shown that both T_4 and T_3 are capable of bidirectional transfer across the blood-brain barrier by way of the CSF and the return of the hormones from CSF to plasma favored T_4 over T_3. Passage of T_3 from blood to CSF was more rapid and complete than T_4, which is contrary to the finding in rats.

It has also been shown that intrathecally injected T_4 influences body temperature regulation in dogs[22] and that intrathecally administered T_3 in rats causes greater heart rate stimulation than intravenous T_3[23]. These studies indicate that thyroid hormones can regulate certain body functions by effects exerted at specific sites within the brain in addition to direct effects on the organs themselves and that these sites can be approached through the CSF.

It is of interest that PA is detected in the tela choroidea, the precursor of the choroid plexus in the 11-day rat fetus[24], and thus could have a role in fetal brain development. Anatomical studies on sheep, pig and human fetuses show that epithelial tight junctions develop early in cerebral endothelium and choroid plexus epithelium[25]. At the 50 to 70-days gestation period in the sheep, and the late fetal and early new-born period in the rat, there is development of the blood-brain barrier mechanisms that decrease penetration of hydrophyllic molecules. At the same time the protein concentration in the CSF decreases.

A further complexity in the distribution of thyroid hormones to the CNS is seen in studies of the equilibration of continuously infused labeled hormones with tissues[26]. The ratio of T_3 derived from injected T_4 to T_3 derived from plasma fails to achieve equilibrium with the plasma and other organs, and the brain maintains a ratio higher than in other tissues. Studies of this kind that focus on specific regions of the brain would be of interest. It is also important to note, especially in the context of this symposium, that hormone accumulation in the brain varies during maturation. Brain uptake of T_4 in 10 day old rats is 3 times higher than in 30 day old rats[27].

An additional point of interest is that iodide ion is specifically transported by the choroid plexus in the direction of CSF to blood[28]. This is responsible for the very low ratio of CSF to serum iodide meas-ured in man[29]. This transport mechanism is genetically related to iodide transport in thyroid cells[30].

As noted earlier, glial cells are a potentially important component of the blood-brain barrier. In our laboratory, we are investigating the uptake of thyroid hormone into cultured cells as models of intracellular uptake in the CNS, and we have included a human glioma cell line in these studies. We have also made a point of looking at the uptake of T_4 as well as T_3 since the major source of intracellular T_3 in nerve cells is from monodeiodination of T_4 after it is taken into the cell. In the glioma cells as well as in human and mouse neuroblastoma cells and human medulloblastoma cells, specific transport of thyroid hormones can be demonstrated at the plasma membrane. Thus, nervous system cells share this property with hepatocytes[31], skeletal myoblasts[32] and several other cell types that have been investigated[31,33].

Our initial studies were done with the mouse neuroblastoma cell line NB41A in the cell culture medium, RPMI 1640. Although 33% of the labeled T_3 accumulated by the cells in 2 hours could be blocked by a saturating concentration of T_3 (10µM), we were unable to demonstrate any saturable uptake of T_4 (Fig 1). When the culture medium was replaced by Hank's physiological salt solution, however, the total uptake of both T_3 and T_4 was 3-fold higher and both T_3 and T_4 uptake was suppressed by 10µM hormone (77% and 25%, respectively). We then showed that the components of RPMI responsible for the interference with thyroid hormone uptake were amino acids of the L- system. As shown in Fig 2, saturable uptake of T_4 was completely blocked by phenylalanine at the concentration present in RPMI, 0.09mM. Isoleucine (0.4mM) and cycloleucine (10mM) were equally effective. Amino acids of the other classes were less inhibitory, as

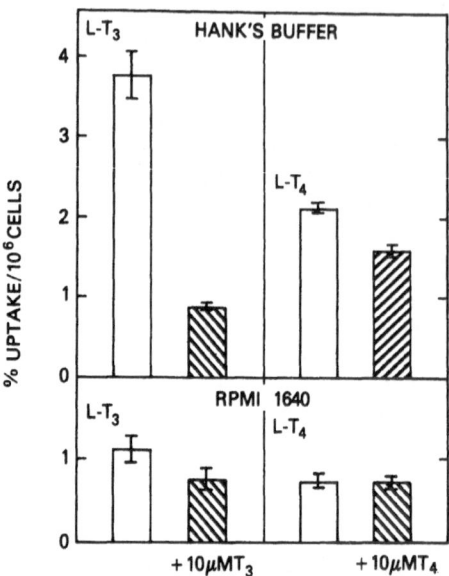

Fig 1. Effect of incubation medium and excess hormone on L-T$_3$ and L-T$_4$ uptake by mouse neuroblastoma cells (NB4IA3). After 45 min preincubation confluent cells were incubated at 37°C for 2 h with 10 pM ^{125}I-labeled thyroid hormone in the absence or presence of 10μM unlabeled hormone. Wells measuring 28.2 cm^2 contained 4 ml of culture medium (RPMI 1640) or Hank's balanced salt solution. After washing with phosphate buffered saline + 0.1% serum albumin, the cells were harvested and their radioactivity measured. Mean ± SEM of 3 experiments performed in duplicate. Experiments were performed under diminished light to avoid spontaneous deiodination especially of T$_4$, in RPMI due to riboflavin and folic acid.

illustrated in Fig 2, and the combined effect of multiple amino acids was required to reach the inhibition of T$_3$ uptake observed in RPMI. It thus appears possible that cellular uptake of T$_3$ and T$_4$ in the CNS in vivo may be affected by variations in amino acid levels.

The accumulation of T$_3$ and T$_4$ by a monolayer culture of human neuroblastoma cells in Hank's buffer is illustrated in Fig 3. Both hormones were rapidly taken up by the cells. The levels reached at 2 hours (13% and 10% of the total) were higher than in the mouse neuroblastoma, and saturation by excess hormone was more complete (90% and 73%). The saturable uptake plateaued after 60 min and was 37% less for T$_4$ than for T$_3$. The saturable uptake is dependent on cell energy as demonstrated by almost complete inhibition in the presence of antimycin or oligomycin, drugs that interfere with ATP production. The dose response to antimycin is shown in Fig 4.

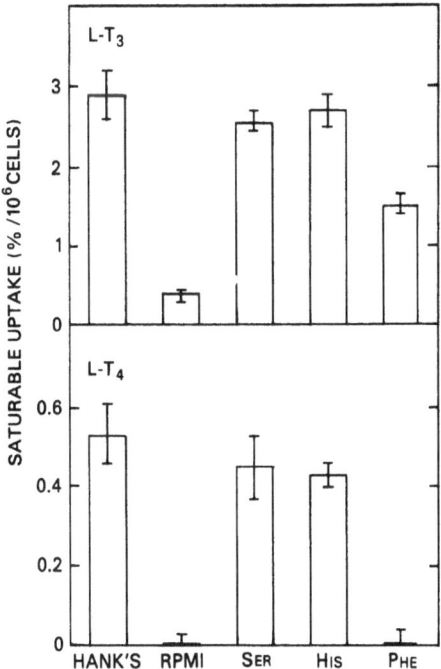

Fig 2. <u>Effect of amino acids on L-T$_3$ and L-T$_4$ saturable uptake by mouse neuroblas toma cells (NB41A3).</u> Confluent cells were preincubated for 45 min and incubated for 2 h in Hank's buffer containing 0.38mM serine, 0.10mM histidine or 0.09mM phenyl alanine. Other conditions as in Fig 1. Saturable uptake is defined as the difference between total uptake and uptake in the presence of excess T$_3$ or T$_4$.

A key question is whether the hormone accumulated by the cells actually crosses the plasma membrane. In fibroblasts, this has been demonstrated directly by fluorescence microscopy[34] showing that rhodamine-labeled T$_3$, after attaching to the cell surface, moves into coated pits and then enters the cytoplasm via coated vesicles. This energy-dependent phenomenon can be blocked by endocytosis inhibitors such as monodansylcadaverine (MDC). We have shown that MDC also inhibits T$_3$ and T$_4$ uptake by mouse neuroblastoma and human medulloblastoma and glioma cells. Some of these data are given in Fig 4. They indicate that thyroid hormone uptake by cells of the CNS probably involves a receptor-mediated endocytosis mechanism.

Another approach to demonstrating intracellular penetration is to examine nuclear uptake in the cells. After incubation of neuroblast monolayers with T$_3$ or T$_4$, the cells are disrupted and the nuclei are isolated and their hormone content measured. We have shown that the accumulation of labeled hormone in nuclei is strongly diminished in the presence of antimycin or MDC. These agents, however, do not inhibit uptake by isolated nuclei. Further evidence is presented in Fig. 5, based on the apparent affinity of nuclear receptors after whole cell incubations compared to the affinity determined with previously isolated nuclei. It can be seen that the receptor affinity for L-T$_3$ and L-T$_4$ measured in intact cells is more than 3 times that in isolated nuclei. This can be attributed to a step-up in the concentration of free hormone in the cytoplasm over the concentration in the medium, the latter value being used in the calculation of Ka in intact cells.

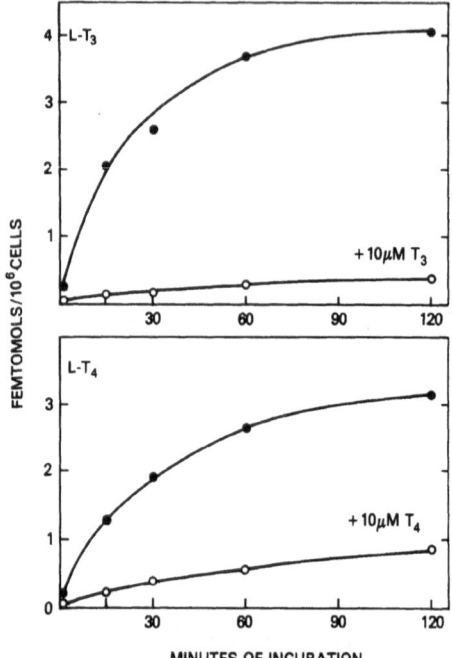

Fig 3. Time course of L-T_3 and L-T_4 total and non-saturable uptake by human neuroblastoma cells (SK-N-SH). Cells were incubated at 25°C in Hank's buffer with 30 pM ^{125}I-labeled hormone. Wells measuring 9.6 cm^2 contained 1 ml of buffer. Other conditions as in Fig 1. Saturable uptake of L-T_3 and L-T_4 were 90% and 73% of total uptake, re spectively, and saturable L-T_4 uptake at 2 h was 37% less than saturable L-T_3 uptake.

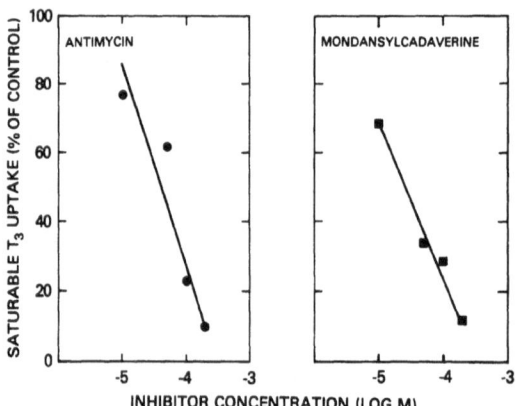

Fig 4. Inhibition of L-T_3 uptake in mouse neuroblastoma cells by antimycin and monodansylcadaverine. Confluent cells were preincubated for 30 min at 37°C in RPMI 1640 containing the inhibitor. Incubation after addition of 25pM L-T_3 was for 30 min. Wells measuring 28.2 cm^2 contained 4 ml of medium. Other conditions as in Fig 1.

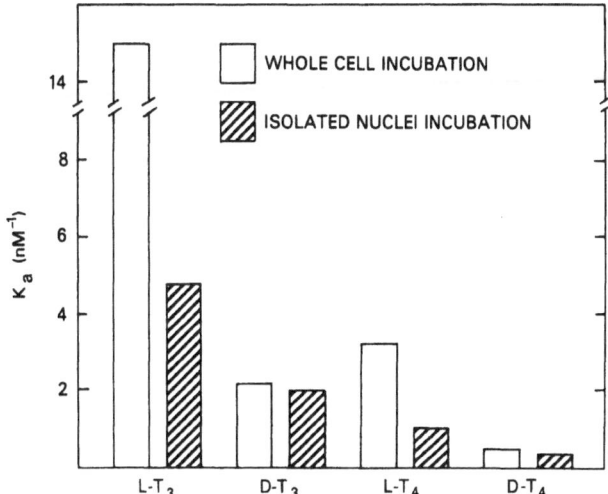

Fig 5. <u>T3 and T4 nuclear receptor affinity in mouse</u>
<u>neuroblastoma (NB41A3)</u>. Intact cells or isolated nuclei
were incubated for 120 min in RPMI or 30 min in 0.25 M
sucrose-25mM Tris-1 mM $MgCl_2$ with 25pM labeled L-T_3, D-T_3,
L-T_4 or D-T_4 and increasing amounts of unlabeled
hormone. Radioactivity in nuclei was counted and Scatchard
analysis of receptor-bound hormone was done using LIGAND
program. Nuclear affinity (Ka) for L-isomers was approximately 2x higher than for D-isomers when measured in isolated
nuclei. With intact cells, Ka was more than 3-fold higher
than with isolated nuclei for both L-isomers, but
remained unchanged for D-isomers.

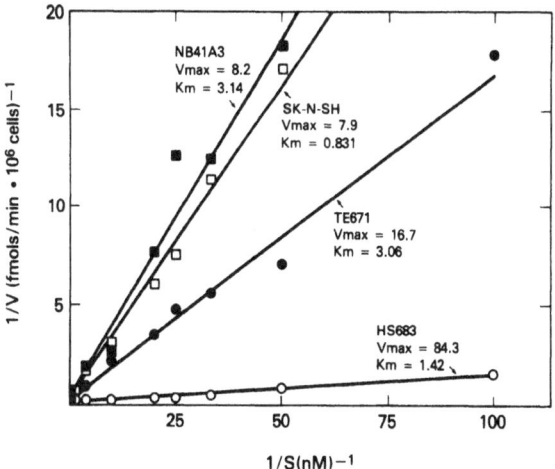

Fig 6. <u>Analysis of the initial L-T4 uptake rate in</u>
<u>mouse neuroblastoma NB41A3, and human neuroblastoma</u>
<u>SK-N-SH, medulloblastoma TE671, and glioma HS683</u>.
Cells were incubated for 1 min at 25°C in Hank's buf-
fer with 30pM L-T_4 and increasing amounts of unlabeled
T_4. Other conditions as in Fig 3.

Fig. 5 also shows that the nuclear receptor affinity for $D-T_3$ and $D-T_4$ measured with intact cells is not different from that determined with isolated nuclei. This indicates that the transport mechanism at the plasma membrane is stereospecific, and that the transport of $L-T_3$ and $L-T_4$ is "active", as defined by accumulation against a concentration gradient.

In further experiments, we measured the initial saturable cell uptake at 1 min as a function of hormone concentration in the medium. The data in Fig 6 are plotted according to Lineweaver and Burk. The Km values for T_4 uptake in all four cell lines were similar, ranging from 0.8 to 3.1nM. Maximum velocities, however, were widely disparate, being lowest for the human neuroblastoma cells (Vmax=7.9 fmols/min/10^6 cells) and highest for the human glioma cells (Vmax=84 fmols/min/10^6 cells).

The strikingly different Vmax obtained with glioma cells suggests a special role for T_4 transport in these cells. Saturable cell uptake (31.6 fmols/10^6 cells at 60 min) was much higher than in human neuroblasts (4.4 fmols/10^6 cells) but the difference in nuclear uptake was much less (0.383 fmols/10^6 cells in glioma cells compared to 0.105 fmols/10^6 cells in neuroblasts). Since antimycin and MDC depressed the specific uptake in glioma cells, we can conclude that the large amount of T_4 that they accumulate is located within the cytoplasm. Further studies are required to determine whether this is simply the result of increased cytoplasmic binding or whether it might be related to a transcellular transport function of these cells. The latter possibility is consistent with the finding that the apparent affinity of the nuclear receptor decreased from 2.5 nM^{-1} to 0.22 nM^{-1} in the presence of antimycin, presumably reflecting a 10-fold decrease in free T_4 concentration within the cytoplasmic compartment. The high free T_4 concentration in the unblocked glial cell could provide a gradient for transport out of the cell if the plasma membrane transport into the cell is polarized.

The cultured cells that we have been studying are immature and may therefore be models for the developing brain rather than the mature brain. The mouse neuroblastoma cells, however, can be induced to differentiate by exposure to 0.5mM sodium butyrate (Fig 7). This causes an increase in cytoplasm: nucleus ratio, neurite outgrowth and induction of tyrosine hydroxylase and other enzymes[35,36]. As shown in Fig 8, butyrate caused an increase in T_3 uptake at 2 hours. The initial (1 min) uptake was also increased. The apparent Ka of nuclear binding in intact cells with the finding that the apparent affinity of the nuclear receptor was decreased by butyrate but the Ka in isolated nuclei was not affected. The apparent inconsistency of an increased intracellular transport but a decreased intracellular free T_4 concentration is unexplained. As described by others in developing rats[37] and in cultured rat glial cells[38] we also found that neuroblast differentiation was accompanied by an increase in nuclear receptor number.

In our studies with CNS cells, and also with myoblasts[32], we have found no evidence for a separate transport mechanism at the nuclear envelope. Oppenheimer and his coworkers, however, have concluded that there is nuclear membrane transport of T_3 in hepatocytes and other tissues, including brain[39,40]. In mature neurons, Dratman and colleagues[41] have shown that intravenously injected [^{125}I]T_3 is concentrated in the synaptosomal fraction of whole rat brain (minus cerebellum) and that the concentration is higher in synaptosomes than in the brain cytosol fraction. They have also shown[42] that [^{125}I]T_4 injection is followed by accumulation of [^{125}I]T_3 in synaptosomes, and that the labeled T_3/T_4

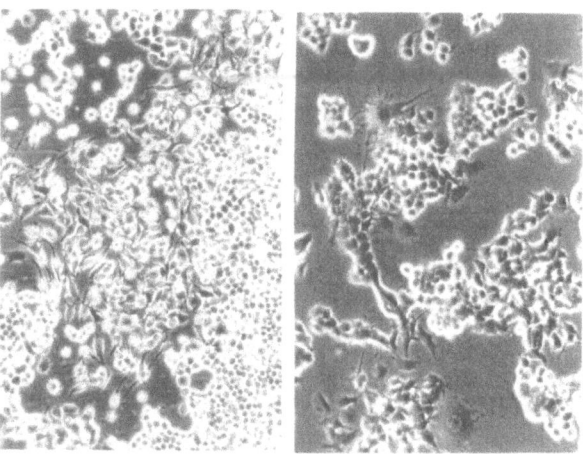

Fig 7. <u>Differentiation of mouse neuroblastoma
NB41A3 cells induced by butyrate</u>. Left panel:
Undifferentiated cells grown for 4 days in RPMI
1640 + 5% fetal bovine serum. Right panel:
Differentiated cells grown for 7 days in the
presence of 0.5 mM sodium butyrate.

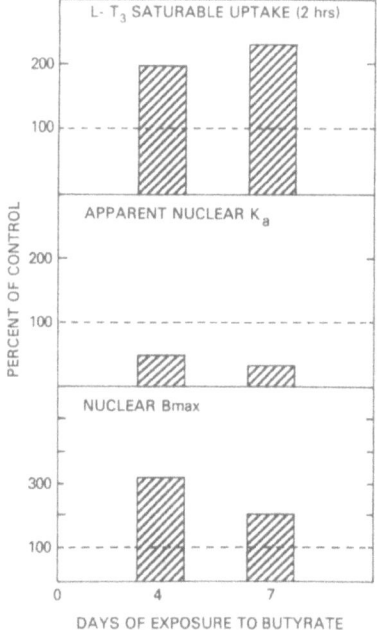

Fig 8. <u>Effect of butyrate on cell
uptake, nuclear receptor affinity (Ka)
and maximal binding capacity (Bmax)
for L-T$_3$ in intact mouse neuroblastoma
(NB41A3)</u>. Cells were grown in RPMI
1640 containing 0.5 mM sodium butyrate
for 4 or 7 days. Then intact cells
were incubated with 25pM [^{125}I]L-T$_3$
and Scatchard analysis performed
(see Legends to Figs 1 & 4).

ratio is higher in synaptosomes than in brain cytosol. No labeled T_3 was detected in plasma. Although they conclude that the data favor the uptake of T_4 by nerve endings, and conversion there of T_4 to T_3, they also raise the possibility that T_3 is generated from T_4 in other compartments of the neuron - or even elsewhere in the CNS - and is subsequently translocated to the nerve ending. This would be consistent with the known transport of metabolites from the nerve cell body peripherally along the axon[43]. It is also possible, however, that hormone entering the synaptosome is transported in retrograde fashion toward the cell body[43].

This report has briefly summarized what we now know about thyroid hormone transport to the central nervous system. The data are still sketchy and much remains to be done. Obviously the brain is a complex organ and major differences in thyroid hormone transport and metabolism are to be expected in its constituent parts, so study of different brain regions as well as different cell types is required. We also need to distinguish between findings in the mature brain and those during fetal and postnatal development. Thyroid hormones play a very different role in these stages of the organism, and possible variations in hormone delivery to cells may contribute to these differences. Finally, in the malnutrition that often accompanies iodine deficiency, we need to ask whether PA synthesis in the choroid plexus is compromised, as it is in the liver. If so, important effects in thyroid hormone delivery to the brain may be expected.

REFERENCES

1. M.W. Brightman, Morphology of blood-brain interfaces, Exp Eye Res (suppl) 25:1-25 (1977).
2. M. Bradbury, The Concept of the Blood-Brain Barrier, John Wiley and Sons, Chichester (1979).
3. E. Levin, Are the terms blood-brain barrier and brain capillary permeability synonomous?, Exp Eye Res (suppl) 25:191-199 (1977).
4. G.A. Hagen and L.A. Solberg, Jr., Brain and cerebrospinal fluid permeability to intravenous thyroid hormone, Endocrinology 95:1398 (1974).
5. W.M. Pardridge, Carrier-mediated transport of thyroid hormones through the rat blood-brain barrier: primary role of albumin-bound hormone, Endocrinology 105:605-612 (1979).
6. T. Terasaki and W.M. Pardridge, Stereospecificity of triiodothyronine transport into brain, liver and salivary gland: role of carrier- and plasma-protein mediated transport, Endocrinology 121:1185-1191 (1987).
7. W.M. Pardridge, Receptor-mediated peptide transport through the blood-brain barrier, Endocrine Rev 7:314-330 (1986).
8. W.M. Pardridge and E.M. Landow, Tracer kinetic model of blood brain barrier transport of plasma protein-bound ligands. Empiric testing of the free hormone hypothesis, J Clin Invest 74:745-752 (1984).
9. J. Robbins and M.L. Johnson, Possible significance of multiple transport proteins for the thyroid hormones, in: "Free Hormones in Blood", A. Albertini and R.P. Ekins, eds., Elsevier Biomedical Press, Amsterdam (1982).
10. J. Prothero and A.C. Burton, The physics of blood flow in capillaries: the nature of the motion, Biophys J 1:565 (1961).
11. C. Crone, The blood-brain barrier: facts and questions, in: "Homeostasis of the Brain", B.K. Seisjo, S.C. Sorensen, eds., Munksgaard, Copenhagen (1971).

12. J. Robbins and J.E. Rall, Proteins associated with the thyroid hormones, Physiol Rev 40:415-489 (1960).

13. G. Hagen and W.J. Elliott, Transport of thyroid hormones in serum and cerebrospinal fluid, J Clin Endocrinol Metab 37:415-422 (1973).

14. K. Felgenhauer, Protein size and cerebrospinal fluid composition, Klin Wochenschr 52:1158-64 (1974).

15. B. Weisner and U. Kanerz, The influence of the choroid plexus on the concentration of prealbumin in CSF, J Neurolog Sci 61:27-35 (1983).

16. G. Schreiber, Synthesis, processing and secretion of plasma proteins by the liver and other organs and their regulation, in: "The Plasma Proteins: Structure, Function & Genetic Control", F.W. Putman, ed., Academic Press, Orlando (1987).

17. P.W. Dickson, A.R. Aldred, P.D. Marley, D. Bannister and G. Schreiber, Rat choroid plexus specializes in the synthesis and the secretion of transthyretin (prealbumin), J Biol Chem 261:3475-3478 (1986).

18. J. Herbert, J.N. Wilcox, K.T.C. Pham, R.T. Fremeau, Jr., M. Zeviani, A. Dwork, D.R. Soprano, A. Makover, de W.S. Goodman, E.A. Zimmerman, J.L. Roberts and E.A. Schon, Transthyretin: a choroid-plexus specific transport protein in human brain, Neurology 36:900-911 (1986).

19. P.W. Dickson, A.R. Aldred, J.G.T. Mentiny, P.D. Marley, W.H. Sawyer and G. Schreiber, Thyroxine transport in choroid plexus, J Biol Chem 262:13907-13915 (1987).

20. H.F. Cserr, Physiology of the choroid plexus, Physiol Rev 51:273-367 (1971).

21. B. Chernow, K.D. Burman, D.L. Johnson, R.A. McGuire, J.T. O'Brian, L. Wartofsky and L.P. Georges, T_3 may be a better agent than T_4 in the critically ill hypothyroid patient: evaluation of transport across the blood-brain barrier, Critical Care Med 11:99-104 (1983).

22. H. Kaciuba-Uscilko, J. Sobocinska, S. Kozlowski and A.W. Ziemba, The effect of intraventricular thyroxine administration on body temperature in dogs at rest and during physical exercise, Experentia 32:351-352 (1975).

23. M. Goldman, M.B. Dratman, F.L. Crutchfield, A.S. Jennings, J.A. Maruniak and R. Gibbons, Intrathecal triiodothyronine administration causes greater heart rate stimulation than intravenously delivered hormone, J Clin Invest 76:1622-1625 (1985).

24. M. Kato, D.R. Soprano, A. Makover, K. Kato, J. Herbert and de W.S. Goodman, Localization of immunoreactive transthyretin mRNA in fetal and adult rat brain, Differentiation 31:228-235 (1986).

25. N.B. Saunders, Ontogeny of the blood-brain barrier, Exp Eye Res (suppl) 25:523-550 (1977).

26. M-J. Obregon, F. Roelfma, G. Morreale de Escobar, F. Escobar del Rey and A. Querido, Exchange of triiodothyronine derived from thyroxine with circulating triiodothyronine as studied in the rat, Clin Endocrinol 10:305-315 (1979).

27. E. Vigouroux, J. Clos and J. Legrand, Uptake and metabolism of exogenous and endogenous thyroxine in the brain of young rats, Horm Metab Res 11:228-232 (1979).

28. J. Wolff, Transport of iodide and other anions in the thyroid gland, Physiol Rev 44:45 (1964).

29. J.B. Alpers and J.E. Rall, The metabolism of iodine in the cerebrospinal fluid, J Clin Endocrinol Metab 15:1482 (1955).

30. J. Wolff, R.H. Thompson and J. Robbins, Congenital goitrous cretinism due to absence of iodide-concentrating ability, J Clin Endocrinol Metab 24:699-707 (1964).

31. E.P. Krenning and R. Docter, Plasma membrane transport of thyroid hormone, in: "Thyroid Hormone Metabolism", G. Hennemann, ed., p. 107-131 (1986).

32. A. Pontecorvi, M. Lakshmanan and J. Robbins, Intracellular transport of 3,5,3'-triiodo-L-thyronine in rat skeletal myoblasts, Endocrinology 121:2145-2152 (1987).

33. S-Y. Cheng, Characterization of binding and uptake of 3,3',5-triiodothyronine in cultured mouse fibroblasts, Endocrinology 112:1754-1762 (1983).

34. S-Y. Cheng, F.R. Maxfield, J. Robbins, M.C. Willingham and I. Pastan, Receptor mediated uptake of 3,3',5-triiodo-L-thyronine by cultured fibroblasts, Proc Natl Acad Sci USA 77:3425 (1980).

35. K.N. Prasad, Differentiation of neuroblastoma cells in culture, Physiol Rev 50:129-265 (1975).

36. K.N. Prasad, Butyric acid: a small fatty acid with diverse biological functions, Life Sciences 27:1351-1358 (1980).

37. T. Valcana and P.S. Timiras, Nuclear triiodothyronine receptors in the developing rat, Molec Cell Endocrinol 11:31-41 (1978).

38. J. Ortiz-Caro, F. Montiel, A. Pascual and A. Aranda, Modulation of thyroid hormone nuclear receptors by short chain fatty acids in glial C6 cells, J Biol Chem 261:13997-14004 (1986).

39. J.H. Oppenheimer and H.L. Schwartz, Stereospecific transport of triiodothyronine from plasma to cytosol and from cytosol to nucleus in rat liver, kidney, brain and heart, J Clin Invest 75:147-154 (1985).

40. A.D. Mooradian, H.L. Schwartz, C.N. Mariash and J.H. Oppenheimer, Transcellular and transnuclear transport of 3,5,3'-triiodothyronine in isolated hepatocytes, Endocrinology 117:2449-2456 (1985).

41. M.B. Dratman, F.L. Crutchfield, J. Axelrod, R.W. Colburn and T. Nguyen, Localization of triiodothyronine in nerve ending fractions of rat brain, Proc Natl Acad Sci USA 73:941-944 (1976).

42. M.B. Dratman and F.L. Crutchfield, Synaptosomal [^{125}I]triiodothyronine after intravenous [^{125}I] thyroxine, Am J Physiol 235:638-647 (1978).

43. B. Grafstein and D.S. Forman, Intracellular transport in neurons, Physiol Rev 60:1167-1283 (1980).

THYROID HORMONE RECEPTORS IN THE DEVELOPING BRAIN

Philippe De Nayer and Béatrice Dozin*

Hormone and Metabolic Research Unit
University of Louvain Medical School
Avenue Hippocrate 75, B-1200 Brussels, Belgium

Introduction

Most if not all the effects of thyroid hormones result from the interaction of the hormone with specific nuclear receptors [1]. The receptors have recently been identified as the product of the c-erb-A proto-oncogene [2,3]. Several sub-types have been characterized [4,5]. Interestingly, the expression of the proto-oncogene may vary in different tissues. The receptor appears to be inserted in the chromatin as a complex with associated proteins [6]. The nature of these nuclear proteins and their possible role in the modulation of the receptor activity are ill-defined. They could be involved in the regulation of the properties of the receptor, its affinity for the chromatin and its affinity for the hormones through phosphorylation, ribosylation, acetylation induced a.o. by thyroid hormones [7-9].

Among the target organs the central nervous system presents the unique feature of being strictly dependent on thyroid hormones for structural organization and function during a well-defined period of development. The mature brain appears to have lost responsiveness to thyroid hormones, as measured by O_2 consumption, mitochondrial α-GPD activity and cytosolic malic enzyme stimulation although the high levels of receptor establish the cerebral tissue as a target for thyroid hormones. These observations lead to the proposal that in brain and liver thyroid hormones may regulate the expression of different genes. The regulation of gene expression in a given cell may differ depending whether the brain is still developing or has acquired maturity; in addition, the response to thyroid hormones may vary according the cell type.

The alterations in morphological organization of the brain resulting from hypothyroidism have been documented [10-13]. Numerous biochemical parameters are affected by altered thyroid states (for review see 14-16). Recent biochemical data on the effect of thyroid hormones on nerve cell differentiation indicate that they regulate microtubule assembly by changing the concentration and/or the activity of MAPS (tau fraction) [17]. The critical period of effectiveness of thyroid hormones in brain maturation raises a special problem. The correct organization of

---

*Present address: Istituto Nazionale per la Ricerca sul Cancro, Genova, Italy.

the neuronal network depends on the precise synchronization of three major steps: cell division, migration and differentiation. The primary effect of thyroid hormones on these processes implies the control of gene expression in different cell types. This control is probably exerted in a specific set of few genes. An approach to gain insight in this regulatory mechanism includes the analysis of the receptors in different cell types, and ideally, in a given cell type at different stages of maturation.

In this review, we will focus on the nuclear receptors. Cytosol binding sites for thyroid hormones have been described in the developing brain. They show high Ka [18,19] or number [20] at birth and a decrease of these parameters during the postnatal period. A marked variation in T_3 binding activity was observed among different brain areas [21]. A cytosolic binding protein for thyroid hormones has been characterized in fetal rat and mouse brain cells in primary cultures [22-23]. The role of these binding proteins in thyroid hormone action is still not clear. They could be involved in transport, storage or regulatory functions of hormone metabolism. The fact that maturational changes do not occur in the liver is worth mentioning.

Topographical Distribution of the Nuclear Receptors in the Brain

The number of binding sites per mg of DNA varies widely from one tissue to another. Of the tissues studied, the anterior pituitary contains the highest density of receptor sites (6,000 per cell) and the testis the lowest (16 per cell). The number of nuclear binding sites in spleen is small, and close to that in circulatory lymphocytes (about 400 per cell). The adult brain contains about 2,000 binding sites per cell. In neonatal brain their number (4,000 per cell) is higher than in adults and close to that in adult liver [24]. The receptor density -expressed in binding capacity (ng T_3/mg DNA)- is not uniform in different brain areas. The anterior pituitary and the cerebral cortex contain more receptors than the cerebellum and the brain stem. The difference spans from 0.5 ng T_3 to 0.05 ng T_3/mg DNA. The Ka of the receptors in different parts of the brain appears quite similar [25-29].

A further analysis of the distribution of T_3 receptors in rat cortical cells suggests a preferential localization in neurons [26,30]. A similar finding was reported in embryonic and neonatal chick brain [31]. Studies with pure cultures of neurons and astrocytes from embryonic and neonatal chick brain have confirmed the predominant localization of the receptors in neuronal nuclei [32,33]. It is clear that thyroid hormone receptors are not restricted to neuronal cells, but also appear in glial cells [34,35] suggesting that both cell types are potential targets for thyroid hormones. A direct effect of T_3 on protein phosphorylation has indeed been demonstrated in neuronal and glial cells derived from rat cortex [36,37].

In adult rats hypothyroidism does not appear to change the binding properties of the hepatic nuclear recetors [38-39] contrary to what is observed in the brain. The effect of thyroid hormone deficiency on brain nuclear receptors is not uniform: in the anterior pituitary the Ka of the receptors increases [25,40] and their number decreases [25]. In the cerebral cortex however the affinity is significantly diminished [25,41] and the number of receptors increased [25,29,41,42]. The observations point to different regulatory mechanisms acting on thyroid hormone receptors in liver and brain, and within different areas of the brain.

T_3 receptors are present in nuclei from 7 weeks old human embryos [43]. In the human fetal brain high affinity thyroid hormone receptors (Ka 2×10^{10} M^{-1} have been detected at 10 weeks of gestation [44]. Their concentration is low (46 fmol T_3/mg DNA) but a tenfold increase is noted at 16 weeks, coincident with the period of neuroblast multiplication. At that period of development, brain is the only organ to contain high concentrations of T_3 [44]. As pointed out by the authors, it is impossible at the present time to conclude that thyroid hormones play a role in neuroblast proliferation and/or differentiation although the presence of receptor and T_3 might suggest that this is the case [43,44]. Interestingly, the receptor appears to be present in the human embryo before onset of the thyroid function [43]. High affinity receptors have been identified in the lung of fetuses at 13 weeks of gestation; the binding capacity increased by 65% between 12-13 weeks and 16-19 weeks of gestation to reach 420 fmol/mg DNA [45]. Besides lung, liver and heart of 16-18 weeks old fetuses contain a significant number of high affinity T_3 receptors [44] (Table 1).

Table 1. T_3 receptors in human fetal tissues (adapted from Bernal and Pekonen, 1984).

Organ	Age (weeks)	Ka (10^{10} M^{-1})	Receptor conc. (fmol T_3/mg DNA)
Brain	10	1.9	46
	12	1.9	264
	16	2.0	479
	17	1.9	390
	18	3.0	344
Liver	16	1.5	352
Heart	16-18	0.9	201
Lung	16	1.8	544

In rat fetal brain T_3 receptors are present at 14 days of gestational age [46]. After birth the number of receptors increases rapidly in the early post-natal period and remains at high level during the first 2-3 weeks, before decreasing to adult values [29,46-50]. In heart and lung, the T_3-binding capacity is low at birth but increases during maturation [46,51,52].

In chick embryo brain the T_3 maximal binding capacity rises between 9 and 12 days, decreases at 17 days and further declines after hatching [53]. Yet, in an another study of T_3 ontogenesis in chick brain embryos the binding capacity of neuronal nuclei was shown to increase with age (7-11 days) whereas the glial receptor remained at low level until late embryogenesis (19 days). In adult brain, the number of T_3 receptors in neuronal nuclei was 8,000 per nucleus as compared to 1,000 in glial nuclei [31].

In fetal lambs the T_3 receptor concentration increases 2.6-fold from the 50th day to the 82d day of gestation. From 82 to 100 days, the concentration remained constant [54]. In these different species, the changes in receptor numbers correspond to critical periods of brain development (neuroblast proliferation, synaptogenesis).

The changes in nuclear receptor properties during development do not seem restricted to changes in the receptor number. Evidence for qualitative modifications in hormone binding have been reported during maturation. In neonatal rat brain the affinity of the receptor increases starting at the end of the first week (0.7×10^9 M^{-1}), reaching maximal values (1.8×10^9 M^{-1}) at day 20 and to decline to adult values after one month [49]. The affinity for T_4 was not modified [55]. Hepatic receptors did not show these changes in affinity for T_3 [49].

In the brain of foetal lambs, the affinity for T_3 increased from 50 days (1.0×10^{10} M^{-1}) to 80 days (1.9×10^{10} M^{-1}) returning to 50 days value at 100 days [54]. Developmental [54-56] and regional [57] differences in T_3, T_4 and analogs binding by nuclear receptors from rat, chick and lamb brain have been reported. The maturational changes in the properties of the receptor raise the possibility of differential expression of receptors sub-types. They could also be due to a control exerted on the receptor by associated proteins or factors [8,58,59]. Tissue specific age dependent chromatin-associated factors interacting with the receptor are able to change the binding properties of salt-extracted nuclear receptors [60] (Table 1). These data suggest that the modulation of the binding properties of the receptor through chromatin-factors may lead to organ specificity for thyroid hormones and may regulate the expression of thyroid-hormone dependent effects in different target tissues.

Table II. Effect of chromatin-associated factors on the affinity
for T_3 of salt-extracted nuclear receptors
(Ka, 10^9 M^{-1}) Mean $\pm$ SD

A. Brain chromatin	Liver receptors
9 days	0.29 ± 0.18
18 days	0.85 ± 0.65
50 days	0.35 ± 0.64
B. Liver chromatin	Brain receptors
9 days	0.33 ± 0.39
18 days	0.31 ± 0.75
50 days	0.32 ± 0.59

Primary culture of brain cells offers an interesting tool to analyze in vitro the complex events occuring in development. In cultures of cells dissociated from cerebral hemispheres of 14 days old embryonic mice the number of receptors increased during 6 days, the period of neuronal proliferation [32]. The number and the Ka of nuclear T_3 receptors in cultured neurons obtained from the cerebral hemispheres of 15-16 old rat embryos increased between 5 to 9 days of culture, indicating that the in vitro model adequately reflects the in vivo situation, and in addition confirming that the receptor increase is mainly occuring in neurons. In astrocytes the increase in receptor number is less marked and happens only after 21 days of culture [33].

Neonatal hypothyroidism increases the number of T_3 receptors in brain [29,48,50,61,62]. A slight decrease in Ka was observed whereas the administration of T_3 enhanced the Ka and diminished the number of receptors. Hepatic receptors did not change in these conditions [60].

Conclusions

The analysis of the properties of the T_3 nuclear receptors during development reveals several interesting features:

a) The number and the affinity of the receptors in different organs do not follow the same maturational pattern. The effect of thyroid hormone deficiency on receptor number and affinity varies in different organs.

b) The distribution of the receptors and the modulation of their properties is not uniform in different brain areas.

c) Neuronal cells contain more receptors than glial cells.

d) In human and in the animal models (rat, chick, lamb) the surge in receptor number and affinity coincides with critical periods in brain maturation (neuroblasts differentiation, synaptogenesis).

e) The properties of receptors in brain may be due to the presence of a specific sub-type of receptor and/or to chromatin-associated factors.

Finally, it should be stressed that receptors do not as such control the expression of thyroid hormone effects. In addition to post-receptor events, the interpretation of in vitro findings should also take into account differences in cellular transport of T_3 from plasma to cytosol, and from cytosol to nucleus in different tissues [63].

Acknowledgements

The financial support of FRSM (Belgium), and the secretarial help of Mrs Th. Lambert are gratefully acknowledged.

References

1. J.H. Oppenheimer, H.L. Schwartz, C.N. Mariash, W.B. Kinlaw, N.C.W. Wong and H.C. Freake. Advances in our understanding of thyroid hormone action at the cellular level. Endocrine Rev. 8:288-308 (1987).

2. J. Sap, A. Munoz, K. Damm, Y. Goldberg, J. Ghysdael, A. Leutz, H. Beug, B. Vennström. The c-erb-A protein is a high-affinity receptor for thyroid hormones. Nature 324:635-640 (1986).

3. C. Weinberger, C.C. Thompson, E.S. Ong, R. Lebo, D.J. Gruol and R.M. Evans. The c-erb-A gene encodes a thyroid hormone receptor. Nature 324:641-646 (1986).

4. C.C. Thompson, C. Weinberger, R. Lebo and R.M. Evans. Identification of a novel thyroid hormone receptor expressed in the mammalian central nervous system. Science 237:1610-1614 (1987).

5. D. Benbrook and M. Pfahl. A novel thyroid hormone receptor encoded by a cDNA clone from a human testis library. Science 238:788-791 (1987).

6. H.H. Samuels, A.J. Perlman, B.M. Raaka and F. Stanley, Thyroid hormone receptor synthesis and degradation and interaction with chromatin components, in: "Molecular basis of thyroid hormone action" J.H. Oppenheimer and H.H. Samuels, eds., Academic Press, New York (1983).

7. V.M. Nikodem, D.R. Huang, B.L. Trus and J.E. Rall. The effects of thyroid hormone on in vitro phosphorylation, acetylation, and ADP ribosylation of rat liver nuclear proteins. Horm. Metab. Res. 15:550-554 (1983).

8. J. Bismuth, A. Anselmet and J. Torresani. Triiodothyronine nuclear receptor and the role of non-histone protein factors in in vitro triiodothyronine binding. Biochim. Biophys. Acta 840:271-279 (1985).

9. J. Ortiz-Caro, F. Montiel, A. Pascual and A. Aranda. Modulation of thyroid hormone nuclear receptors by short-chain fatty acids in glial C6 cells. J. Biol. Chem. 261: 13997-14004 (1986).

10. J. Legrand. Hormones thyroïdiennes et maturation du système nerveux. J. Physiol. Paris 78:603-652 (1983).

11. J. Legrand. Effects of thyroid hormones on central nervous system development, in: "Neurobehavioral Teratology", J. Yanai, ed., Elsevier Science Publ., Amsterdam (1984).
12. J.M. Lauder and M.C. Bohn. Thyroid hormones and corticosteroids as temporal regulators of postnatal neurogenesis in the cerebellum and hippocampus, in: "Progress in Psychoneuroendocrinology", F. Brambilla, G. Racagni and D. de Wied, eds., Elsevier Biomedical Press, Amsterdam (1980).
13. A. Seiger and A.C. Granholm. Thyroid hormone dependency of the developing brain, in: "Neurohistochemistry: Modern Methods and Applications", Alan R. Liss, Inc. (1986).
14. D.H. Ford and E.B. Cramer. Developing nervous system in relation to thyroid hormones, in: "Thyroid Hormones and Brain Development", G.D. Grave, ed., Raven Press, New York (1977).
15. J.H. Dussault and J. Ruel. Thyroid hormones and brain development. Ann. Rev. Physiol. 49:321-334 (1987).
16. J. Nunez. Thyroid hormones, in: "Handbook of Neurochemistry", Vol. 8, A. Lajtha, ed., Plenum Publ. Corp., New York (1985).
17. J. Nunez. Microtubules and brain development: the effects of thyroid hormones. Neurochem. Int. 7:959-968 (1985).
18. S. Geel. Development-related changes of triiodothyronine binding to brain cytosol receptors. Nature 269:428-430 (1977).
19. B. Dozin-Van Roye and Ph. De Nayer. Triiodothyronine binding to brain cytosol receptors during maturation. FEBS Lett. 96:152-154 (1978).
20. A.M. Lennon, J. Osty and J. Nunez. Cytosolic thyroxine-binding protein and brain development. Molec. Cell. Endocrinol. 18:201-214 1980.
21. S.E. Geel, L. Gonzales and P.S. Timiras. Properties of triiodothyronine binding sites in cerebral cortical cytosol. Endocrine Res. Commun. 8:1-18 (1981).
22. A.M. Lennon, F. Chantoux, J. Osty and J. Francon. A high affinity thyroid hormone binding protein in the cytosol of embryonic rat brain cells in primary cultures. Biochem. Biophys. Res. Commun. 116:901-908 (1983).
23. G. Shanker, N.R. Bhat and R.A. Pieringer. Investigations on myelination in vitro: thyroid hormone receptors in cultures of cells dissociated from embryonic mouse brain. Biosci. Rep. 1:289-297 (1981).
24. J.H. Oppenheimer. The nuclear receptor-triiodothyronine complex: relationship to thyroid hormone distribution, metabolism and biological action, in: "Molecular Basis of Thyroid Hormone Action", J.H. Oppenheimer and H.H. Samuels, eds., Academic Press, New York (1983).
25. B. Dozin and Ph. De Nayer. Triiodothyronine receptors in adult rat brain: topographical distribution and effect of hypothyroidism. Neuroendocrinol. 39:261-266 (1984).
26. J. Ruel, R. Faure and J.H. Dussault. Regional distribution of nuclear T3 receptors in rat brain and evidence for preferential localization in neurons. J. Endocrinol. Invest. 8:343-348 (1985).
27. H.L. Schwartz and J.H. Oppenheimer. Nuclear triiodothyronine receptor sites in brain: probable identity with hepatic receptors and regional distribution. Endocrinology 103:267-273 (1978).
28. N.L. Eberhardt, T. Valcana and P.S. Timiras. Triiodothyronine nuclear receptors: an in vitro comparison of the binding of triiodothyronine to nuclei of adult rat liver, cerebral hemisphere, and anterior pituitary. Endocrinology 102:556-561 (1978).
29. T. Valcana. The role of triiodothyronine (T3) receptors in brain development, in: "Neural Growth and Differentiation", E. Meisami and M.A.B. Brazier, Raven Press, New York (1979).
30. J.M. Kolodny, P.R. Larsen and J.E. Silva. In vitro 3,5,3'-triiodothyronine binding to rat cerebrocortical neuronal and glial nuclei suggests the presence of binding sites unavailable in vivo. Endocrinology 116:2019-2028 (1985).

31. M.A. Haidar, S. Dube and P.K. Sarkar. Thyroid hormone receptors of developing chick brain are predominantly in the neurons. Biochem. Biophys. Res. Commun. 112:221-227 (1983).
32. A. Pascual, A. Aranda, V. Ferret-Sena, M.M. Gabellec, G. Rebel and L.L. Sarlìeve. Triiodothyronine receptors in developing mouse neuronal and glial cell cultures and in chick-cultured neurones and astrocytes. Dev. Neurosci. 8:89-101 (1986).
33. M. Luo, R. Faure and J.H. Dussault. Ontogenesis of nuclear T3 receptors in primary cultured astrocytes and neurons. Brain Res; 381:275-280 (1986).
34. F. Courtin, F. Chantoux and J. Francon. Thyroid hormone metabolism by glial cells in primary culture. Molec. Cell. Endocrinol. 48:167-178 (1986).
35. J. Ortiz-Caro, B. Yusta, F. Montiel, A. Villa, A. Aranda and A. Pascual. Identification and characterization of L-triiodothyronine receptors in cells of glial and neuronal origin. Endocrinology 119:2163-2167 (1986).
36. J. Ruel, J.M. Gavaret, M. Luo and J.H. Dussault. Regulation of protein phosphorylation by triiodothyronine (T3) in neural cell cultures. Part I: Astrocytes. Molec. Cell. Endocrinol. 45:223-232 (1986).
37. J. Ruel, J.M. Gavaret, M. Luo and J.H. Dussault. Regulation of protein phosphorylation by triiodothyronine (T3). Part II: Neurons. Molec. Cell. Endocrinol. 45:233-240 (1986).
38. B.J. Spindler, K.M. McLeod, J. Ring and J.D. Baxter. Thyroid hormone receptors. Binding characteristics and lack of hormonal dependency for nuclear localization. J. Biol. Chem. 250:4113-4119 (1975).
39. J. Bernal, A.H. Coleoni and L.J. DeGroot. Thyroid hormone receptors from liver nuclei: characteristics of receptors from normal, thyroidectomized and triiodothyronine-treated rats; measurement of occupied and unoccupied receptors, and chromatin binding of receptors. Endocrinology 103:403-413 (1978).
40. K. von Overbeck and Th. Lemarchand-Béraud. Modulation of thyroid hormone nuclear receptor levels by L-triiodothyronine (T3) in the rat pituitary. Molec. Cell. Endocrinol. 33:281-292 (1983).
41. S. Hamada and Y. Yoshimasa. Increases in brain nuclear triiodothyronine receptors associated with increased triiodothyronine in hyperthyroid and hypothyroid rats. Endocrinology 112:207-211 (1983).
42. C.L. Thrall and T. Yanagihara. Alterations of nuclear thyroid hormone receptors in cerebral cortex in vivo. J. Neurochem. 38:669-674 (1982).
43. J. Bernal, K. Liewendahl and B.A. Lamberg. Thyroid hormone receptors in fetal and hormone resistant tissues. Scan. J. Lab. Invest. 45:577-583 (1985).
44. J. Bernal and F. Pekonen. Ontogenesis of the nuclear 3,5;3'-triiodothyronine receptor in the human fetal brain. Endocrinology 114:677-679-679 (1984).
45. L.W. Gonzales and P.L. Ballard. Identification and characterization of nuclear 3,5,3'-triiodothyronine-binding sites in fetal human lung. J. Clin. Endocrinol. Metab. 53:21-28 (1981).
46. A. Perez-Castillo, J. Bernal, B. Ferreiro and T. Pans. The early ontogenesis of thyroid hormone receptor in the rat fetus. Endocrinology 117:2457-2461 (1985).
47. H.L. Schwartz and J.H. Oppenheimer. Ontogenesis of 3,5,3'-triiodothyronine receptors in neonatal rat brain: dissociation between receptor concentration and stimulation of oxygen consumption by 3,5,3'-triiodothyronine. Endocrinology 103:943-948 (1978).
48. T. Valcana and P.S. Timiras. Nuclear triiodothyronine receptors in the developing rat brain. Molec. Cell. Endocrinol. 11:31-41 (1978).
49. B. Dozin-Van Roye and Ph. De Nayer. Nuclear triiodothyronine receptors in brain during maturation. Brain Res. 177:551-554 (1979).

50. K. Ishiguro, Y. Suzuki and T. Sato. Effect of neonatal hypothyroidism on maturation of nuclear triiodothyronine (T3) receptors in developing rat brain. Acta Endocrinol. 95:495-499 (1980).
51. L.J. DeGroot, M. Robertson and P.A. Rue. Triiodothyronine receptors during maturation. Endocrinology 100:1511-1515 (1977).
52. P. Coulombe, J. Ruel and J.H. Dussault. Analysis of nuclear 3,5,3'-triiodothyronine-binding capacity and tissue response in the liver of the neonatal rat. Endocrinology 105:952-959 (1979).
53. D. Bellabarba, S. Bédard, S. Fortier and J.G. Lehoux. 3,5,3'-triiodothyronine nuclear receptor in chick embryo. Properties and ontogeny of brain and lung receptors. Endocrinology 112:353-359 (1983).
54. B. Ferreiro, J. Bernal and B.J. Potter. Ontogenesis of thyroid hormone receptor in foetal lambs. Acta Endocrinol. 116:205-210 (1987).
55. B. Dozin-Van Roye and Ph. De Nayer. Evidence for different binding sites for T4 and T3 in rat liver and brain. VIth Intern. Congress of Endocrinology, abstract 142, p. 280 (1980).
56. D. Bellabarba and J.G. Lehoux. Binding of thyroid hormones by nuclei of target tissues during the chick embryo development. Mech. Ageing Dev. 30:325-331 (1985).
57. M. Margarity, N. Matsokis and T. Valcana. Characterization of nuclear triiodothyronine (T3) and tetraiodothyronine (T4) binding in developing brain tissue. Molec. Cell. Endocrinol. 31:333-351 (1983).
58. Ph. De Nayer and B. Dozin-Van Roye. Effect of chromatin associated factors on the activity of thyroid hormone receptors in rat liver and brain. Biochem. Biophys Res. Commun. 98:1-6 (1981).
59. J.W. Apriletti, Y. David-Inouye, N.L. Eberhardt and J.D. Baxter. Interactions of the nuclear thyroid hormone receptor with core histones. J. Biol. Chem. 259:10941-10948 (1984).
60. Ph. De Nayer and B. Dozin. Thyroid hormones and brain development: modulation of the binding activity of the T3 nuclear receptor by chromatin-associated factors. Molec. Physiol. 7:303-310 (1985).
61. B. Dozin and Ph. De Nayer. Nuclear receptors and cytosolic binding sites for triiodothyronine in rat brain and liver during maturation. Effects of neonatal hypothyroidism, in: "Neuropeptides and Psychosomatic Processes", Endroczi et al., eds., The Hungarian Academy of Sciences, Budapest (1982).
62. D. Bellabarba, S. Fortier, S. Bélisle and J.G. Lehoux. Triiodothyronine nuclear receptors in liver, brain and lung of neonatal rats. Biol. Neonate 45:41-48 (1984).
63. J.H. Oppenheimer and H.L. Schwartz. Stereospecific transport of triiodothyronine from plasma to cytosol and from cytosol to nucleus in rat liver, kidney, brain and heart. J. Clin. Invest. 75:147-154 (1985).

THYROID HORMONE REGULATION OF SPECIFIC mRNAS IN THE DEVELOPING BRAIN

S.A. Stein[1], D.R. Shanklin[4], P.M. Adams[2], G.M. Mihailoff[1,3], M.B. Palnitkar[1], and B. Anderson[1]

[1]Depts. of Neurology, [2]Psychiatry, and [3]Cell Biology
University of Texas Southwestern Medical Center, Dallas, Tx.
[4]Depts. of Pathology and Obstetrics and Gynecology
University of Tenn.-Memphis, Memphis, Tenn.

Supported by NIMH/ADAMHA # MH42469(SAS), NIMH/ADAMHA #MH43017(SAS) and United Cerebral Palsy Research and Educational Foundation #377-87(SAS)

Thyroid hormones, T3 and T4, have been shown to play significant but poorly understood roles in development and differentiation of rodent and human brain (1-7). In the human, disorders of maternal and fetal thyroid function include maternal and secondary fetal iodine deficiency, maternal hypothyroidism or hyperthyroidism, as well as disorders related to deficient fetal autonomous thyroid hormone secretion, i.e., goiter or sporadic congenital hypothyroidism. These disorders are identifiable causes of mental retardation (4, 8, 9, 10), cerebral palsy (11, 12), and other significant neurological abnormalities (5, 6, 11).

Significant and unresolved issues concerning the neurological impairments and their relationship to alterations in iodine and thyroid hormone in the fetal brain include the facts that:

1) Iodine replacement prior to pregnancy in iodine deficient women does not completely prevent mental or motor abnormalities (8);

2) Treatment of sporadic congenital hypothyroidism in the first postnatal month has significantly reduced the incidence of severe mental retardation (4). Despite this fact, evidence exists that mild mental retardation, learning disabilities, and mood motor and neurological abnormalities do occur (5, 6);

3) Maternal thyroid hormone aberration during pregnancy represents an identifiable cause of cerebral palsy (12) and lowered IQ (10) in children;

4) The neurological signs and neuropathological evaluation, particularly in cerebral cortex may be temporally related to alteration of fetal nervous system development as early as the first trimester (9, 11, 12, 14, 15, 16).

To assist us in understanding the pathophysiology and treatment of these disorders, a multidisciplinary approach to animal models of the different disorders is required.

Thyroid Hormone and Human and Rodent Brain Development

Of the animal models of hypothyroidism, the fetal and neonatal hypothyroid rat (1, 2), and progeny of the maternally iodine deficient (1, 15) or hypothyroid rodent (49) are reliable models of the human conditions (1, 4, 8, 11, 12)[The rodent at 16-19 days of gestation corresponds to the human fetus at about 12 weeks of gestation](1). The human and rodent (mouse and rat) are similar with respect to timing of functional and anatomical development of the fetal thyroid gland, thyroid hormone secretion, thyroid hormone receptors, detection of brain T3 and T4, and the maturation of the hypothalamo-pituitary-thyroid axis (1, 16, 17).

The type and localization of pathological changes in human(12) and hypothyroid rat brains (1,7), and electrophysiological (1), behavioral and learning abnormalities (1, 26, 27) are similar. Just as with early T4 in the human (4), T4 given to rodents before 10 days of postnatal life may(1) or may not (27, 28) normalize the behavioral, cerebral cortical, or biochemical abnormalities of hypothyroidism. Unfortunately, no animal model can totally simulate the human.

The human neurological (5, 6, 9, 11) and rodent behavioral signs (1-3, 18, 26, 27) reflect on anatomic abnormalities in specific regions of the cerebral cortex including the motor cortex and its corticospinal projections, parietal lobes, temporal lobe associative cortex, as well as hippocampus and brainstem.

When considered in terms of the normal sequence of fetal brain development which is similar in the rodent and human, these human neurological and rodent behavioral signs and neuropathology can be related to specific periods of fetal brain development and to specific changes in thyroid hormone secretion. Prior to 8-10 weeks of gestation in the human and 15 d pc in the rodent, the development of major brain structures occurs(1, 14, 20, 21). This temporally correlates with potential dependence on maternal thyroid hormones transported through the placenta (1, 3). The time of onset of rodent thyroid function(15-19 days of gestation) occurs immediately after brainstem neurogenesis (8-13 days post conception[d pc] (20) and the onset (13-20 days postconception) of forebrain neurogenesis (21); this timing correlates with the differentiation of these neurons[15 d pc to 4 days after birth(ab)], primarily in cerebral cortical layers V and VI (1, 22, 23). A similar temporal relationship for neuronal differentiation and thyroid hormone is seen in the human motor cerebral cortex (14, 16) between 9 and 24 weeks of gestation which corresponds with fetal thyroid hormone secretion.

In the rodent, the period from 17 d pc to 10 d ab is associated with the final stages of cerebral cortex neuronal migration (Layers III and IV), initial dendritic and axonal outgrowth of cerebral cortical layers I,IV,V, and VI (22, 23) ; afferent projections to the cerebral cortex (22) appear as does synaptogenesis (24) in the cerebral cortex. Further increases in T3 and T4 from 10 d ab to 45 d ab to peak values correlate with maximal rates of glial cell replication (25), increased dendritic branching and synaptogenesis, continued neurogenesis and differentiation in other brain regions, and myelination (1, 2). A similar sequence occurs in human brain from the second trimester to 2 years of age (14, 16).

In the human with congenital hypothyroidism or iodine deficiency and the fetal and neonatal hypothyroid rodent (1,13), each of the prenatal neuroanatomical events including neurogenesis, migration, axonal outgrowth, dendritic process ontogeny, synaptogenesis, gliogenesis, and myelinogenesis may be prolonged or changed in a permanent or T4 reversible fashion in the hypothyroid brain(1,7, 20, 28). Alteration of the duration of the sequence of brain development as well as individual developmental events within the sequence is important in understanding the rodent and human behavioral disorders of fetal thyroid hormone. These manifestations of these disorders differ in severity related to the timing, duration, and extent of maternal and fetal thyroid hormone or iodine abnormality in utero.

Although the rodent is particularly useful as a model, most of the behavioral and anatomical data have come from study of postnatal rodent hypothyroidism, resulting from surgically produced (1,3) or pharmacologically induced (20) hypothyroidism. These treatments may have independent pharmacological effects as well as morbidity and mortality and both maternal and fetal hypothyroidism (1, 3, 15). The effects of maternal and fetal iodine deficiency and maternal hypothyroidism encompass stages of development from neurogenesis to differentiation to synaptogenesis. Postnatal hypothyroidism interferes with cerebral cortex dendritic development and axonal outgrowth(1).

However, a definition of the effects of congenital hypothyroidism related to the timing of fetal thyroid hormone secretion and early cerebral cortical neuronal differentiation requires the production (1) of in utero late gestation hypothyroidism. A reproducible model of late fetal hypothyroidism would allow the dissection of maternal from fetal effects and isolate a narrow period of critical fetal brain development in the rodent and human with congenital hypothyroidism occurring after autonomous fetal thyroid hormone secretion.

The hyt/hyt congenitally hypothyroid mouse(29,30) provides a model for partial but severe inherited congenital hypothyroidism. The hyt/hyt hypothyroid mouse is advantageous because:

1) The primary inherited hypothyroidism characterized by significantly reduced T4 levels(32), reduced thyroid gland T4 (30), reduced thyroid gland iodine uptake (30), and elevated serum TSH-like activity (31);

2) Hypothyroidism is produced without artificial intervention;

3) It is distinct from postnatal hypothyroid rats and other inbred hypothyroid mice, with secondary hypothyroidism or other associated abnormalities (31) because of the fetal timing and primary nature of the hypothyroidism,

4) The hypothyroidism results from an autosomal recessive genetic defect(29).

5) Relatively equivalent numbers of hypothyroid (hyt/hyt) and euthyroid heterozygote controls(hyt/+) are produced within a litter which allows valid comparisons:

6) The hypothyroidism occurs at a time of autonomous fetal thyroid hormone secretion(after 15 days post conception).

7) The timing of hypothyroidism in the hyt/hyt mouse corresponds temporally with cerebral cortex(CC) and total brain neuronal differentiation, particularly of CC layer V and thalamocortical afferents to CC layer IV (22).

8) The timing of the hypothyroidism also corresponds with new fetal total brain and cerebral cortex mRNA synthesis (33).

The hyt/hyt mice and their hyt/+ littermates permit testing of a model for thyroid hormone deficiency and action in utero(Figure 1).

REDUCED THYROID HORMONE (T3/T4)

↓

ABNORMAL BRAIN AND CEREBRAL CORTEX
GENE EXPRESSION

↓

ABNORMAL NEUROANATOMICAL DEVELOPMENT

↓

ABNORMAL REFLEXIVE AND ADAPTIVE BEHAVIOR

Fig. 1. Model of Thyroid Hormone Action in Developing Brain.

Reduced or altered thyroid hormone levels may lead to alteration of fetal and neonatal brain and cerebral cortex mRNA synthesis. Some of these mRNA abnormalities may contribute to abnormal neuroanatomical development and subsequent abnormal behavior(Figure 1).

Behavioral, Neuroanatomical, and Endocrinological Evaluation of the hyt/hyt Mice

The purposes of our studies were to utilize the inherited hyt/hyt congenitally hypothyroid mouse to characterize:

1)The neuroanatomical and reflexive behavioral abnormalities that occur in the early neonatal period in hypothyroidism;

2)The specific brain and cerebral cortex molecular biological abnormalities that occur in the late gestation and neonatal period related to thyroid hormone deficiency that might have functional significance; and

3)The sensitivity of fetal brain to thyroid hormone(Scheme 1).

Our colony of genetically hypothyroid mice was established from breeders (C.RF-hyt, N10F13-16) obtained from Jackson Laboratory(Bar Harbor, ME), and was maintained by mating hyt/+ sisters(T4 > 6 ug/dl) and T4 supplemented hyt/hyt(hypothyroid) brothers(serum T4 levels < 1.1 ug/dl)(36); these matings produced litters of hyt/hyt and hyt/+ offspring for study.

Day of birth hyt/hyt animals were distinguished from hyt/+ littermates by serum thyroxine level (T4) (34), and histological and ultrastructural evaluation of the thyroid gland(Figure 2). In normal mice, autonomous secretion of T4 begins about 15 d pc (30). In the hyt/hyt mouse, deficient T4 production is associated with delayed and disorganized fetal thyroid gland histological development (30); the

Scheme 1. GENERAL EXPERIMENTAL PROTOCOL.

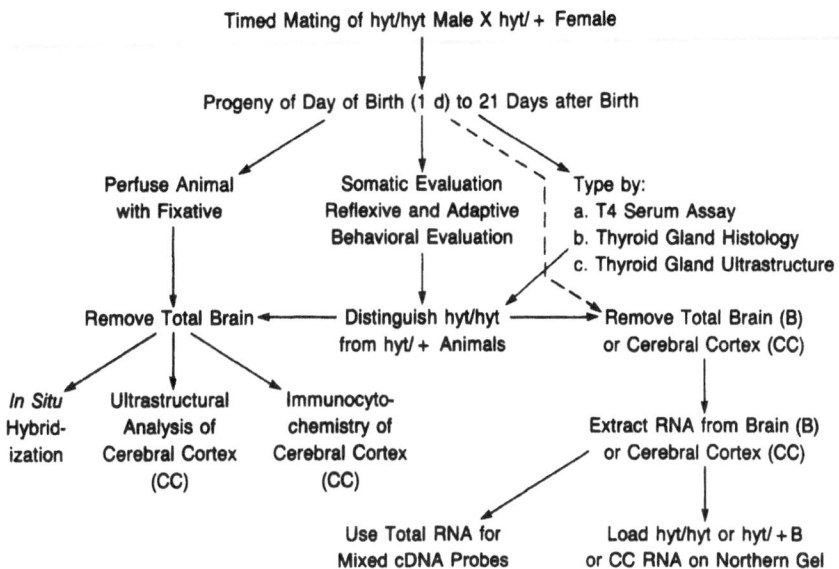

hyt/hyt thyroid gland has reduced or absent colloid, a reduced number of smaller follicles(30) and an increase in the number of follicular cell nuclei as early as 17 d pc. Our histological distinction of the thyroid glands from 1 day hyt/hyt and hyt/+ littermates was based on the following criteria: 1)The percentage of follicles which contained colloid (0-100%);. 2)The size of the follicles(+ to +++); and 3)The size of intrafollicular epithelial gaps(+ to ++++)(Figure 2, Panels A and B).

The initial step in this analysis was to determine a sample size which would in cross section fill the microscopic field. This was required because hyt/hyt glands are hypoplastic compared to the normal. In the hyt/hyt animals, less than 50% of the follicles contained colloid. This reduction in follicles with colloid correlated with the functional status of the gland. The hyt/+ normal gland had 75% or more colloid containing follicles. In the hyt/hyt thyroid gland, gaps between epithelial cells were rare(0-1+), while gaps were readily apparent and numerous in the hyt/+ glands(3+).

The critical features for differentiation on electronmicroscopy were graded on a scale from 1+ to 3+ and included: 1)The arrangement and distance between nuclei; 2)The amount of cytoplasm and colloid. There was abundant cytoplasm in the normal gland with occasional long gaps between nuclei(Figure 2, Panels C and D). By contrast, hyt/hyt follicles have nuclei very close together with reduced cytoplasm and colloid. Within an individual litter, distinction of 1 day(day of birth) hyt/hyt(T4 < 1 ng/ml) and hyt/+(> 4.0 ng/ml) mice could be made by T4 levels. The anatomical and T4 samples were analyzed in coded fashion to prevent investigator bias. The anatomical indices corresponded exactly with the prediction of a hyt/hyt versus a hyt/+ mouse by T4 levels. Based on the use of these distinctions, hyt/hyt or hyt/+ animals were divided for the neuroanatomical or molecular biological studies(Scheme 1).

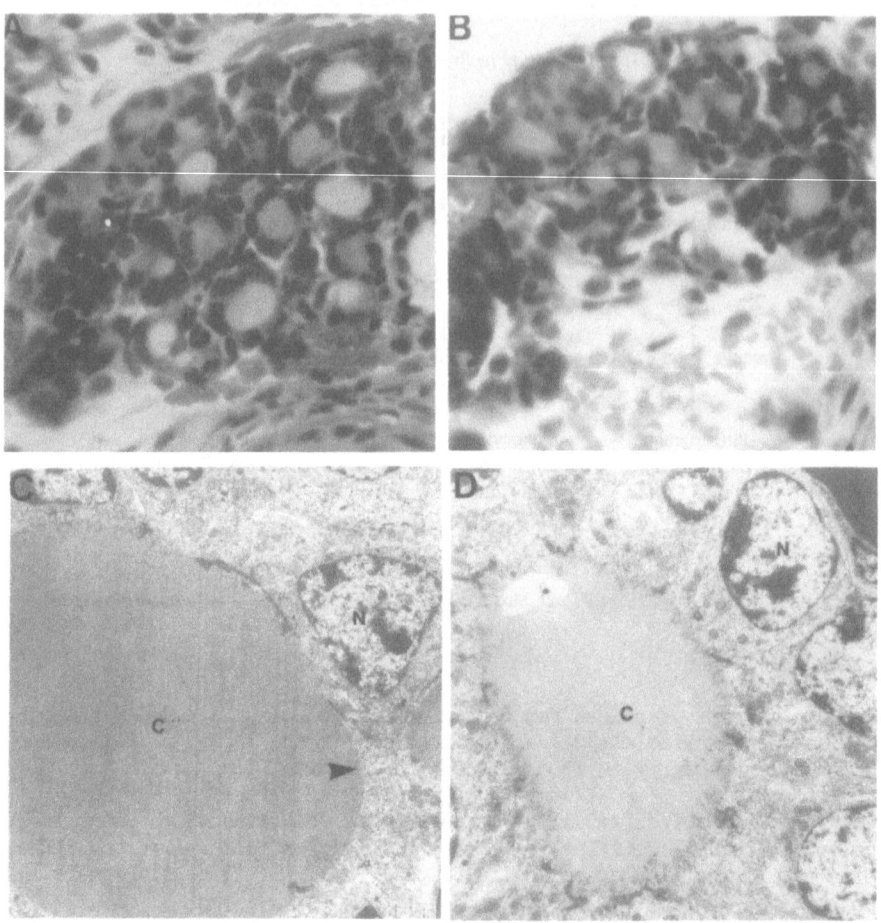

Fig. 2. PHOTOMICROGRAPHS OF HISTOLOGICAL SECTIONS AND ELECTRONMICROGRAPHS OF 1 DAY OLD hyt/hyt and hyt/+ MOUSE THYROID GLANDS. Panel A: Histology of hyt/+ Thyroid Gland. Follicle development with varied cross sectional profiles was observed. The hyt/+ glands had 75% or more colloid containing follicles compared to less than 50% in the hyt/hyt glands(Panel B). Several gaps were present in the epithelial rim of the follicles (arrows). Hematoxylin and eosin. Original magnification 400X. Panel B: hyt/hyt Photomicrograph. Tighter, seemingly more cellular follicles with closely packed epithelial nuclei. Gaps were absent. Hematoxylin and eosin. Orig. magnifi. 400X. Panel C: Electronmicrograph(EM) of hyt/+ Thyroid Gland: Follicle with smoothed out colloid boundary and fairly dense granular colloid was seen. Epithelial gaps with spreading of nuclei including back to back follicles (arrow) were present. Microvilli were inconspicuous. Print magnifi. 9100X. Panel D: EM of hyt/hyt Gland: Small Follicle with a regular cytoplasmic-colloid boundary in prominent microvilli. Thyroid cells had less colloid and cytoplasm per nucleus and nuclei were more closely applied. Print magnifi. 9100X.[Reprinted with permission(63)].

Preweaning Somatic and Behavioral Reflexive, Motor and Adaptive Evaluations(Scheme 1, Figure 3)

Hyt/hyt and hyt/+ mice were specifically distinguished (d ab) by T4 levels at 21 days after birth(d ab)[hyt/hyt= 0.58 +/- 0.28 ugm/dl (n=13), hyt/+=6.04 +/- 1.72 ugm/dl(n=17)(T=11.37, 27 df, p<.01)](32).

The somatic and behavioral studies on these 29 mice established that the inherited hyt/hyt abnormality caused severe neonatal hypothyroidism that was related to significant reductions in body length, body weight, and significant delays in age of ear raising and eye opening. There was delayed age of onset of normal reflexive behaviors and impaired performance of other later adaptive behaviors including swimming escape and locomotor activity. As with T4 levels, these measures can also be used to distinguish the hyt/hyt mouse from the hyt/+ mouse in the early neonatal period. The hypothyroidism in the hyt/hyt animals impairs somatic and brain function at least prior to 5-15 days after birth.

Previous research on the appearance and subsequent disappearance of reflexive behaviors provided the foundation for normal motor

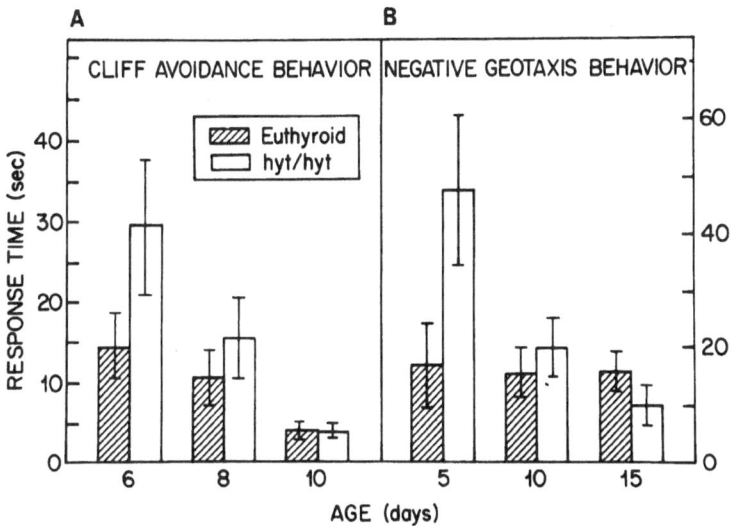

Fig. 3(Panels A and B). REFLEXIVE BEHAVIORAL EVALUATION OF hyt/hyt AND hyt/+ NEONATAL ANIMALS. The reflexes evaluated for each mouse were cliff avoidance and negative geotaxis. The days of age and the methods have been previously reported (32). In the normal hyt/+ mice, the negative geotaxis and cliff avoidance response had developed by 5-8 days after birth. The hyt/hyt neonates were found to be abnormally slow to develop on these behavioral reflexes when compared to the hyt/+ mice. For cliff avoidance (Panel A), the hyt/hyt animals were significantly delayed at 6 days of age when compared to the normal animals(t=2.06, p < .05, 27 df). For negative geotaxis (Panel B), a similar significant delay was seen for the hyt/hyt animals versus the normal hyt/+ littermates at 5 days of age(t=2.07, p <.05, 27 df). As the hyt/hyt animals grew older, the time required to make the criterion response became similar to the normal animals and was not delayed[Reprinted with Permission(32)].

development in rodent and human (18, 35). Cliff avoidance, surface righting, and negative geotaxis were among the normal behaviors that ontogenically develop in the rodent soon after birth (32, 35). The time of their appearance and disappearance have been suggested as an an indicator of the normal development of the rodent nervous system and the subsequent appearance of more complex behaviors (18). Demonstration of delays in cliff avoidance and negative geotaxis reflexive behavior is similar to what had been observed in the fetal and neonatal hypothyroid rat (18). The development of rooting and suckling behavior preceded normal neonatal feeding behavior(18) and other reflexes including the placing response and body righting (2) were delayed in the hypothyroid rodent. These reflexes and early behaviors may be sensitive to the function and anatomical development of the corticospinal tracts and to developing cerebellum, brainstem, and vestibular system (18,35). Delayed and abnormal anatomical development in these areas have been observed in other hypothyroid neonatal rodents (1,2,7). Ultrastructural evaluation of the neonatal hyt/hyt and hyt/+ cerebral cortex demonstrated delayed development of 8 day hyt/hyt cerebral cortical neurons which are less differentiated than the 1 day old hyt/+ neurons (19). Despite eventual normal performance, the behavioral and anatomical delay may be particularly important because nervous system development involves multiple and complex events that are precisely timed. Alteration in the anatomical events underlying the reflexive behavior may lead to longterm impairment of adaptive behavior.

Swimming escape and locomotor activity represent more complex behaviors. These are built on the normal performance of early reflexes as well as coordination of multiple motor and adaptive behaviors. Since the significant abnormalities that we have observed in swimming escape, locomotor activity, and weight occur at 30 to 40 days of age (data not shown), this suggested that the delays observed in early neonatal reflex and somatic measures may predict or be correlated with later behavioral abnormalities and perhaps abnormal neuroanatomical development(32).

Control of Gene Expression by Thyroid Hormone and The hyt/hyt Mouse

Thyroid hormone may affect either an increase or decrease in the levels of the components of the translational apparatus (ribosomal RNA, t RNA, small or large ribosomal subtypes, endoplasmic reticulum) (36, 37) or of the transcriptional apparatus(RNA polymerase I,II, and III) (36, 37). The effects of thyroid hormone may also be limited to variable increases or decreases or the turning on or turning off of individual mRNAs or subsets of specific mRNAs, at particular developmental stages and in specific cells or tissues (37). In adult rat brain and liver, thyroid hormone may significantly affect the abundance of 2-3% of the total poly A+ mRNAs (38). The changes in mRNA levels of these and other mRNAs (36) have been attributed to altered transcription (36), altered mRNA stability (36) or altered processing (36). These changes follow the poorly understood interaction of T3 and T4 with: 1)The putative thyroid hormone nuclear receptor (49); 2)Tissue specific trans-acting factors; and 3)5' and 3' flanking DNA sequences of the regulated genes.

Since the hyt/hyt animals are deficient in thyroid hormone starting during late gestation, these animals afford a means of looking at the effects of thyroid hormone deficiency during a critical period of molecular biological and neuroanatomical development of brain with particular relevance for cerebral cortex. These animals may be used to evaluate the presence and potential effects of specific brain mRNA changes related to fetal and neonatal thyroid hormone deficiency. Neuronal differentiation, particularly of cerebral cortical layers V and VI is temporally correlated with general brain (20,000-30,000 mRNAs) and cerebral cortex specific synthesis of new poly A+ mRNAs (17 d pc to birth) and the onset of fetal thyroid function (33). This also correlates with the normal ontogeny of important brain specific protein and mRNA synthesis such as tubulin, actin, neurofilaments, somatomedin, and glial fibrillary acidic protein. The purposes of our molecular studies were to determine:

1)Which of these mRNAs were expressed in mouse total fetal and neonatal brain and cerebral cortex (Figure 4, Panels A,B, and C);

2)When and where these mRNAs were expressed (Figures 5,6,7 and 8);

3)Whether any of the mRNAs that were expressed in fetal brain were regulated by thyroid hormone (Figures 6 and 7).

The first purpose was accomplished by utilizing cDNAs homologous to specific proteins and mRNAs: 1)That had been shown to be present in the brains of fetal and neonatal animals or neuronal tissue culture (40, 41, 43); 2)That had important functions based on anatomical or biochemical studies in the developing nervous system (40, 42, 43, 49). 3)That were regulated by thyroid hormone in other tissues or adult brain (37, 38, 42). The 75 mRNAs that were chosen included cytoskeletal mRNAs (40) and trophic substances (42, 45)(Figure 4).

The presence of these mRNAs was initially assessed by cDNA-DNA slot blot analysis(Figure 4, Panels A, B and C). Common, known mRNAs, tubulin, IGF1 and other mRNAs of potential functional importance were detected in normal brain and cerebral cortex at birth. The pattern of gene expression differed in early neonatal total brain versus cerebral cortex in normal animals. As seen in Figure 4, the level of gene expression for individual mRNAs also differed between cerebral cortex and total brain.

The presence and ontogeny of the mRNAs detected on the slots blots was confirmed by northern gel hybridizations to total RNA from fetal and neonatal total brain(14 d pc to 6 mos.) or cerebral cortex(1-7 d pc) (Figures 5,6, and 7). Actin, neurofilament 69kd cDNA protein, and tubulin were all detected in 1 day hyt/+ cerebral cortex and total brain. The mRNA that may be homologous with the neural thyroid hormone receptor was detected as early as 14 d pc in total brain and 1 day cerebral cortex.

Regulation of Specific Gene Expression by Thyroid Hormone in Mouse Brain

The next set of issues on the detected mRNAs utilizing the hyt/hyt mice were: 1)To identify brain and cerebral cortex mRNAs possibly regulated by thyroid hormone in fetal and neonatal brain; and 2)To try to determine why some mRNAs were specifically altered. To do this, zeta probe sheets for northern gel hybridizations were prepared from 1, 5 and 7 day cerebral cortex, brain remainder, and total brain RNA from hyt/hyt and hyt/+ littermates. These sheets were hybridized to cDNA probes reflecting mRNAs that had been detected in cerebral cortex and total brain(Figures 4 and 5).

Thyroid hormone does not cause global changes in all mRNAs but may act to regulate the gene expression of specific mRNAs in specific brain

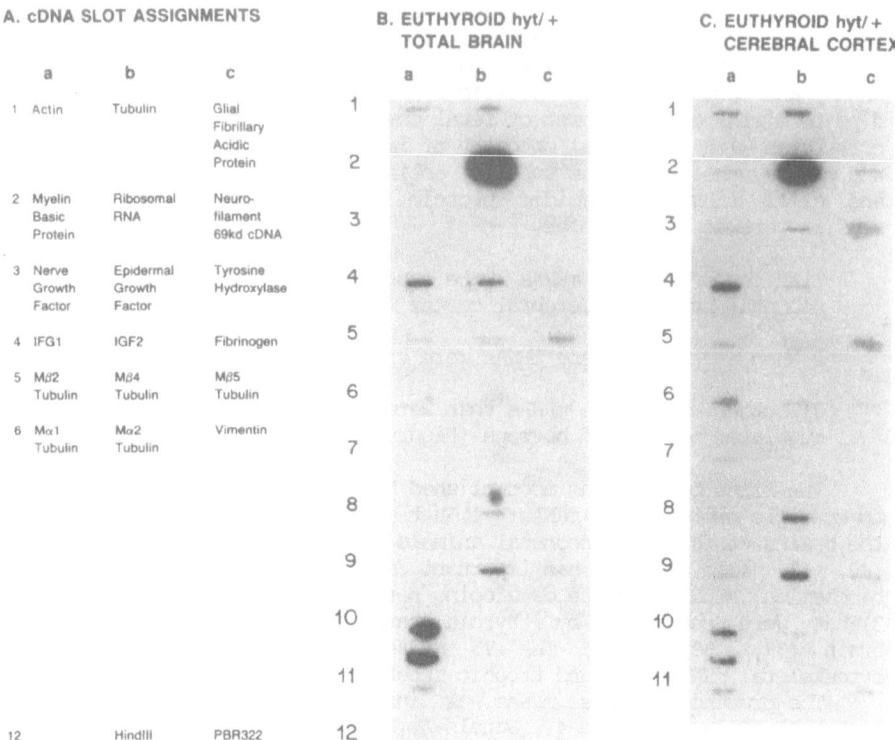

Fig. 4. cDNA-DNA SLOT BLOT HYBRIDIZATIONS. The cDNA probes were loaded in equal amounts on zeta probe paper with a slot blot apparatus and hybridized with equal amounts of mixed cDNA probes. The mixed cDNA probes were transcribed from total RNA from 1 day old hyt/+(euthyroid) total brain and cerebral cortex(CC). The lane assignments for individual cDNAs are shown in Panel A. The abundance of the mRNAs in 1 day old euthyroid total brain and cerebral cortex was reflected by the autoradiographic density of the zeta probe sheets following hybridization. For 1 day old euthyroid hyt/+ total brain(Panel B), the mRNAs that were detected were actin, tubulin, somatomedins IGF1 and IGF2, and the tubulin isotypes Mβ2, Mβ 4, Mβ5, Mα1, and Mα2. The most abundant mRNAs were ribosomal RNA, actin, tubulin, IGF1, IGF2, and Mβ5 tubulin. The specificity of the hybridization was suggested by the fact that no signals were noted for the non-specific controls(PBR322 and the Hind III Digest of lambda) or for fibrinogen, which was specifically made in the liver. No signals were detected for myelin basic protein and glial fibrillary acidic protein, which are not made until two to three days after birth(45). In comparing hyt/+ total brain to 1 day old euthyroid hyt/+ cerebral cortex(Panel C), the number of different mRNA synthesized at detectable levels was greater in CC as was general mRNA abundance. The mRNAs that were present included: actin, tubulin, neurofilament 69 KD cDNA, epidermal growth factor, IGF1, IGF2 and all of the tubulin isotypes. In CC, higher abundances of Mα1 and Mβ5 tubulins, IGF1, and neurofilament 69KD cDNA were noted in CC versus total brain(Lanes 7-11 = Unknown adult cDNAs homologous to common mRNAs).

areas at times that are correlated with neuronal differentiation and other events in the developing brain. In hyt/hyt versus hyt/+ mice, specifically, Epidermal Growth Factor (EGF) mRNA and Mβ5 and Mα1 tubulin isotype mRNAs were altered in abundance in neonatal brain, cerebral cortex, and brain remainder(total brain minus cerebral cortex)(Figures 6 and 7), while other mRNAs were unchanged in abundance i.e. ribosomal RNA(rRNA), total tubulin, and other tubulin isotypes. Reduced thyroid hormone, as reflected by the hyt/hyt mouse, led to changes in abundance of some of these total brain and cerebral cortex mRNAs as early as the day of birth. The differences in specific mRNA abundance between the hyt/+ and hyt/hyt animals may reflect on thyroid hormone regulation of late gestation and early neonatal mRNA synthesis after the onset of autonomous fetal thyroid hormone secretion.

The presence of these mRNAs in day of birth(1 day) brain demonstrated fetal brain and cerebral cortex synthesis of these specific mRNAs. The change of abundance in the day of birth hyt/hyt versus hyt/+ mice suggests that the fetal brain is sensitive to thyroid hormone deficiency. Not only is this in keeping with the neurological, neuroanatomical, and behavioral data in rodents, sheep, and human (1), but also with biochemical data in the progeny of maternally hypothyroid rodents (3). In the offspring of these rats, the pervasive defects in protein synthesis and rRNA synthesis that begin in midgestation (44) may be related also to effects of thyroid hormone deficiency on gene expression.

Thyroid Hormone, Epidermal Growth Factor(EGF), and Brain Development

The presence of EGF mRNA in fetal and neonatal cerebral cortex has not been shown previously. However, EGF has distinct effects on the neonatal nervous system(47,48). Our results suggest that thyroid hormone level may determine or regulate the abundance of the EGF mRNA in the cerebral cortex (Figure 6). EGF mRNA and protein were regulated primarily or secondarily by thyroid hormone in neonatal skin (46) and other rodent adult tissues (47). In combination with our findings, the regulation of EGF suggests that the EGF gene may contain thyroid responsive elements that extend across tissues and developmental age or that common tissue trans-acting thyroid responsive elements are present for EGF. The mRNA for EGF demonstrated higher abundance in the late fetal and early neonatal period and a marked decline between 6-16 days (Figure 5).

Brain constitutive gene expression as revealed by mRNA ontogeny data is potentially an important mechanism for the turning on and turning off of mRNAs in late gestation and the early neonatal period. Thyroid hormone regulation must be superimposed on this constitutive expression.

Although the potential role of fetal and neonatal EGF is poorly understood, EGF stimulates astroglial proliferation (47) and oligodendroglial differentiation as reflected by the synthesis of myelin related enzymes and myelin basic protein (50) which are required for normal myelination. Abnormal anatomical indices of myelination are noted in young hyt/hyt cerebral cortex and corpus callosum(48). Alternatively, EGF promotes neonatal neuronal process outgrowth and neuronal survival in certain discrete areas of subcortex and cortex(49). Therefore, given our results, thyroid hormone may promote the fetal and early neonatal synthesis of EGF mRNA as a prelude to late neonatal EGF protein synthesis. EGF may then play specific roles in postnatal(25) glial proliferation and myelination (47,50) and neuronal differentiation (49).

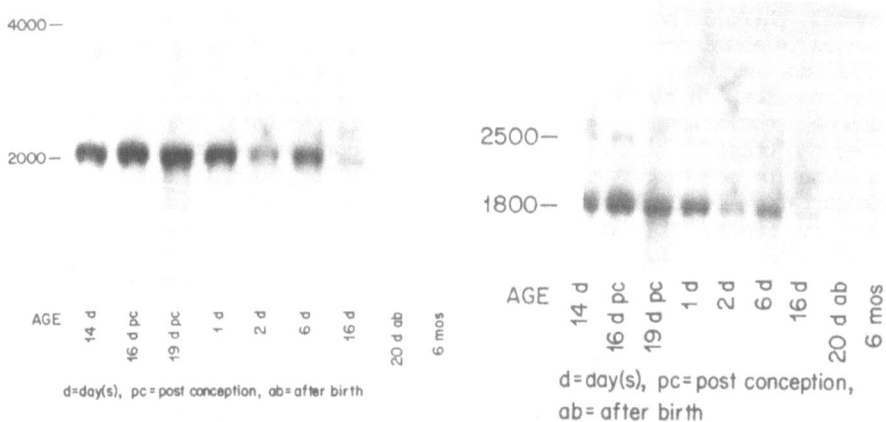

A DEVELOPMENTAL ABUNDANCE
OF EGF
IN hyt/+ NORMAL TOTAL BRAIN

B DEVELOPMENTAL ABUNDANCE
OF Mβ5
IN hyt/+ NORMAL TOTAL BRAIN

d=day(s), pc=post conception, ab=after birth

d=day(s), pc=post conception,
ab=after birth

Fig. 5. mRNA ABUNDANCE ONTOGENY of EPIDERMAL GROWTH FACTOR(EGF) mRNA(Panel A) and Mβ5 TUBULIN(Panel B) ISOTYPE mRNA. This utilized northern gel hybridizations with total brain RNA from 14 d pc (days, post-conception) to 6 months of age. Ten ugm of total RNA from hyt/+ or +/+ euthyroid brain was used and hybridized with random primer labeled cDNA probes. PANEL A: EGF. EGF was detected as early as 14 d pc in normal mouse total brain and 1 day of age(day of birth) in the cerebral cortex. EGF was present in high abundance in total brain in late gestation and the early neonatal period and then fell significantly in abundance after 6 days after birth(d ab). Total brain and cerebral cortex EGF mRNA unlike its peripheral counterpart that is 4750 b.p (45), reflects both 4000 and 2000 b.p. mRNAs; the 2000 bp mRNA is significantly more abundant in the late fetal and neonatal brain. PANEL B: Mβ5 TUBULIN ISOTYPE mRNA. Similar to EGF, the Mβ5 isotype mRNA and the Mα1 isotype mRNA(data not shown) of tubulin demonstrated high abundance in the late fetal and early neonatal brain with a rapid fall in abundance after 6 d ab.

Microtubules, Tubulin, Tubulin Isotypes, and Thyroid Hormone

Similarly, the results with the tubulin isotypes have demonstrated changes in abundance of the Mβ5 and Mα1 isotype mRNAs in the hyt/hyt versus the hyt/+ animals at different neonatal ages and in different brain regions at different ages. Since tubulin comprises 25% of neonatal neuronal protein, changes in tubulin mRNAs may have important biochemical consequences (51). Thyroid hormone regulates the synthesis of fetal (52) and neonatal brain tubulin protein (53). As reported by others for fetal tubulin protein (52), a differential regional sensitivity to thyroid hormone as well as increases or decreases for specific tubulin isotype mRNAs in response to thyroid hormone was suggested by our results in cerebral cortex, total brain, and brain remainder.

Given the anatomical events that are occurring during this time and the ontogeny of these tubulin isotype mRNAs, the alteration of tubulin abundance may be of potential functional and anatomical

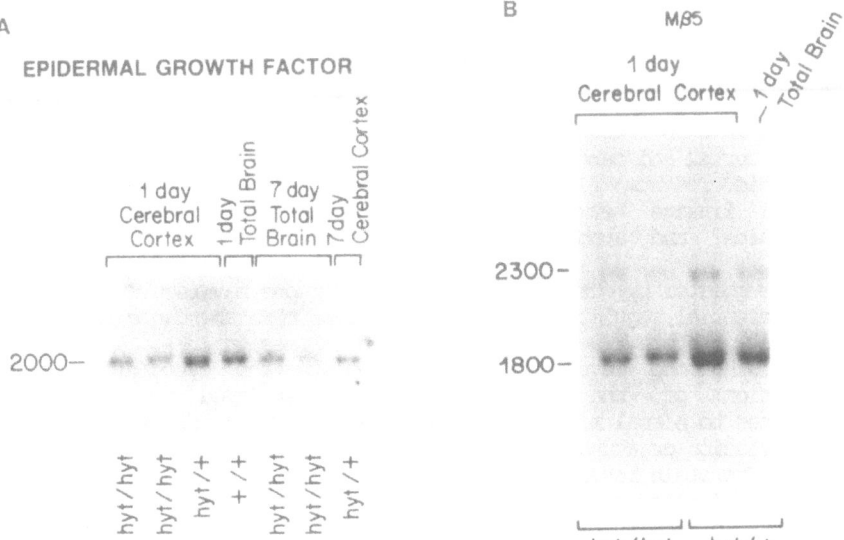

Fig. 6. PANEL A: NORTHERN GEL HYBRIDIZATION OF EGF mRNA IN TOTAL BRAIN AND CEREBRAL CORTEX FROM 1 AND 7 DAY OLD hyt/hyt and hyt/+ MICE. EGF mRNA was depressed in abundance in the 1 day hyt/hyt cerebral cortex versus the 1 day hyt/+ cerebral cortex. PANEL B: NORTHERN GEL HYBRIDIZATION of Mβ5 ISOTYPE mRNA in 1 DAY hyt/hyt AND hyt/+ CEREBRAL CORTEX, The abundance of the Mβ5 mRNA was significantly reduced in the hyt/hyt CC versus the hyt/+ normal CC at 1 day of age. The abundance of Mβ5 mRNA in hyt/hyt and hyt/+ total brain was not significantly different. However, the abundance of Mβ5 mRNA was depressed in the 1 day hyt/+ versus hyt/hyt brain remainder(data not shown); the brain remainder represented all of the brain minus the cerebral cortex from the same animal. Therefore at 1 day, Mβ5 mRNA is depressed in hyt/hyt cerebral cortex but increased in hyt/hyt brain remainder versus hyt/+ remainder. No abundance changes were noted for the tubulin isotypes Mβ4 or total tubulin mRNAs in CC.

significance. Tubulins and the multiple tubulin isotypes, neurofilaments, and actin form the cytoskeletal scaffolding for the normal nervous system and form the basis for microtubules, neurofilaments, and microfilaments respectively. Microtubules play a significant role in determining the size and shape of neurons (51). Microtubules contribute to neurogenesis, migration, fast axoplasmic transport, neurite extension, growth cone movement, and the morphology and polarity of axons and dendrites (51). Microtubules are made up of complexes of heterodimers of α and β tubulin isotypes and microtubule associated proteins(MAPs). MAPs may bind the individual tubulins and stabilize the microtubule (51) as well as assisting in tubulin assembly for microtubules (58). Multiple tubulin isotypes have been described which may be ubiquitous or specific for brain (40) or other tissues (40, 51).

Similar to EGF, the ontogeny of these mRNAs in our studies and those of others demonstrated a higher abundance of some mRNAs, i.e. tubulin isotypes(Figure 5) in the late fetal and early neonatal period and the subsequent rapid decline for these mRNAs. The constitutive stockpiling of tubulin mRNA likely precedes neuronal process outgrowth and the decrease in these sequences may correspond with the onset of tubulin protein synthesis and terminal neuronal differentiation (43).

In neonatal hypothyroidism, a severe reduction in dendritic arborization and in axonal process outgrowth in neonatal hypothyroidism (54) may be related to the observed significant decrease in microtubule density compared to normal animals (55). This may help to explain the reduction in axonal and dendritic processes, and dendritic branching in cerebral cortex and cerebellum that has been repeatedly observed in hypothyroid rodents (7). These data and our data also suggest a potential linkage between abnormal process development, abnormal microtubules, and abnormal tubulin synthesis, possibly of specific isotypes.

As suggested by our results, thyroid hormone may regulate tubulin by differential regulation of tubulin isotype mRNA abundance; this is consistent with other work on tubulin mRNA expression (57). Alteration of thyroid hormone levels may also regulate the cytoskeleton by: 1)Regulation of the tau microtubule associated protein which contributes to normal microtubule assembly (58); 2)Alteration in the phosphorylation or abundance of neurofilaments (56); or 3)Alteration in other MAP protein levels (59). An additional role for thyroid hormone regulation of MAPs and tubulin isotypes may be in its contribution to functional diversity (51) of microtubules within neurons and between neurons in the same and different brain regions. A single neuron as well as axonal subpopulations may make different MAPs with distinct intracellular localizations (60) and multiple tubulin isotypes (61).

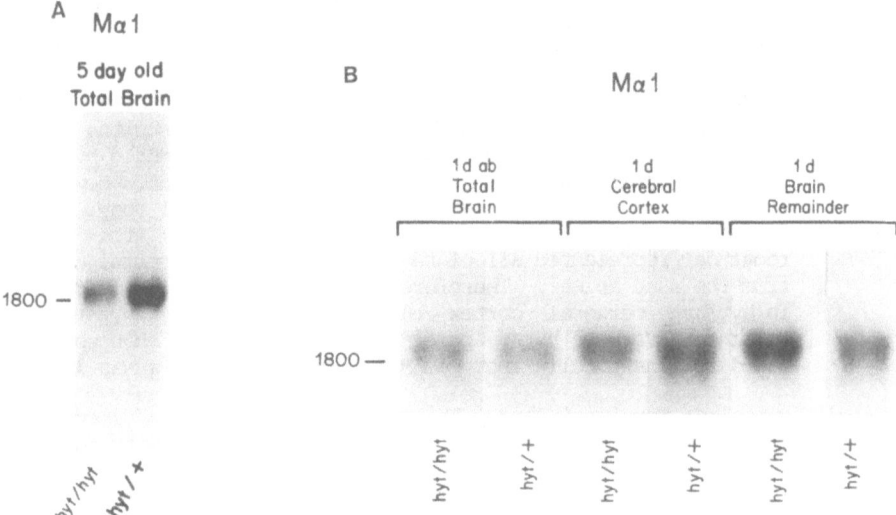

Fig. 7. NORTHERN GEL HYBRIDIZATION OF Mα1 TUBULIN ISOTYPE mRNA IN 1(Panel B) and 5 DAY(Panel A) TOTAL BRAIN, CEREBRAL CORTEX(CC), and BRAIN REMAINDER. PANEL A: 5 DAY TOTAL BRAIN. Mα1 tubulin isotype showed a significantly lower abundance in the 5 day old hyt/hyt total brain versus the 5 day old hyt/+ euthyroid brain. Panel B:1 DAY BRAIN. The abundance of this mRNA was not significantly changed in abundance in 1 day total brain or 1 day cerebral cortex. However, in brain remainder similar to Mβ5, Mα1 isotype mRNA was depressed in the hyt/+ animals. In 1 and 7 Day Old Cerebral Cortex, rRNA was not altered in abundance in total brain or cerebral cortex in the hyt/hyt versus the hyt/+ animals(Data not shown).

Different microtubules may be comprised of distinct tubulin isotypes (51). Presently, although the different isotypes are made from different genes with distinct protein products, the specific functions of different isotypes in the nervous system are still unknown.

Given the potential thyroid hormone regulation and the developmental and brain regional abundance differences of the tubulin isotypes i.e. Mα1 and Mβ5 and the role of microtubules and microtubular assembly in neurite extension (51), the purposes of the next set of studies were: 1)To determine the brain cellular localizations of the tubulin isotype mRNAs by in situ hybridization; and 2)To try to relate this localization to specific brain and cerebral cortex developmental or differentiation events occurring from 17 d pc to day of birth. On in situ hybridization (62), we found the 1 day old mouse brain to have abundant Mα1 mRNA which was distributed ubiquitously to cells throughout the hemispheres. The Mα1 mRNA showed greatest abundance in cerebral cortex where it could be detected easily in cells of both the cortical plate and sub-plate layers in multiple regions including frontal and parietal cortex(Figure 8). Since this was done in day of birth animals, the fetal mRNA for this isotype was made in these specific cellular regions of cerebral cortex.

The Mα1 mRNA was also easily detected in the differentiating pyramidal cells of the hippocampus and the stratum granulosum of the developing dentate gyrus and to a lesser extent compared to cerebral cortex in the basal ganglia and the white matter around the ventricles. The Mβ5 isotype mRNA was also abundant in all regions of the cerebral cortex of 1 day old animals(data not shown).

The localization of the potentially thyroid hormone regulated Mβ5 and Mα1 tubulin was primarily to cerebral cortex, a site of active axonal outgrowth and differentiation as well as other brain regions. The cellular sites of this localization and the temporal correlation of this localization with active differentiation in some of these brain areas may suggest a role for these isotypes and thyroid hormone in late fetal and early neonatal neuronal process growth. These also reflect areas of the brain that have shown specific neuropathological abnormalities in the hypothyroid rodent and the human(1,7,13). The localization and abundance of the Mβ5 and Mα1 isotype mRNAs will be looked at in the future in the hyt/hyt brain.

SUMMARY

The hyt/hyt mouse provides a useful model of human sporadic congenital hypothyroidism. This mouse demonstrates molecular, somatic, neuroanatomical, and behavioral abnormalities and emphasizes the utility of an integrated approach in problems related to thyroid hormone. Since rodent and human tubulin isotypes have structurally conserved regions (40,57) and may have distinct neuroanatomical functions, the study of some of the tubulin isotype mRNAs may be useful in understanding the mechanism(s) of action of thyroid hormone on the developing brain and the effects of thyroid hormone deficiency in human congenital hypothyroidism. Our observations suggest that the fetal brain is sensitive at a molecular level to thyroid hormone and that early neonatal behavioral and anatomical events are altered in disorders of autonomous thyroid hormone secretion as in the hyt/hyt mouse. Our work provides support for the idea that late gestational alteration in thyroid hormone levels contributes to abnormal brain and

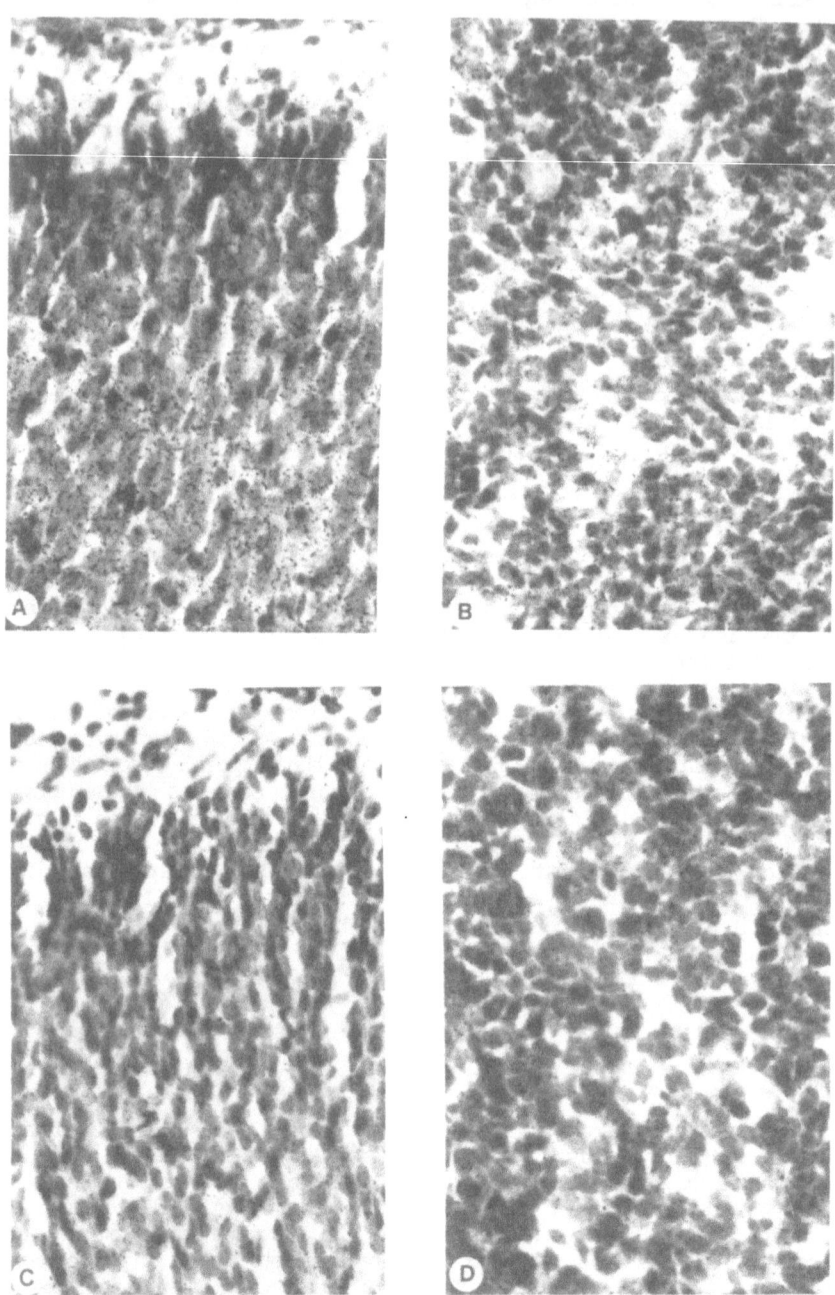

Fig. 8. Panels A-D. IN SITU HYBRIDIZATION OF Mα1 TUBULIN ISOTYPE mRNA
WITH 1 DAY OLD MOUSE BRAIN. Illustrated here are a series of
autoradiographs in which the density of silver grain
accumulation indicates the relative presence or absence of mRNA
for Mα1 tubulin. Substantial labeling was observed in the
developing cerebral cortex (as seen here in the superficial
layers) (Panel A), while slightly less dense labeling could be
seen in the caudate-putamen (Panel B). Panels C (cerebral
cortex) and D (caudate-putamen), which are adjacent sections
treated with RNAase, exhibit minimal labeling and thus suggest
that the silver grain accumulation illustrated in Panels A and B
is specific for Mα1 tubulin.

cerebral cortex gene expression of a small group of the total mRNAs. i.e. tubulin isotypes and EGF, and to subsequent abnormal anatomical and behavioral development.

ACKNOWLEDGEMENTS

The authors thank Dr. V. A. Galton for performing the T4 assays and Drs. V.A. Galton, S. Brady, and G. Bloom for critical review of the manuscript. The authors are grateful to Phillip Gibson, Amy Anthony, Phillip Toy, Ken Bourell, Frank Moretta, and Debra Hollis for technical assistance and Desiree Beard for typing this manuscript. The donation of the T4 antibody(Dr. P. Reed Larsen) and cDNAs for IGF1, NGF, IGF2(Dr. Graham Bell), for the tubulin isotypes, GFAP, and neurofilament(Dr. Nicholas Cowan), tubulin and actin(Dr. Donald Cleveland), tyrosine hydroxylase(Dr. Dona Chikairaishi), and for the neural thyroid hormone receptor(Dr. Ron Evans) is appreciated.

REFERENCES

1. G.M. De Escobar, F. Escobar del Rey, Thyroid Hormone and the Developing Brain, in: "Congenital Hypothyroidism", J.H. Dussault, P. Walker, ed., Academic Press, New York, 85-127 (1983).
2. J.T. Eayrs, W.A. Lishman, The maturation of behavior in hypothyroidism and starvation, Brit. J. of Anim. Behav. 3:17-24 (1955).

3. C.E. Hendrich, W.J. Jackson, S.P. Porterfield, Behavioral testing of progenies of Tx(Hypothyroid) and growth hormone treated Tx rats: an animal model for mental retardation, Neuroendo. 438:429-437 (1984).
4. R.Z. Klein, Infantile hypothyroidism then and now: the results of neonatal screening, Curr. Prob. Ped. 15:1-58 (1985).
5. J. Birrell, G.J. Frost, and J.M. Parkin, The development of children with congenital hypothyroidism, Dev. Med. Child Neuro. 25:512-519 (1983).
6. J.F. Rovet, D. Westbrook, R.M. Ehrlich, Neonatal thyroid deficiency: early temperamental and cognitive characteristics, J Am. Acad. Child. Psych. 23:10-22 (1984).
7. J. Legrand, Hormones thyroidiennes et maturation du systeme nerveux, J. Physiol., Paris 78:603-652 (1982-1983).
8. R. Fierro-Benitez, Iodized Oil and Mental Development, in:"Iodine Deficiency Disorders and Congenital Hypothyroidism", G. Medeiros-Neto, R.M.B. Maciel, A. Halpern, ed., Ache, Sao Paolo, Brasil, 120-126 (1986).
9. R. MacFaul, S. Dorner, E.M. Brett, D.B. Grant, Neurological abnormalities in patients treated for hypothyroidism from early life, Arch. Dis. Child. 53:611-619 (1978).
10. E.B. Mann, W.S. Jones, R.H. Holden, E.D. Mellits, Thyroid function in human pregnancy VIII: retardation of progeny aged 7 Years; relationships to maternal age and maternal thyroid function, Amer. J. of Obstet. Gynecol., 111:905-916 (1971).
11. G.R. DeLong, J.B. Stanbury, R. Fierro-Benitez, Neurological signs in congenital iodine-deficiency disorder (endemic cretinism), Dev. Med. & Child Neur. 27:317-324 (1985).
12. K.B. Nelson, J.H. Ellenberg, Antecedents of cerebral palsy: I. univariate analysis of risks, Amer. J. of Dis. Child. 139:1031-1038 (1985).

13. N.P. Rosman, "Neurological and muscular aspects of hypothyroidism in childhood", ed. by A.L. Prensky, The Pediatric Clinics of North America, Saunders, Philadelphia, 575-594 (1975).

14. R.L. Sidman and P. Rakic, Development of the human central nervous system, in: "Histology and Histopathology of the nervous system", W. Haymaker and R.D. Adams, ed., Thomas, Springfield, Il., 1:3-145, (1974).

15. L. Jianquin, W. Xin, Y. Yuquin, W. Kewei, Q. Dakai, X. Zhenfu, and W. Jun, The effects on fetal brain development in the rat of a severely iodine deficient diet derived from an endemic area: Observations on the First Generation, Neuropath. and App. Neurobio. 12:261-276 (1985).

16. M. Marin-Padilla, Prenatal and early postnatal ontogenesis of the human motor cortex: a Golgi study. I. The sequential development of the cortical layers, Brain Res. 23:167-191 (1970).

17. D.A. Fisher, J.H. Dussault, J. Sack, and I.J. Chopra, Ontogenesis of hypothalamic-pituitary-thyroid function and metabolism in man, sheep, and rat, Rec. Progr. Horm. Res. 33:59-107 (1977).

18. C.H. Narayanan, Y. Narayanan, R.C. Browne, Effects of induced thyroid deficiency on the development of suckling behavior in rats, Physiol. Behav. 29:361-370 (1982).

19. S.A. Stein, P.M. Adams, Neurobiological and molecular biological abnormalities in the hyt/hyt hypothyroid mouse, American Thyroid Assoc., Abstract #108, Washington, D.C. (1987).

20. C.H. Narayanan, Y. Narayanan, Cell formation in the motor nucleus and mesencephalic nucleus of the trigeminal nerve of rats made hypothyroid by propylthiouracil, Exp. Brain. Res. 59:257-266 (1985).

21. M. Berry, Development of the cerebral neocortex of the rat, in: "Aspects of Neurogenesis", G. Gottlieb, ed., Academic Press, New York, 3-68 (1974).

22. S.P. Wise, J.W. Fleshman and E.G. Jones, Maturation of pyramidal cell form in relation to developing afferent and efferent connections of rat somatic sensory cortex, Neurosci. 4:1275-1297 (1979).

23. J.M. Donatelle, Growth of the corticospinal tract and the development of placing reactions in the postnatal rat, J. Comp. Neurol., 175:207-232 (1977).

24. D.A. Kristt, Neuronal differentiation in somatosensory cortex of the rat. I.relationship to synaptogenesis in the first postnatal week, Brain Res. 150:467-486 (1978).

25. R.R. Sturrock, Histogenesis of the anterior limb of the anterior commissure of the mouse brain, I. a quantitative study of changes in the glial population with age, J. Anat. 117:17-35 (1974).

26. J.W. Davenport, Perinatal hypothyroidism in rats. persistent motivational and metabolic effects, Dev. Psychobiol. 9:67-82 (1986).

27. J.W. Davenport, L.M. Gonzalez, R.S. Hennies, and W.W. Hagquist, Severity and timing of early thyroid deficiency as factors in the induction of learning disorders in rats, Horm. Behav. 7:139 (1976).

28. T. Noguchi, T. Sugisaki, I. Satoh and M. Kudo, Partial restoration of cerebral myelination of the congenitally hypothyroid mouse by parenteral or breast milk administration of thyroxine, J Neurosci, 45:1419-1426 (1985).

29. W.G. Beamer, E.M. Eicher, L.J. Maltais and J.L. Southard,

Inherited primary hypothyroidism in mice, Science 212:61-62 (1981).

30. W.G. Beamer, L.A. Cresswell, Defective thyroid ontogenesis in fetal hypothyroid (hyt/hyt) Mice, Anat. Rec. 202:387-393 (1982).

31. W.G. Beamer, M.C. Wilson, E.H. Leiter, Endocrinology, In: "The mouse in biomedical research", H.L. Foster, J.D. Small, J.G. Fox, ed. III, Academic Press, New York, 165-245 (1983).

32. P.M. Adams, S.A. Stein, M. Palnitkar, A. Anthony, L. Gerrity, Evaluation and characterization of the hyt/hyt hypothyroid mouse I: somatic and behavioral studies, Neuroendocrinology (in press) (1988).

33. N. Chaudhari and W.E. Hahn, Genetic expression in the developing brain, Science 220:924-928 (1983).

34. D.L. St. Germain, V.A. Galton, Comparative study of pituitary-thyroid hormone economy in fasting and hypothyroid rats, J. Clin. Invest. 75:679-688 (1983).

35. J. Altman, K. Sudarshan, Postnatal development of locomotion in the laboratory rat, Animal Behav. 23:896-920 (1975).

36. J.R. Tata, The action of growth and developmental hormones, Biol. Rev. 55:285-319 (1980).

37. H.C. Towle, "Effects of thyroid hormones on cellular RNA metabolism," J.H. Oppenheimer and H. Samuels, ed., Academic Press, New York, 179-213 (1983).

38. S.A. Stein, Thyroid hormone control of gene expression in Sprague-Dawley rat brain and liver, Ann. of Neuro. 18:385 (1985).

39. C.T. Thompson, C. Weinberger, R. Lebo, R.M. Evans, Identification of a novel thyroid hormone receptor expressed in the mammalian central nervous system, Science 237:1610-1614 (1987).

40. S.A. Lewis, M. G. Lee and N.J. Cowan, Five mouse tubulin isotypes and their regulated expression during development, J. Cell Bio. 101:852-861 (1984).

41. S.A. Lewis and N.J. Cowan, Temporal expression of mouse glial fibrillary acidic protein mRNA studied by a rapid in situ hybridization procedure, J. Neurochem. 913-919 (1985).

42. J.A. Fagin, S. Sianina and S. Melmed, Triiodothyronine stimulates rat pituitary insulin-like growth factor-I gene expression in vitro and in vivo, American Thyroid Association, Abstract #94, (1986).

43. J. Bond, and S. Farmer, Regulation of tubulin and actin mRNA production in rat brain: expression of a new B-tubulin mRNA with development, Molec. Cell Bio. 3:1333-1342 (1983).

44. C.E. Hendrich, W. Ocasio-Torres, J. Berdecia-Rodriquez, S.P. Porterfield, Brain and liver ribosomal protein synthesis and profiles in hypothyroid mothers and their progenies, American Thyroid Assoc., Abstract #106 (1987).

45. J. Scott, M. Urdea, M. Quiroga, R. Sanchez-Pescador, N. Fong, M. Selby, W.J. Rutter, G.I. Bell, Structure of a mouse submaxillary epidermal growth factor and seven related proteins, Science 221:236-240 (1983).

46. S.B. Hoath, J. Lakshmanan, S.M. Scott, D.A. Fisher, Effect of thyroid hormones on epidermal growth factor concentration in neonatal mouse skin, Endocrin. 112:308-314 (1983).

47. R.M. Gubits, P.A. Shaw, E.W. Gresik, A. Onetti-Muda, T. Barka, Epidermal growth factor gene expression is regulated differently in mouse kidney and submandibular gland, Endocrin. 119:1382-1387 (1986).

48. D.L. Simpson, R. Morrison, J. de Vellis, H.R. Herschman, Epidermal growth factor binding and mitogenic activity of

purified populations of cells from the central nervous system, *J. Neurosci. Res.* 8:453-462 (1982).

49. R.S. Morrison, H. I. Kornblum, F. M. Leslie, R. A. Bradshaw, Trophic stimulation of cultured neurons from neonatal rat brain by epidermal growth factor, *Science*, 238:72-75(1987).

50. G. Almazan, P. Honegger, J.M. Matthieu and B. Guentert-Lauber, Epidermal growth factor and bovine growth hormone stimulate differentiation and myelination of brain cell aggregates in culture, *Dev. Brain Res.* 21:257-264 (1985).

51. S.T. Brady, Cytotypic specialization of the neuronal cytoskeleton and the cytomatrix: implications for neuronal growth and regeneration, in: "Cellular and Molecular Aspects of Neural Development and Regeneration", A. Goria, et. al, ed., Springer-Verlag, New York, (1988).

52. T. Takahashi, Transplacental effects of 3,5-dimethyl-3'-isopropyl-1-thyronine on tubulin content in fetal brains in rats, *Japanese Jrnl. of Physio.* 34:365-368 (1983).

53. S. Chaudhury, D. Chatterjee, P.K. Sarkar, Induction of brain tubulin by triidothyronine: dual effect of the hormone on the synthesis and turnover of the protein, *Brain Research* 339:191-194 (1985).

54. Ch. Marc, A. Rabie, Microtubules and neurofilaments of the sciatic nerve fibers of the developing Rat: effects of thyroid deficiency, *Int. J. Dev. Neurosci.*, 3:353-358 (1985).

55. Ch. Faivre, C. Legrand, A. Rabie, Effects of thyroid deficiency and corrective effects of thyroxine on microtubules and mitochondria in cerebellar purkinje cell dendrites of developing rats, *Dev. Brain Res.* 21-30 (1983).

56. Ch. Marc, M. Clavel, A. Rabie, Non-phosphorylated and phosphorylated neurofilaments in the cerebellum of the rat: an immunocytochemical study using monoclonal antibodies. development in normal and thyroid-deficient animals, *Dev. Brain Res.* 249-260 (1986).

57. D. W. Cleveland, K.F. Sullivan, Molecular biology and genetics of tubulin, *Ann. Rev. Biochem.* 54:331-365 (1985).

58. J. Francon, A. Fellous, A. Lennon and J. Nunez, Is thyroxine a regulatory Signal for neurotubule assembly during brain development? *Nature* 266:188-190 (1977).

59. A. Hargreaves, B. Yusta, A. Aranda, J. Avila, A. Pascual, Triiodothyronine (T3) induces neurite formation and increases synthesis of a protein related to MAP1B in cultured cells of neuronal origin, *Dev. Brain Res.*, 141-148 (1988).

60. I. Gozes, K. Sweadner, Multiple tubulin forms are expressed by a single neurone, *Nature* 294:477-480 (1981).

61. R. Cumming, R.D. Burgoyne, N.A. Lytton, Axonal sub-populations in the central nervous system demonstrated using monoclonal antibodies against a-Tubulin, *Eur. J. of Cell Bio.* 31:241-248 (1983).

62. J.N. Wilcox, C.E. Gee, J.L. Roberts, In situ cDNA:mRNA hybridization: development of a technique to measure mRNA levels in individual cells, *Meth. in Enzym.* 124:510-533 (1986).

63. S.A. Stein, D.R. Shanklin, A. Taurog, M.G. Roth, L. Krulich, C.M. Chubb, P.M. Adams, Evaluation and characterization of the hyt/hyt hypothyroid mouse II: abnormalities of TSH and the thyroid gland, *Neuroendocrinology* (in press) (1988).

THYROID INFLUENCES ON THE DEVELOPING CEREBELLUM

AND HIPPOCAMPUS OF THE RAT

Jean M. Lauder

Dept. of Cell Biology and Anatomy
University of North Carolina
Chapel Hill, N.C., 27599, USA

INTRODUCTION

Although cretinism was already described by the turn of
the century, the involvement of thyroid deficiency in this
malady was not reported until the 1930's (Kerley, 1936). Soon
thereafter, animal studies were carried out which made clear
the necessity of thyroid hormones for general somatic and
neural development during the neonatal and early postnatal
period in the rat, which is roughly equivalent to the first
year of life in the human (Salmon, 1936; Scow and Simpson,
1945; Eayrs and Taylor, 1951; Eayrs and Horn, 1955; Hamburg
and Vicari, 1957; Eayrs, 1961).

In 1965, Legrand and co-workers reported effects of
neonatal hypothyroidism on cerebellar development in the rat,
which suggested that thyroid hormones are required for the
normal rate and amount of cell acquisition from an actively
proliferating germinal zone, the external granular layer.
This report prompted us to choose the rat cerebellum to study
the effects of hypo- and hyperthyroidism on germinal cell
proliferation and neuronal differentiation (including axonal
growth and synaptogenesis) in the developing brain (Nicholson
and Altman, 1972a,b,c; Lauder, et al., 1974; Lauder, 1977,
1978, 1979). Later, we examined the effects of altered thyroid states on another postnatally developing brain region,
the hippocampus (Lauder and Mugnaini, 1977, 1980; Lauder and
Ingraham, in preparation). These studies are discussed below,
together with related work from other laboratories, in an
attempt to provide an overview of the types of roles thyroid
hormones may play in brain development, based on studies in
the rodent cerebellum and hippocampus. The reader is also
directed to a more comprehensive review by Legrand (1984).

CELL PROLIFERATION AND DIFFERENTIATION IN THE CEREBELLUM

In the newborn rat, the external granular layer (EGL),
located on the surface of the cerebellum, consists of a

proliferating population of cells which increase in number during the first postnatal week. This is followed by a progressive cessation of cell proliferation and onset of neuronal differentiation, followed by the movement of these cells out of the EGL as they actively migrate (granule cells) or are passively displaced (basket and stellate cells) to reach their final destinations (Altman, 1972a,b; Rakic, 1971, 1972). Thus, the area and width of the EGL reaches a peak after the first postnatal week, then declines, until it disappears at about three weeks of age. In altered thyroid states occurring during this period, the timecourse of EGL growth and decline is changed (Legrand, 1965; Nicholson and Altman, 1972a; Lewis et al., 1976; Lauder, 1977; Legrand et al., 1976; Patel et al., 1979) suggesting effects of thyroid hormones on 1) the rate of cell proliferation within the EGL or its cessation, and/or 2) the rate of movement of differentiating cells out of this germinal zone. In our own studies, we have examined the effects of neonatal hypo- and hyperthyroidism on the rate of cell proliferation in the EGL (Lauder, 1977), the time of cessation of proliferation of specific neuronal precursors (Nicholson and Altman, 1972a), and the rate of movement of neurons out of the EGL (Lauder, 1979), discussed below.

The **methods** used in all of our studies were as follows (Nicholson and Altman, 1972a). Animals were made hypothyroid by daily injection with 0.05 ml of 0.2% propylthiouracil (PTU) in 1% carboxymethyl cellulose on postnatal(P) days 0 (birth)- 10 (P0-10); 0.1 ml 0.2% PTU on P11-20 and 0.1% 0.4% PTU on P21- 30. Thyroids from these animals were monitored histologically at all ages. Animals were judged to be hypothyroid on the basis of lack of colloid and a hyperplastic follicular epithelium. Hyperthyroid animals were injected with 1 ug L-thyroxine (T_4) in 0.1 ml physiological saline on P0-7; 2 ug on P8-14; 3 ug on P15-21; and 5 ug on P22-30, according to the tolerance schedule devised by Hamburgh et al. (1964).

Cell proliferation in the external granular layer

In **hypothyroidism** (Ho) it was found that the EGL proliferates for a longer period of time, as judged by the prolonged presence of cells which can be labelled with ^{3}H-thymidine, a marker of DNA synthesis. However, no effect on the length of the cell cycle could be demonstrated using the percent labelled mitoses method of analysis, even though the percentage of cycling cells (growth fraction) was decreased and the doubling time increased (Lauder, 1977; Lewis et al., 1976). This was indeed puzzling since it was clear that cell acquisition was retarded and prolonged in the Ho EGL. However, upon closer examination we found that the absolute number of labelled mitotic figures increased with time after ^{3}H-thymidine injection, suggesting a prolonged mitotic period, which could not be detected when the data were expressed as a percentage of total mitoses. This observation was confirmed by the finding of increased numbers of cells in the last phases of mitosis (anaphase-telophase), suggesting an inhibition of cleavage itself. We concluded, therefore, that the effects of

Ho on cell proliferation in the EGL occur at the level of mitosis, in particular the cleavage stimulus.

The effects of **hyperthyroidism** (Hr) on cell proliferation in the EGL were easier to understand, since we found that the length of the cell cycle was signficantly shortened following treatment of neonates with thyroxine, due mainly to a decrease in the length of the G_1 phase. We concluded that the reason the EGL disappeared earlier in Hr was in part due to the fact that cells proliferated at an accelerated rate and reached the point of differentiation prematurely, since they were destined to complete a predetermined number of cycles before reaching this point. This hypothesis was supported by the finding that some neurons derived from the EGL were formed earlier than normal in Hr animals, implying premature cessation of proliferation of their precursor cells (Nicholson and Altman, 1972a). A stimulation of cell proliferation in the EGL by thyroxine was also reported by Weichsel (1974), Legrand et al. (1976), Seress (1978) and Rabie et al.(1979). A trend towards a shortened cell cycle and G_1 phase was also found in a study by Patel et al. (1979), although they did not consider these effects to be statistically significant.

The question of whether the effects of altered thyroid states on cell proliferation in the EGL are direct or indirect still must be answered. In favor of a direct effect is our finding that excess thyroxine affects the length of the cell cycle by shortening G_1. This is the phase of the cycle which is involved in the decision to differentiate or complete another round of proliferation (Fox and Pardee, 1971). Moreover, it is this same part of the cycle which is shortened when thyroid hormones are presented to cell lines in vitro , where effects are presumably direct (Burki and Tobias, 1970; Defesi and Surks, 1981; Defesi et al., 1985). In primary cultures from developing rat or mouse cerebellum no such effects on proliferation of EGL cells has been demonstrated (Messer et al. 1984, 1985), although effects on the number of glial cells were found. However, it is not clear whether proliferation of EGL cells occurs at sufficient levels in these cultures to show effects even if they were present. Thus, this issue must await further clarification. Also there is some question as to whether all postnatal germinal zones respond to thyroid hormones in the same way (Seress, 1977, 1978). However, the presence of specific nuclear receptors for thyroid hormones during brain development, (see Chapter by DeNayer) is consistent with the possibility that effects of thyroid hormones on cell proliferation, where they do exist, could occur by direct interactions with the genome.

Granule cell migration and parallel fiber development

Granule cells, quantitatively the largest contingent of neurons formed from the external granular layer (EGL), initiate neurite outgrowth after permanently withdrawing from their division cycle, which takes place in the proliferative zone located in the outer part of the EGL. These cells then pass to the inner part of the EGL (the subproliferative zone), where they orient themselves in the transverse plane. Two axonal processes are first extended parallel to the long axis of the folia (parallel fibers), followed by the emergence of a third, vertically oriented process which descends into the underlying molecular layer and grows towards the internal granular layer (the final destination of the granule cells). The cell body of the granule cell then migrates through the molecular layer to the internal granular layer, presumably using cellular substrates for guidance, such as the Bergmann glia (Del Cerro and Swarz, 1976; Rakic, 1971) and/or Purkinje cell dendrites (Das et al., 1974). The granule cells accomplish this migration either by actively moving in an amoeboid-like fashion, while trailing the descending axon behind (Rakic, 1971) or by translocation of the cell body through this process (Altman, 1975).

We examined the effects of altered thyroid states on granule cell migration at 10 days postnatal using multiple survival ^{3}H-thymidine autoradiography to label cells during their proliferative cycle and then follow their movements out of the EGL, and through the molecular layer to the internal granular layer (Lauder, 1979). Animals were treated as described above. We also studied the effects of hypo- and hyperthyroidism on the growth of parallel fibers using a lesioning and degeneration staining method which allowed us to measure the length of longitudinally oriented bundles of these axons during the first three postnatal weeks (Lauder, 1978).

In these studies it was found that **hypothyroidism** significantly reduced the rate of granule cell migration through the molecular layer, although the movement of these cells within the EGL and their rate of exit from it were unaffected. Likewise, this treatment significantly retarded the growth of parallel fibers, resulting in a permanent deficit in their length.

Hyperthyroidism accelerated the exit of cells from the EGL into the molecular layer as well as the rate of migration of these cells to the internal granular layer. However, the movevment of cells from the proliferative to subproliferative zone within the EGL was retarded. (This explains the reduced percentage of cycling cells in the EGL, decreased growth fraction, since these postmitotic cells dilute the proliferating cell population). Parallel fiber growth was significantly accelerated in these animals, an effect which was proportional to the increased rate of granule cell migration. These effects are in agreement with the findings of Rabie et al. (1979) who reported that thyroxine "induced migration of newly formed granule cells".

The results of these studies emphasize the close relationship between parallel fiber growth and granule cell migration and suggest that changes in the rate of migration observed in altered thyroid states could be secondary to effects on the rate of axonal growth in the developing cerebellum.

Synaptogenesis and granule cell death

Hypothyroidism significantly retards the developmental increase in both the density and total number of synapses after the first two weeks postnatal, a deficit which remains into adulthood (Nicholson and Altman, 1972b,c; Rebiere and Dainat, 1976; Rebiere and Legrand, 1972a). Parallel fibers, which constitute the vast majority of axons in the molecular layer, exhibit deficits in the number of synaptic varicosities located along their length both in terms of the number of sites per unit length and the total number of sites per parallel fiber (Lauder, 1978). Moreover, Purkinje cell dendrites and dendritic spines, with which these axons synapse, are hypoplastic (Brown et al., 1976; Rebiere and Legrand, 1972b).

Deficits in the amount of synaptosomal protein in animals made Ho from birth can be restored by replacement therapy with thyroxine if given early enough (Rabie and Legrand, 1973). A "critical period" for such recovery clearly exists, since if replacement is delayed until after the first 4 weeks of postnatal life, this deficit cannot be ameliorated (Rebiere and Legrand, 1972a).

Alterations in the synaptic organization of the Ho cerebellum have also been noted (Hajos et al., 1973; Crepel, 1974; Rebiere and Dainat, 1976; Rabie et al., 1977). These include: 1) the persistence of basket cell axon-Purkinje cell somatic spine synapses, which are normally transient; 2) decreased density of basket cell axon terminals; 3) increased numbers of Purkinje cell axon collaterals and climbing fibers; 4) the persistence of transient climbing fiber synapses on Purkinje cell bodies; and 5) retarded development of glomeruli (sites of synapses between mossy fibers, granule cell dendrites and Golgi cell axons). Together with our own evidence for synaptic deficits discussed above, it is clear that lowered levels of thyroid hormones during the postnatal period can have lasting effects on circuitry development in the cerebellum.

The synaptic alterations and deficits found in Ho animals is especially interesting in light of reports that such animals also exhibit an increased amount of cell death in the internal granular layer, which may be related to the reduced capacity of granule cells to make the correct number of synapses with their main synaptic targets, the Purkinje cells (Lewis et al., 1976; Rabie et al., 1979a).

In **hyperthyroidism**, both synaptogenesis and parallel fiber growth are accelerated in the molecular layer during the first 3 weeks after birth (Nicholson and Altman, 1972b,c; Lauder, 1978). In fact, a dramatic increase in the final length of parallel fibers is apparent in such animals. The net result is a normal density of synapses in the molecular layer by 24 days, despite the fact that there is a profound deficit in the number of granule cells contributing to this neuropil (Nicholson and Altman, 1972a). There is no effect of excess thyroxine on the number of synaptic varicosities per unit length of parallel fiber (Lauder, 1978), yet because of the greater length of these axons, each parallel fiber contains more than the normal number of these synaptic sites. However, the molecular layer has a drastic reduction in total numbers of synapses due to the fact that the area of this layer is greatly decreased, in part due to the granule cell deficit.

The dramatic stimulation of parallel fiber growth by hyperthyroidism may have functional consequences for the cerebellum, since longer parallel fibers will excite Purkinje cells more laterally placed than these same fibers would normally contact. Since the projections of Purkinje cells to specific deep cerebellar nuclei differ depending on where the Purkinje cells are placed along the medial-lateral axis of the folia (Bell and Dow, 1967; Palay and Chan-Palay, 1974), additional Purkinje cells may be excited by these longer parallel fibers, which will then influence inappropriate deep nucei, leading to abnormal functioning of this circuitry.

Thus, both **hypo- and hyperthyroidism** lead ultimately to significant reductions in total synapses in the cerebellum, and probably to functional miswiring of developing neuronal circuitry.

CELL PROLIFERATION AND DIFFERENTIATION IN THE HIPPOCAMPUS

The hippocampus, like the cerebellum, is a structure which undergoes a large part of its development postnatally. Therefore, this brain region has also been studied with regard to the effects of altered thyroid states during the neonatal period. There are both similarities and differences in the effects of neonatal hypo- and hyperthyroidism on the hippocampus compared to the cerebellum, which may reflect the specific differences in ongoing developmental processes and their regional specificities.

Various aspects of hippocampal development have been exmained, including general areal and volumetric growth, DNA, RNA and protein content (Rabie et al., 1979b), proliferation and migration of precursor cells (Rami et al., 1986a), time of granule cell formation (Lauder and Ingraham, in preparation), dendritic development of granule and pyramidal cells (Rami et al., 1986b), and growth of granule cell axons, the mossy fibers (Lauder and Mugnaini, 1977, 1980).

Cell proliferation and granule cell genesis

Hypothyroidism appears to produce relatively severe effects on the volumetric, longitudinal and areal growth of the hippocampus in contrast to the forebrain where such effects are smaller. Thus there is a desynchronization of coordinated growth between these two brain regions (Rabie et al., 1979). Such a mismatch could lead to miswiring of connecting circuitry and/or behavioral deficits.

These hippocampal deficits are accompanied by a significant decrease in the rate and final amount of cell acquisition (Rabie et al., 1979). According to Rami et al. (1986a), the effects of Ho on the number of proliferating granule cell precursors in the polymorph and granule cell layers are largely non-significant, although it appears from their Figure 3 that there is a significant decrease in the number of proliferating precursors in the polymorph layer at the earliest age examined (5 days). In contrast, they found a highly significant deficit in the rate of cell migration from the polymorph layer to the granular layer. The results of our own study on the timecourse of granule cell genesis (Lauder and Ingraham, in preparation; see Figure 1 below), indicate that Ho leads to an increase in the number of granule cells formed on day 2, followed by a reduction in the number of cells generated at 5 days and thereafter. This is consistent with the results of Rami et al. (1986a).

In **hyperthyroidism**, Seress (1978) found a slight stimulation of cell proliferation in the polymorph and granular layers at 2 days postnatal, followed by a highly significant decrease thereafter. This is of interest since we have found that Hr produces a significant increase in the number of granule cells formed at this time, whereas the number produced thereafter is reduced (Fig. 1).

It appears, therefore, that both **hypo- and hyperthyroidism** have effects on the proliferation of granule cell precursors which influence the time of cessation of their division cycle. In both cases there seems to be an early contingent of granule cells which begin their differentiation prematurely, followed by a deficit in the number of granule cells formed thereafter. In Ho, there is also a reduction in the rate of migration of these cells from their germinal zone in the polymorph layer to their final destination in the granular layer.

Granule and pyramidal cell differentiation

The development of the granule cell-pyramidal cell pathway appears to be particularly vulnerable to altered thyroid hormone levels during the postnatal period. For example, Rami et al. (1986b) reported that the dendritic arborizations of both of these cell types are hypoplastic in **hypothyroid** animals, and that the severity of the deficit in pyramidal cell dendrites is found in that population of cells which receive input from granule cell axons, the mossy fibers. This same population of cells is the most responsive to replacement therapy with thyroxine, according to these authors.

In a study of the effects of **hyperthyroidism** on the developing granule cell-pyramidal cell pathway (Lauder and Mugnaini, 1977, 1980) we found that thyroxine could stimulate the growth of the mossy fibers, in a dose-dependent manner, causing them to sprout collaterals which grew down into the infrapyramidal region and synapsed with the basal dendrites of pyramidal cells. Normally, the mossy fibers innervate only the apical pyramidal cell dendrites where they synapse on

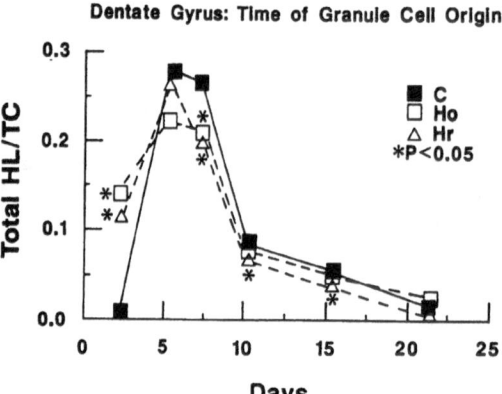

Dentate Gyrus: Time of Granule Cell Origin

Figure 1. Effects of hypo- (Ho) and hyperthyroidism (Hr) on granule cell genesis (time of origin) in the rat hippocampus. Cells were labelled with ^{3}H-thymidine (6.7 Ci/mM, 10 uCi/g bw) on the days indicated and animals were allowed to survive to 60 days of age prior to sacrifice. Heavily labelled cells (HL) are assumed to have stopped proliferating on the day of thymidine injection since they have not undergone label dilution. Animals were made Ho and Hr as described in the text. Counts of HL and total cells (TC) were made in the dentate gyrus using matched horizontal sections. Data are expressed as HL/TC. ■---■ : controls; □---□ : hypothyroid; △---△ : hyperhtyroid. * = p <.05 (Fisher multiple F test). From Lauder and Ingraham, in preparation.

large dendritic spines ("excrescences"). In Hr animals these characteristic spines were also found on the basal dendrites of pyramidal cells where they received synaptic input from the ectopic infrapyramidal mossy fibers. This suggests that Hr can lead to changes in axonal growth, and as a consequence can also cause dendritic spines and synapses to form on an inappropriate part of a target cell. Clearly such miswiring of developing hippocampal circuitry could have longlasting functional and behavioral consequences.

CONCLUSIONS

Investigations of the effects of altered thyroid states on postnatal development of the rat brain have demonstrated the importance of proper levels of thyroid hormones for all phases of neural ontogeny, from germinal cell proliferation to axonal growth and synaptogenesis. Although the underlying mechanisms for these growth regulatory functions are not well understood, it is possible that a common final pathway may be found, perhaps at the level of nuclear receptors and gene regulation. Molecular biologic approaches to this question will hopefully uncover the fundamental actions of these important hormones and further clarify their roles in development of the nervous system.

REFERENCES

Altman, J., 1972a, Postnatal development of the cerebellar cortex in the rat I. The external germinal layer and the transitional molecular layer, J. Comp. Neurol., 145:353.

Altman, J., 1972b, Postnatal development of the cerebellar cortex in the rat II. Maturation of the components of the granular layer, J. Comp. Neurol., 145:399.

Altman, J., 1975, Postnatal development of the cerebellar cortex in the rat. IV. Spatial organization of bipolar cells, parallel fibers and glial palisades, J. Comp. Neurol., 163:427.

Bell, C.C., and Dow, R.S., 1967, Cerebellar circuitry, Neurosci. Res. Prog. Bull., 5:177.

Brown , W.J., Verity, M.A., and Smith, R.L., 1976, Inhibition of cerebellar dendrite development in neonatal thyroid deficiency, Neuropath. appl. Neurobiol., 2:191.

Burki, H.J., and Tobias, C.A., 1970, Effect of thyroxine on the cell generation cycle parameters of cultured human cells, Exp. Cell Res., 60:445.

Crepel, F., 1974, Excitatory and inhibitory processes acting upon cerebellar Purkinje cells during maturation in the rat; influence of hypothyroidism, Exp. Brain Res., 20:403.

Das, G.D., Lammert, G.L., and McAllister, J.P., 1974, Contact guidance and migratory cells in the developing cerebellum, Brain Res., 69:13.

Defesi, C.R., and Surks, M. I., 1981, 3,5,3'-Triiodothyronine effects on the growth rate and cell cycle of cultured GC cells, Endocrinology, 108:259.

Defesi, C.R., Fels, E.C., and Surks, M.I., 1985, L-Triiodothyronine (T_3) stimulates growth of cultured GC cells by action early in the G_1 period: Evidence for mediation by the nuclear T_3 receptor, Endocrinology, 116:2062.

Del Cerro, M.P., and Swarz, J.R., 1976, Prenatal development of Bergmann glial fibres in rodent cerebellum, J. Neurocytol., 5:669.

Eayrs, J.T., 1961, Protein anabolism as a factor ameliorating the effects of early thyroid deficiency, Growth, 25:175.

Eayrs, J.T., and Horn, G., 1955, The development of cerebral cortex in hypothyroid and starved rats. Anat. Rec., 121:53.

Eayrs, J.T., and Taylor, S.H., 1951, The effect of thyroid deficiency induced by methyl thiouracil on maturation of the central nervous system, J. Anat. (Lond.), 85:350.

Fox, T.O., and Pardee, A.B., 1971, Proteins made in the mammalian cell cycle, J. Biol. Chem., 246:6159.

Hajos, F., Patel, A.J., and Balazs, R., 1973, Effect of thyroid deficiency on the synaptic organization of the rat cerebellar cortex, Brain Res., 50:389.

Hamburgh, M., and Vicari E., 1957, Effect of thyroid hormone on nervous system maturation, Anat. Rec., 127:302.

Hamburgh, M., Lynn, E., and Weiss, E.P., 1964, Analysis of the influence of thyroid hormone on prenatal and postnatal maturation of the rat, Anat. Rec., 150:147.

Kerley,G.C., 1936, Childhood myxedema: observations through infancy, childhood and early life. Endocrinology, 20:611.

Lauder, J.M., and Mugnaini, E., 1977, Early hyperthyroidism alters the distribution of mossy fibers in the rat hippocampus, Nature (Lond.), 268:335.

Lauder, J.M. and Mugnaini, E., 1980, Infrapyramidal mossy fibers in the hyperthyroid hippocampus: A light and electron microscopic study in the rat, Dev. Neurosci., 3:248.

Lauder, J.M., 1977, The effects of early hypo- and hyper-thyroidism on the development of rat cerebellar cortex. III. Kinetics of cell proliferation in the external granular layer, Brain Res., 126:31.

Lauder, J.M., 1978, Effects of early hypo- and hyperthyroid-ism on development of rat cerebellar cortex. IV. The parallel fibers, Brain Res., 142:25.

Lauder, J.M., 1979, Granule cell migration in the developing rat cerebellum. Influence of neonatal hypo- and hyper-thyroidism, Dev. Biol., 70:105.

Lauder,J.M., Altman, J. and Krebs, H., 1974, Some mechanisms of cerebellar foliation: Effects of early hypo- and hyperthyroidism, Brain Res., 76:33.

Legrand, J., 1984, Effects of thyroid hormones on central nervous system development, in: "Neurobehavioral Teratology", J. Yanai, ed., Elsevier, Amsterdam, p 331.

Legrand, J., 1965, Influence de l'hypothyroidisme sur la maturation du cortex cerebelleux. C.R. Acad. Sci. (Paris), 261:544.

Legrand, J., Selme-Matrat, M., Rabie, A., Clos, J., and Legrand, C., 1976, Thyroid hormone and cell formation in the developing rat cerebellum, Biol. Neonate, 29:368.

Lewis, P.D., Patel, A.J., Johnson, A.L. and Balazs, R., 1976, Effect of thyroid deficiency on cell acquisition in the postnatal rat brain: A quantitative histological study, Brain Res., 104:49.

Messer, A., Maskin, P., and Snodgrass, G.L., 1984, Effects of triiodothyronine (T$_3$) on the development of rat cerebellar cells in culture. Int. J. Devl. Neurosci., 2:277.

Messer, A., Snodgrass, G.L. and Maskin, P., 1985, Timecourse of effects of triiodothyronine on mouse cerebellar cells cultured by two different methods, Int. J. Devl. Neurosci., 3:291.

Nicholson, J.L. and Altman, J., 1972a, The effects of early hypo- and hyperthyroidism on the development of the rat cerebellar cortex. I. Cell proliferation and differentiation, Brain Res., 44:13.

Nicholson, J.L. and Altman, J., 1972b, The effects of early hypo- and hyperthyroidism on the development of the rat cerebellar cortex. II. Synaptogenesis in the molecular layer, Brain Res., 44:25.

Nicholson, J.L. and Altman, J., 1972c, Synaptogenesis in rat cerebellum: Effects of early hypo- and hyperthyroidism, Science, 176:530.

Palay, S.L. and Chan-Palay, V., 1974, "Cerebellar Cortex, Cytology and Organization", Springer-Verlag, New York, pp 4-5; 63-71.

Patel, A.J., Lewis, P.D., Balazs, R., Bailey, P. and Lai, M., 1979, Effects of thyroxine on postnatal cell acquisition in the rat brain. Brain Res., 172:57.

Rabie, A. and Legrand, J., 1973, Effects of thyroid hormone and undernourishment on the amount of synaptosomal fraction in the cerebellum of the young rat., Brain Res., 61:267.

Rabie, A., Favre, C., Clavel, M.C., and Legrand, J., 1977, Effects of thyroid dysfunction on the development of the rat cerebellum, with special reference to cell death within the internal granular layer, Brain Res., 120:521.

Rabie, A., Favre, C., Clavel, M.C., and Legrand, J., 1979a, Sequential effects of thyroxine on the developing cerebellum of rats made hypothyroid by propylthiouracil, Brain Res., 161:469.

Rabie, A., Patel, A.J., Clavel, M.C., and Legrand, J., 1979b, Effect of thyroid deficiency on the growth of the hippocampus in the rat, Dev. Neurosci., 2:183.

Rakic, P., 1971, Neuron-glia relationships during granule cell migration in developing cerebellar cortex. A Golgi and electron microscopic study in Macacus rhesus, J. Comp. Neurol., 141:283.

Rakic, P., 1972, Extrinsic cytological determinants of basket and stellate cell dendritic pattern in the cerebellar molecular layer, J. Comp. Neurol., 146:335.

Rami, A., Rabie, A., and Patel, A.J., 1986a, Thyroid hormone and development of the rat hippocampus: Cell acquisition in the dentate gyrus, Neuroscience, 19:1207.

Rami, A., Patel, A.J. and Rabie, A., 1986b, Thyroid hormone
    and development of the rat hippocampus: Morphological
    alterations in granule and pyramidal cells,
    Neuroscience, 19:1217.
Rebiere, A. and Dainat, J., 1976, Repercussions de l'hypo-
    thyroide sur la synaptogenese dans le cortex cerebel-
    leux du rat, Acta Neuropathol., 35:117.
Rebiere, A. and Legrand, J., 1972a, Donnees quantitatives sur
    la synaptogenese dans le cervelet du rat normal et
    rendu hypothyroidien par le propylthiouracyle, C.R.
    Acad. Sci (Paris), 274:3581.
Rebiere, A. and Legrand, J., 1972b, Comparative effects of
    underfeeding, hypothyroidism and hyperthyroidism on
    the histological maturation of the molecular layer of
    the cerebellar cortex of the young rat, Arch. Anat.,
    mic. morphol. Exp., 61:105.
Salmon, T.N., 1936, Effect of thyro-parathyroidectomy in
    newborn rats, Proc. Soc. Exp. Biol. Med., 35:489.
Scow, R.O. and Simpson, M.E., 1945, Thyroidectomy in the
    newborn rat. Anat. Rec., 91:209.
Seress, L., 1977, The postnatal development of rat dentate
    gyrus and the effect of early thyroid hormone treat-
    ment, Anat. Embryol., 151:335.
Seress, L., 1978, Divergent responses to thyroid hormone
    treatment of the different secondary germinal layers
    in the postnatal rat brain, Hirnforschung, 19:395.
Weichsel, M., 1974, Effect of thyroxine on DNA synthesis and
    thymidine kinase activity during cerebellar develop-
    ment, Brain Res., 78:455.

QUANTITATIVE STUDIES OF THE EFFECTS OF HYPOTHYROIDISM ON THE DEVELOPMENT

OF THE CEREBRAL CORTEX

Antonio Ruiz-Marcos

Unidad de Neuroanatomia
Instituto Cajal
C.S.I.C., Madrid

INTRODUCTION

It is very well known for many years that neonatal hypothyroidism deranges the development of the cerebral cortex (C.C.), producing profound and often irreparable alterations in its structure[4,7].

It seems therefore important to study how this derangement is produced, trying to establish where, inside the structure of the C.C., the damage is more pronounced, and how it evolves once the hypothyroid condition is established.

The extreme complexity of the neuropil which forms the microstructure of the C.C. imposes some limitations to these types of studies as, due to this complexity, the differences induced in it by any pathological condition is, in most cases, far from being evident by a simple inspection at the microscope of a section of the C.C. properly stained.

To look at a microscopic section of the C.C. stained by any silver procedure is, in some way, similar to looking at a screen where many pictures are projected simultaneously. If we do not have the proper filters which allow us to see these pictures individually we reach the apparent paradox that, because of the excess of information present on the screen, we do not get the information contained in it. In this respect mathematical models and some special mathematical algorithms are, in a sense, one of the types of "filters" which can be used to obtain the information contained in one section of the C.C. stained according to the rapid Golgi procedure. This makes possible the study of some of the properties of the structure of the neuropil of the C.C., and how this structure is affected by some pathological conditions, such as hypothyroidism.

Eayrs[4] was one of the first to apply quantitative techniques to the study of the effects that hypothyroidism has on cortical neurons. Although he deserves the credit of having been a pioneer in these types of studies at a time when sophisticated computer techniques were not available, his results, probably because of the lack of technical facilities, were not conclusive. Thus, he already mentioned that this pathological condition affects the density of the dendritic arborization of cortical neurons, but did not mention if this effect was produced with preference in some

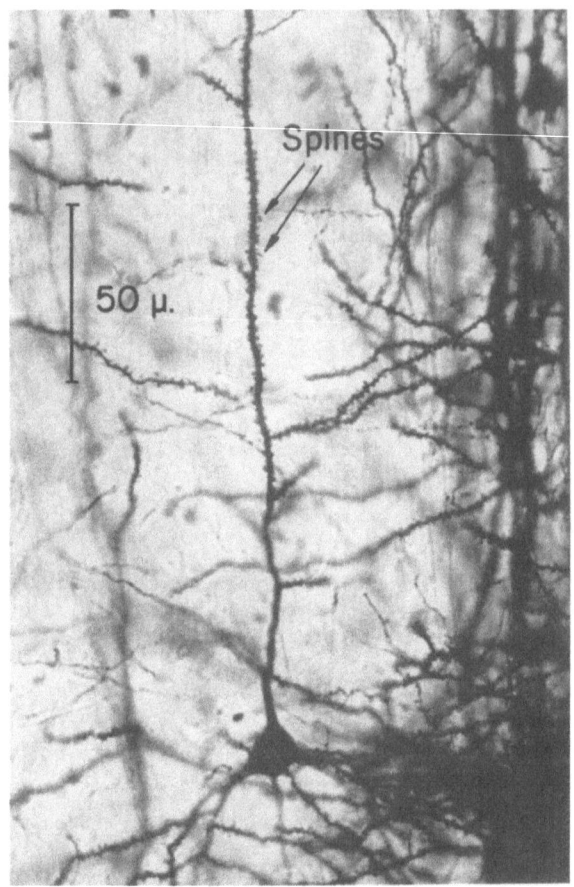

Fig. 1. Microphotography of a portion of the
apical shaft of a pyramidal neuron of
the layer V of the cerebral cortex of
a rat 30 days old, showing dendritic
spines distributed along it. Rapid
Golgi method.

specific region of this structure[5]. This result leads, further, to the
idea, generally accepted, that hypothyroidism affects the morphology of
cortical neurons as a whole making them hypoplastic[6,14]. Nevertheless, the
mathematical study of the distribution of dendritic spines along the apical
shafts of pyramidal neurons of layer V, and the density of the dendritic
arborization of pyramidal cells of layer III of the C.C., has shown that
hypothyroidism induced by surgical thyroidectomy performed on Wistar rats
at neonatal[10,23,27] or adult ages[22] affects more profoundly the region of
these neurons located in the superficial layers of the C.C. than elsewhere.

    The simple microscopic observation of the sequence of dendritic spines
along the apical shafts of pyramidal neurons (Fig. 1), leads us to think
they are distributed at random along them. Nevertheless, in 1969, a

mathematical study of the number of spines, counted in portions of 50 μ. along the apical shafts of pyramidal neurons of layers III and V of the visual cortex of the mouse, allowed us to demonstrate that the spines are not distributed at random along these shafts, but according to a very well defined distribution[30] similar to the ones shown on the left-hand panel of Fig. 2. The properties of this distribution and of its evolution with the age of the animal were, further, described by a set of equations or mathematical model[28]. This model was found to be valid to define the distribution of spines along the shafts of pyramidal neurons of cortices other than the visual and for species other than the mouse, such as rat, hamster, cat, monkey and man. With it it was possible to describe the changes induced by total darkness in the maturation of the mouse visual cortex[28].

The importance of dendritic spines in the establishment of neuronal connections[3], and the results obtained from the above mentioned study of the effect of darkness on the development of the mouse visual cortex, lead us to think that this mathematical model could be used, as a tool, to study the effects of hypothyroidism on the development of the cerebral cortex.

STUDIES OF THE DISTRIBUTION OF DENDRITIC SPINES

In order to study, with the help of the above mentioned mathematical model and some mathematical algorithms derived from it, how neonatal hypothyroidism affects the development of the C.C., 5 groups of 8 Wistar rats each were surgically thyroidectomized when they were 10 days old (T). All these groups of animals and their age-paired controls (C) were killed at 20, 25, 30, 40 and 80 days of age. The portion of their brains containing the primary visual and auditory areas of their C.C. were stained according to the rapid Golgi procedure, slightly modified by us[25], and sections 200μ thick of these areas were prepared and mounted. The number of dendritic spines were counted on consecutive segments 50μ long, along 20 apical shafts of pyramidal neurons of layer V of the cerebral cortex of each group of rats, and the data obtained were stored on the permanent magnetic memory of a PDP 11/40 computer for further study.

The results obtained from this study showed that T, performed on rats at 10 days of age, produces, from 10 to 30 days, a smaller increase in the total number of spines counted along the apical shafts than during normal development. From 30 days of age onwards T produces an arrest of the increment of the total number of dendritic spines along the shafts. Furthermore, while it was possible to fit with the mathematical model the distribution of dendritic spines along the apical shafts of pyramidal cortical neurons of T rats 10, 20 and 30 days old, it was no longer possible to find this fitting for T animals older than 30 days. Therefore, T performed on rats at 10 days of age produces, not only an arrest in the production of dendritic spines in their C.C., but also (and what probably is more important) distortion of the distribution of these elements along the apical shafts of pyramidal neurons of layer V of the cerebral cortex[24,25].

Considering that these shafts run through the whole depth of the cortex, and that the dendritic spines of these shafts receive 80% of the specific afferent fibers reaching the C.C.[3], these results indicate that T, induced in the neonatal period in rats, produces a general derangement of the connective properties of the primary areas of the C.C.

Nevertheless, as a consequence of the results obtained from the study of how different T4 treatments, applied to T rats, could restore the above described damage[24,27], it was found that the derangement produced by neonatal T on the connectivity of layer V cortical pyramidal neurons is not

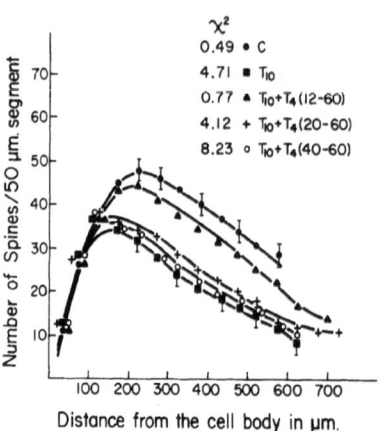

Distributions of Dendritic Spines
Visual Cortex: 60 days old rats

χ^2

0.49 • C
4.71 ■ T_{10}
0.77 ▲ $T_{10}+T_4$ (12-60)
4.12 + $T_{10}+T_4$ (20-60)
8.23 ○ $T_{10}+T_4$ (40-60)

Number of Spines/50 μm. segment

Distance from the cell body in μm.

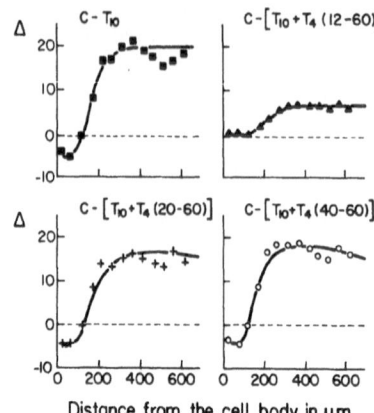

Differences (Δ) in the mean number
of Spines/50 μm., of T and treated
rats versus C (60) rats.

Distance from the cell body in μm.

Fig. 2.   Effect of thyroidectomy at 10 days of age (T_{10}), and of
T_4 treatments, started at 12,20 and 40 days of age, on
the distribution of spines along the apical shafts of
pyramidal neurons. The left-hand panel shows the
experimental distributions corresponding to C, T and
treated rats.   Vertical bars represent 95% confidence
intervals.   The values of X^2 indicate the degree of fit
of the distributions to the mathematical model, a
value >4.5 indicating lack of fit. The right-hand panel
shows the differences between the number of spines of
equivalent segments of the distribution of T and treated
rats and those of C animals.   Data are from Ruiz-Marcos
et al.[27].

uniform along the whole depth of the C.C., being stronger on the more
superficial region.

     The left-hand panel of Fig. 2 shows the experimental distributions of
dendritic spines, measured along the apical shafts of layer V cortical
pyramidal neurons of five different groups of rats, all of them studied at
the same age, 60 days. One of the groups (T_{10}) was thyroidectomized when
the animals were 10 days old and remained untreated throughout their life.
Three of them, also T at 10 days, received a daily injection of 1.5 μg.
thyroxine (T_4)/100 g. of body weight, starting the treatment for each of
these groups when they were 12, 20 and 40 days old, respectively. The
fifth group (C) was sham operated and served as control. The values of the
X^2 parameter, shown on the figure, obtained from the fitting between the
experimental distributions and the mathematical model showed that, while
the T_4 treatments started 2 and 10 days after T could restore the
distribution of dendritic spines to the normal condition corresponding to C
animals, the treatment started 30 days after T had no effect on the
restoration of the damages produced by T on the distribution of dendritic
spines.

     In the right-hand panel of Fig. 2 are represented the differences
between the number of dendritic spines, of equivalent segments, of the
distribution of T and treated rats and those of C animals. The figure
shows that these differences increase with the distance to the cell body,

reaching their maximum value on the region of the shafts located in the more superficial layers of the C.C. Concerning these differences, it seems important to notice that the effect of T on the initial segments of the apical shafts, placed in layers V-IV of the cortex, is opposite to the one produced on the distal parts of these shafts, located in the superficial layers of the C.C. Although it still remains to find an explanation of this opposite effect of T on both regions of the apical shafts, the results obtained indicate that this last effect of T on the proximal part of the shafts is not as strong as the one produced on the distal part, as it can be completely reversed by T_4 treatment started 2 days after T. Furthermore, the fact that all these differences have been studied in animals of the same age, 60 days, and that they decrease as the onset of the treatment is closer to the day on which T was performed, shows that they are a direct consequence of T and not due to some other causes[8].

The direct conclusion drawn from these results is that, at least concerning the dendritic spines distributed along the apical shafts of pyramidal neurons of layer V of the C.C., the effect of T performed at 10 days of age is not uniform along the length of these shafts, being stronger in the region of these shafts located in the more superficial layers of the C.C. At this point the question remains whether this effect is only dependent on the distance to the cell body, or is due to a general derangement of the region of the C.C. where it is produced, something which was somewhat clarified by further experiments.

STUDIES OF DENDRITIC DEVELOPMENT

Obviously, the connectivity of the neurons does not depend only on the total number of dendritic spines present along their dendrites, but also, among other aspects, on the general structure of their dendritic arborization.

In order to study the effect that T could have on the structure of the dendritic arborization of pyramidal neurons, a total of 18 such neurons, chosen at random from layer III of the visual area of the C.C. of 10 control Wistar rats 80 days old, and an equal number of neurons of the same layer of the C.C. of 10 rats of the same age, T when they were 10 days old, were drawn using a camera lucida with a total magnification of 500X. Following a procedure described in detail elsewhere[19,23] the 3 spatial coordinates of the most important points of the dendritic structure, i.e. origin of each dendrite at the soma, inflection points, bifurcation points and end points, were codified and stored in the permanent magnetic memory of a PDP 11/40 computer, according to the instructions of a special program made by us, named ADQUI, for further quantitative study.

The program ADQUI makes use of a sonic digitizer, directly connected to the computer. With this digitizer it is possible to introduce into the computer memory the values of the three spatial coordinates of the selected points to which the preceding set of coordinates corresponds. The values of the two planar (X,Y,) coordinates are transferred by simply touching the corresponding point with the pen of the digitizer. The third coordinate (Z) of these points is measured at the time the drawing of the neuron is made using a sensor (Millitron) attached to the fine focus of the microscope. These values were transferred to the computer memory by touching with the digitizer pen one special scale printed for this purpose on the digitizer board.

As proof that the machine is able to interpret coherently all information as a neuron, and to handle it properly, another program DYNFOT was written in FORTRAN language. Following the instructions of this program the computer is able to reconstruct the whole dendritic

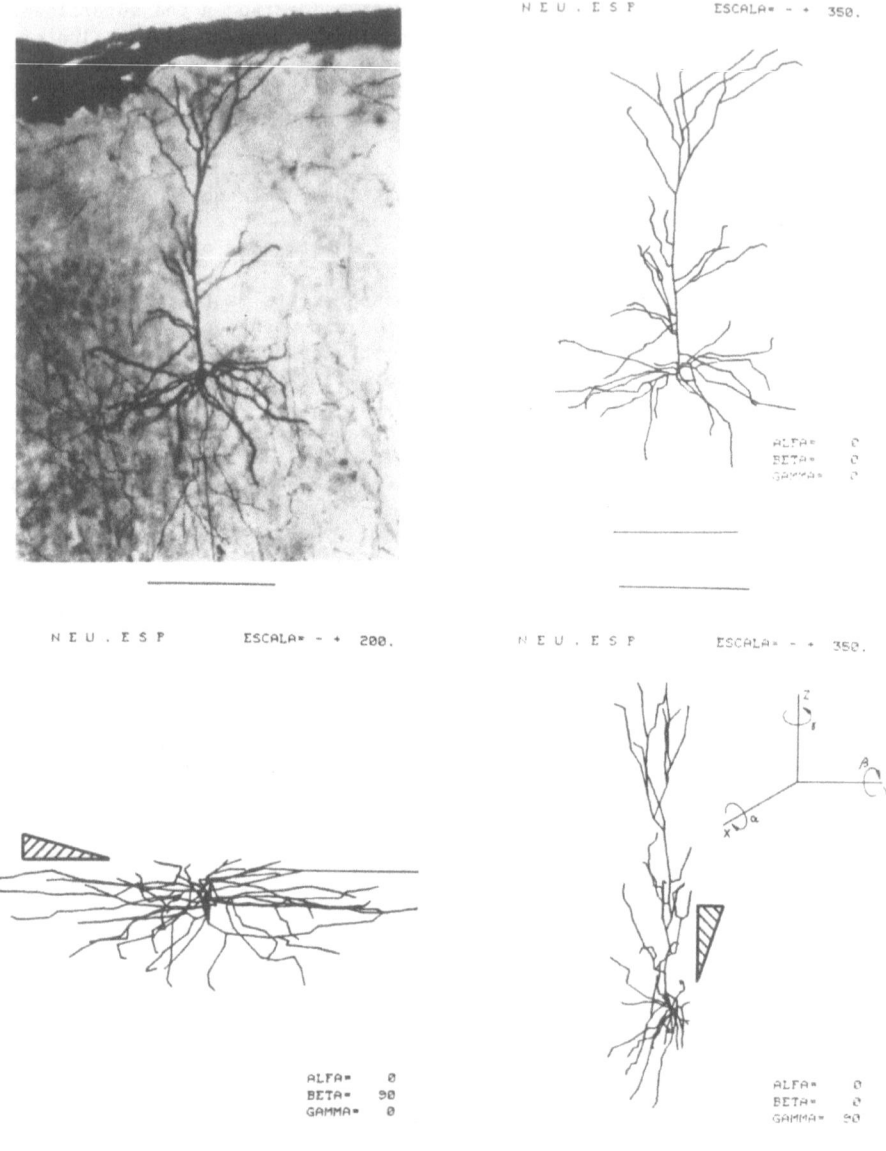

Fig. 3.   Computer reconstruction of the dendritic structure of a
           pyramidal neuron. In the upper-left panel is shown a
           microphotograph of the neuron. In the upper-right the
           reconstruction, made by the computer, in its original
           position. On the lower-left, rotated 90° around the Y axis
           (as seen from below), and on the lower-right, rotated 90°
           around the Z axis (as seen from the side). In these two
           last representations, it can be observed how the basal
           dendrites of the neuron have been cut by the knife of the
           microtome, given the superficiality of the neuron in the
           section. Data are from Ruiz-Marcos[21].

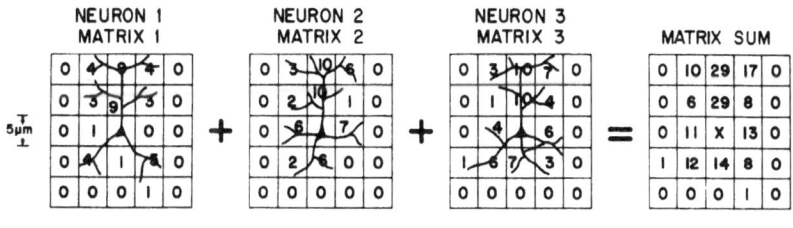

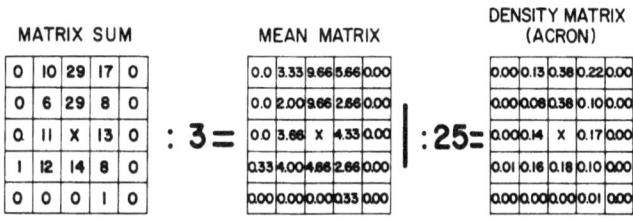

Fig. 4.   Procedure followed by the computer to calculate the mean
           dendritic density matrix corresponding to three hypothetical
           neurons. The neurons are projected on a grid, annotating on
           each individual square the total dendritic length crossing
           it. The numerical matrices corresponding to each individual
           neuron are added, and the matrix sum obtained is normalized.
           Reprinted from Ruiz-Marcos and Ipiña[23], by courtesy of
           Elsevier Biomedical Press, Amsterdam. The Netherlands.

arborization of the neuron showing it on its vectored screen. As the
machine has the spatial information corresponding to each point of the
neuron, making use of the rotational matrix it can recalculate the position
of each point of the dendritic structure, after rotating it in any of the
three angles of space, and once these positions have been calculated, to
show the rotated structure on its screen. Fig. 3 shows an example of the
reconstruction, made by the computer, of a particular neuron shown in the
microphotograph in the upper-left hand panel of the figure[20,21].
Nevertheless, it is to be pointed out that, although this rotation
algorithm and its consequent program is useful to study the shape of the
dendritic structure of a neuron, it is useless to detect the influence that
any given pathological condition, such as hypothyroidism, could· have on the
dendritic structure of a neuronal population.

        With the aim of studying the effect that any given pathological
condition has on the dendritic structure of a set of N neurons, Ruiz-
Marcos and Valverde[29] developed a special algorithm which was later
perfected by Ruiz-Marcos[19].

        According to the revised version of the algorithm and following the
instructions of a new program named ACRON, the computer creates a virtual
grid on its memory, "projects" each individual neuron over the grid (making
the soma of the neuron to coincide with the center of the grid) and
measures the total length of dendrites crossing each individual square
forming the grid (Fig. 4). According to this procedure the computer
transforms one neuron into a numerical matrix. Once the individual
numerical matrices, corresponding to a homogeneous group of neurons, (i.e.,
neurons belonging to the same area and layer of the C.C. of animals of the
same age, raised under the same condition) have been calculated, the

computer proceeds further, adding together all these matrices, finding the matrix sum corresponding to the N original neurons of the group.

The value obtained by the individual elements of this matrix sum clearly depends on the total number of neurons (N) entering the computation and on the area of the individual squares forming the grid. In order to normalize this matrix sum, thus making it independent of these two factors, each individual element of it is divided by the number N of original neurons entering the computation, and by the area of each individual square. As a final result we obtain one normalized mean matrix or ACRON (from Average Computed neuRON). The elements of this ACRON represent the mean dendritic densities of the group of neurons entering the study, at a certain distance from their cell body, defined by the position of each square inside the grid.

Simultaneously with the calculation of the ACRON corresponding to a group of neurons, the computer calculates and stores in its memory the corresponding variance-covariance matrices. This last calculation allows us to proceed further, comparing the ACRONs corresponding to two homogeneous groups of neurons.

In order to find possible existing differences between two ACRONs (or density matrices), the differences between the homologous elements (those which have the same position in the ACRONs) are calculated and, according to a specific statistical procedure, the computer finds the level of statistical significance of these differences. The machine is then instructed to print a sign +, on those places of the ACRONs where the differences (C-T) are positive, and statistically significant beyond a certain level of significance given to the computer as one of the initial data. On those places where the differences are negative, and statistically significant, the computer prints a sign -, and prints blanks on the remaining places where the differences are zero or not statistically significant.

Fig. 5 shows a superimposition of the schemes of three layer III cortical pyramidal neurons and the computer results, obtained from the comparison of the two already mentioned samples of pyramidal neurons, of layer III of the C.C. of C and T rats. It is to be noticed that similar results have been obtained from the comparison of two independent neuronal samples of C and T animals, using two different statistical criteria, described in detail by Ruiz-Marcos[19] and Ruiz-Marcos and Ipiña[23]. According to these results, the region of the dendritic arborization of the pyramidal neurons more affected by T is the one located in the more superficial region of the C.C., and corresponds to the apical tuft of these neurons.

Further studies made by Ruiz-Marcos and Ipiña[23] of the intensity of the effect of T on the dendritic density of pyramidal neurons showed that this intensity increases as the distance from the cell increases along a line parallel to the apical shafts of the pyramidal neurons, reaching a maximum at the upper most part of the neuron. This last result was similar to that already mentioned, concerning the effect of T on the number of dendritic spines along the apical shafts of pyramidal neurons of layer V of the C.C.[22,27].

To find out which of the different possible parameters (such as: number of dendrites, dendritic length, index of ramification, etc.) which define the dendritic structure of a neuron could be more affected by T, Ipiña and Ruiz-Marcos[10] defined the whole dendritic structure of a pyramidal neuron by a set of 10 such variables. The results obtained by these authors from the multivariate analysis of the values attained by these 10 variables in layer III pyramidal neurons of C and T rats, showed that the decrease of dendritic density is due to a failure of the development of the dendrites of the apical tuft region of pyramidal neurons of T animals which, as a consequence, became shorter.

Furthermore, the results obtained by Berbel et al.[1,2], from the studies of the effect of neonatal hypothyroidism on the microtubule density and on their arrangement inside the apical shafts of pyramidal cells of layer V of the C.C. of rats, and on the density of myelinated profiles in different layers of their cortex, showed that these neuronal elements also were more affected by hypothyroidism in the more superficial layers of the cortex than in the rest.

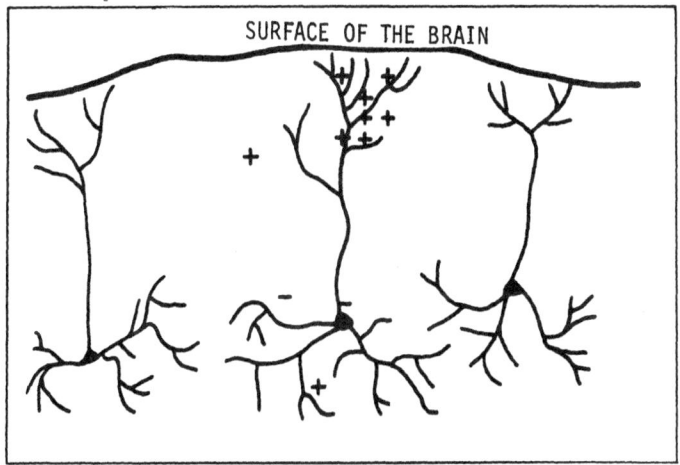

Fig. 5.- Superimposition of the schematic drawings of three pyramidal neurons, and the results obtained from the comparison between the mean dendritic density matrices, corresponding to two neuronal samples of C and T rats. Signs + indicating C>T, with P<0.05. Data are from Ruiz-Marcos and Ipiña[23].

CONCLUSION AND DISCUSSION

The similarity of all these results, and the fact that the more distal region of the basal dendrites are not affected by T, seems to indicate that the more pronounced effect of T on the more distal region of the apical shafts of pyramidal neurons could be due to a general derangement of the more superficial layers of the C.C., and not only to an effect due to the distance to the cell body.

Taking into account the inside-out theory of cortical plate and neuronal development formulated by Rakic[16,17], according to which the more superficial layers of the C.C. are the first to mature, we could think that this non-uniform derangement produced by neonatal hypothyroidism on the development of the different layers of the C.C. is due, at least in part, to the different time and rate of their maturation. However, the study of the relative reduction of dendritic spines along the apical shaft of pyramidal neurons of layer V of the C.C. of rats 90 days old, T in adulthood at 40 days of age[22], showed again that this reduction is greater in the superficial layers of the C.C. than in the rest, in a form similar to the one shown on Fig. 2. As the neurons were already matured at the age of 40 days, when the thyroidectomy was performed on these rats, this effect cannot be attributed only to early maturation, making it necessary to think of other causes besides development by which hypothyroidism affects more profoundly the superficial region of the C.C.

The results obtained by several authors according to which the development of the different subregions of the pyramidal neuronal dendritic arborizations[11,15] and the production of dendritic spines[9,18,31] is influenced by the afferent axonal system to the C.C. lead Ipiña and Ruiz-Marcos[10] to interpret the results described in the present work as being a consequence of the possible damage, induced by T, on the afferent systems to the plexiform cortical layers and, therefore, on the neurons where they originate. These neurons, according to the hypotheses suggested by Marin-Padilla, are the ones located in the mesencephalic nuclei of the reticular formation, as well as the Cajal-Retzius and Martinotti cells[12,13].

Concerning these last considerations, although some of the effects of T performed in adulthood are similar to those produced by neonatal T it is important to point out that, contrary to the findings after early hypothyroidism[25], T performed at adult age does not produce any distortion of the distribution of spines along the apical shaft of layer V cortical pyramidal neurons[22,26]. These last results can be taken as an indication that, although hypothyroidism may somehow affect the structure of the C.C. even if it starts at an adult age, the effect of this condition on the afferent systems to the apical shafts of the pyramidal neurons, and therefore on the neurons where they originate, should be if anything less intense than that produced by hypothyroidism induced early during the life of the animal.

ACKNOWLEDGEMENTS

This work has been supported by grants n° 88/1650 from the Fondo de Investigaciones Sanitarias (FIS) and n° 154/1985 from the Comision Asesora para la Investigacion Cientifica y Tecnica (CAICYT) to Dr. Antonio Ruiz-Marcos. The author wants to express his gratitude to Miss M. E. Fdez. de Molina for typing the manuscript, to Mr. A. Hurtado for his art work and help in the preparation of the figures, and to Mrs. M. C. Gabriel Alvarez for her help with the computer work.

REFERENCES

1. P.J. Berbel, F. Escobar del Rey, G. Morreale de Escobar and A. Ruiz-Marcos, Effect of neonatal hypothyroidism on the microtubule density and arrangement in apical dendrites of pyramidal cells of the rat visual cortex, Neurosc. Lett. S-26 (1986).

2. P.J. Berbel, F. Escobar del Rey, G. Morreale de Escobar and A. Ruiz-Marcos, Differential effect of neonatal hypothyroidism on myelinated profiles in different layers of the cerebral cortex of the rat, Trab. del Inst. Cajal. Vol. LXXV: 37 (1984).

3. M. Colonnier, Synaptic patterns of different cell types in the different laminae of the cat visual cortex: an electron microscopic study, Brain Res. 9: 268-287, 1968.

4. T.J. Eayrs, Thyroid hypofunction and the development of the central nervous system, Nature (London) 172:403-405, 1953.

5. T.J. Eayrs, Influence of the thyroid in the central nervous system, Br. Med. Bull. 16: 122-127 (1960).

6. T.J.Eayrs, Thyroid and the developing brain: anatomical and behavioral effects, in: "Hormones in Development", M. Hamburgh and B.J. Barrington, ed., Appleton-Century-Crofts, New York (1971).

7. D.H. Ford and E. Cramer, Developing system in relation to thyroid hormones, in: "Thyroid Hormones and Brain Development", G.D. Grave, ed., Raven Press, New York (1977).

8. W.T. Greenough, J.M. Juraska and F.R. Volkman, Maze training effects on dendritic branching in occipital cortex of adult rats. Behav. Neurol. Biol. 26: 287-297 (1979).

9. J. Hamory, The inductive role of presynaptic axons in the development of postsynaptic spines, Brain Res. 62:337-344 (1973).

10. S.L. Ipiña and A. Ruiz-Marcos, Dendritic structure alterations induced by hypothyroidism in pyramidal neurons of the rat visual cortex, Dev. Brain Res. 29: 61-67 (1986).

11. D.A. Krisst, Neuronal differentiation in somatosensory cortex of the rat. I. Relationship to synaptogenesis in the first postnatal week, Brain Res. 150: 467-487 (1978).

12. M. Marin-Padilla, Early prenatal ontogenesis on the cerebral cortex (neocortex) of the cat (felix domestica): a Golgi study. I. The primordial neocortical organization, A. Anat. Entwickl. Gesch. 134: 117-145 (1971).

13. M. Marin-Padilla, Neurons of layer I. A developmental analysis, in: "Cerebral Cortex, Vol. 1: Cellular components of the cerebral cortex", A. Peters and E.G. Jones, ed., Plenum Press, New York and London (1984).

14. M.S. Mistkevich and G.N. Moskovkin, Some effects of thyroid hormone on the development of the cerebral nervous system on early ontogenesis, in: "Hormones and Development", M. Hamburg and B.J. Barrington, ed., Appleton-Century-Crofts, New York (1971).

15. M.C. Pinto-Lord and V.S. Caviness, Determinants of cell shape and orientation: a comparative Golgi analysis of cell axon interrelationships in the developing neocortex of normal and reeler mice. J. Comp. Neurol. 187: 49-69 (1979).

16. P. Rakic, Neurons in rhesus monkey visual cortex: systematic relation between time of origin and eventual disposition. Science. 183: 425-427 (1985).

17. P. Rakic, Developmental events leading to laminar and real organization of the neocortex, in: "The Organization of the Cerebral Cortex". F.O. Schmitt, F.G. Worden, G. Adelman and S.G. Dennis, ed., The MIT Press, Cambridge, Massachusetts and London (1981).

18. W. Richter, Neurologische und morphometrische Untersuchungen der Ontogeneses der regio cingularis mesoneocorticalis der Ratte, J. Hirnfors. 21: 53-87 (1980).

19. A. Ruiz-Marcos, Mathematical models of cortical structures and their application to the study of pathological situations, in: "Ramon y Cajal's Contribution to the Neurosciences", S. Grisolia, C. Guerri, F. Samson, S. Norton and F. Reinoso-Suarez, ed., Elsevier Science Publ., Amsterdam, New York, London (1983).

20. A. Ruiz-Marcos, Modelos Matematicos de sistemas neuronales, Invest. y Ciencia (Spanish edition of Scientific American), 93: 98-108 (June 1984).

21. A. Ruiz-Marcos, Mathematical models and algorithms used to study the effect of some pathological conditions such as hypothyroidism or iodine deficiency on the development of the cerebral cortex, in: "Iodine Nutrition Thyroxine and Brain Development", N. Kochupillai, M.G. Karmarkar and V. Ramalingaswami, ed., Tata McGraw-Hill, New Delhi (1986).

22. A. Ruiz-Marcos, P. Cartagena Abella, M.A. Garcia-Garcia, F. Escobar del Rey and G. Morreale de Escobar, Rapid effects of adult-onset hypothyroidism on dendritic spines of pyramidal cells of the rat cerebral cortex, Exp. Brain Res. (Accepted for publication) (1988).

23. A. Ruiz-Marcos and S.L. Ipiña, Hypothyroidism affects preferentially the dendritic densities on the more superficial region of pyramidal neurons of the rat cerebral cortex, Dev. Brain Res. 28: 259-262 (1986).

24. A. Ruiz-Marcos, J. Salas, F. Sanchez-Toscano, F. Escobar del Rey and G. Morreale de Escobar, Effects of neonatal and adult onset hypothyroidism on pyramidal cells of the rat auditory cortex, Dev. Brain Res. 9: 205-213 (1983).

25. A. Ruiz-Marcos, F. Sanchez-Toscano, F. Escobar del Rey and G. Morreale de Escobar, Severe hypothyroidism and the maturation of the rat cerebral cortex, Brain Res. 162: 315-329 (1979).

26. A. Ruiz-Marcos, F. Sanchez-Toscano, F. Escobar del Rey and G. Morreale de Escobar, Reversible morphological alterations of cortical neurons in juvenile and adult hypothyroidism in the rat, Brain Res. 185: 91-102 (1980).

27. A. Ruiz-Marcos, F. Sanchez-Toscano, M.J. Obregon, F. Escobar del Rey and G. Morreale de Escobar, Thyroxine treatment and recovery of hypothyroidism-induced pyramidal cell damage. Brain Res. 239: 559-574 (1982).

28. A. Ruiz-Marcos and F. Valverde. The temporal evolution of the distribution of dendritic spines in the rat visual cortex of normal and dark raised mice, Exp. Brain Res. 8: 284-294 (1969).

29. A. Ruiz-Marcos and F. Valverde, Dynamic architecture of the visual cortex, Brain Res. 19:25-39 (1970).

30. F. Valverde and A. Ruiz-Marcos, Dendritic spines in the visual cortex of the mouse: introduction to a mathematical model, Exp. Brain Res. 8: 268-283 (1969).

31. S.P. Wise, W.J. Fleshman and E.G. Jones, Maturation of pyramidal cells form in relation to developing afferent connections of rat somatosensory cortex, Neurosci. 4: 1257-1297 (1979).

MICROTUBULE ASSEMBLY: REGULATION BY THYROID HORMONES

J. Nunez, D. Couchie and J.P. Brion

INSERM U 282 - CNRS, Hôpital Henri Mondor

94010 Créteil, France

INTRODUCTION

The pioneer work of Eayrs (1) and Legrand (2) have established, several years ago, that the most significant of the abnormalities seen in the brain of a hypothyroid animal is a "hypoplastic neuropil". Eayrs (1), for instance, reported that thyroid hormone deficiency, when established at or before birth, reduces, in the cerebral cortex, the length and the branching of the dendrites of the pyramidal neurons, the density of the terminals and the number of spines. Legrand (2) also reported a permanent and dramatic reduction in the arborization of the dendritic tree of the Purkinje cells in the cerebellum. Lauder (see another chapter of this book) noticed that the average length of the parallel fiber (i.e. the axons of the granule cells) is shorter in the hypothyroid cerebellum. Thus, at the morphological level, both axonal and dendritic outgrowth seems to be impaired by early thyroid hormone deficiency. Several other abnormalities (see 3 for a review), i.e. deficit in cell acquisition, retardation in the timing in cell migration, increase in cell death and in glial cell proliferation, decrease in myelination and synaptogenesis might be secondary to a reduced rate of neurite outgrowth. However thyroid hormones might exert more than one effect on the different neuronal developmental events. Moreover the glial cells (the oligodendrocyte for instance) seem to contain receptors for these hormones and to be their targets (see the article of Sarlieve in this book).

Little is known of the mechanism by which neurite outgrowth begins and how this event is regulated during the early stages of neuronal differentiation. What is known is that massive microtubule assembly is required both "in vivo" and in cultured neurons during neurite outgrowth. Microtubules are good markers of neurites and of neurite outgrowth since: 1) in cell culture the extension of cell processes is very efficiently inhibited by the same antimitotic drugs which inhibit microtubule assembly (4-7); 2) they are the major linear structure of the axons and the dendrites; 3) tubul.. represents 70% of the proteins of the neuronal processes; 4) the ratio between free tubulin subunits and tubulin assembled into microtubules decreases during neurite outgrowth.

Microtubules can be assembled "in vitro" from crude brain supernatants (7) and purified (8) by cycles of polymerization at 37°C and depolymerisation at 0-4°C. These purified microtubules are long helical structures made up of tubulin (see 9 for a review), a protein of 110

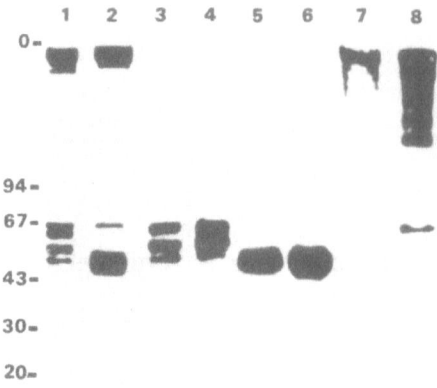

Figure 1. Protein immunoblot of juvenile and adult MAPs. Lanes 1,2: Coomassie blue staining of adult (1) and juvenile (2) thermostable MAPs. Lanes 3-6: protein immunoblot of adult (3,4) and juvenile (5,6) MAPs with a polyclonal (3,5) or a monoclonal (4,6) anti-Tau antibodies. Lanes 7,8: protein immunoblot of adult (7) and juvenile (8) MAPs with an anti-MAP2 polyclonal antibody.

kilodaltons molecular weight, which represents 80% of the polymer, and of several microtubule-associated proteins (MAPs). Up to 35 MAPs copolymerize with microtubules during "in vitro" assembly (10). Among these entities two major groups of MAPs i.e. a group of high-molecular weight proteins (10,11) known as MAP 1 (350 kilodaltons) and MAP 2 (280 kilodaltons), and a group of proteins known as Tau (50-70 kilodaltons) (12) has been purified. All these MAPs are able to promote the assembly of purified tubulin. Pure tubulin very poorly polymerizes in the absence of MAPs.

DIFFERENTIAL EXPRESSION OF MAPs DURING BRAIN DEVELOPMENT

Quantitative and qualitative changes in the expression of MAPs occur during brain development. Changes in composition of Tau proteins were the first to be documented (13,14): two proteins of 48 and 65 kilodaltons are present in the Tau region of the SDS-polyacrylamide gels at immature stages whereas in adulthood the Tau complex is composed of 4-5 entities of 50-70 kilodaltons. Recent data (15) showed however, that only the juvenile entity of 48 kilodaltons is immunologically related to the Tau family (Fig.1). The juvenile 65 kilodaltons protein belongs immunologically to the MAP2 family (15,16) ("small MAP2"). Differential expression of the high-molecular weight MAP2 has also been reported (17): one entity is present at early developmental stages (MAP2b) and two in adulthood (MAP2 a and b). Finally the composition of the MAP1 family (350 kilodaltons) also changes during brain development (16).

MICROTUBULE ASSEMBLY DURING BRAIN DEVELOPMENT; EFFECTS OF THYROID HORMONE DEFICIENCY

The tubulin which is present in the supernatant prepared from euthyroid fetal or new born brain barely polymerizes "in vitro" (18). The rate of assembly increases with age reaching maximal values in adulthood (Fig.2). When this work was begun several years ago, few assumptions were made to explain why tubulin polymerizes so differently depending on the stage of development. The only one which was apparently supported by the experimental data was that the MAPs were present in limiting concentrations at immature stages: adding the mixture of mature MAPs or only one of them, adult Tau for instance, to the juvenile preparation increased the rate of assembly up to that measured in adulthood. However the discovery that juvenile Tau and MAP2 are less active in promoting microtubule assembly than the corresponding adult MAPs (14) raised the possibility that the qualitative changes in MAPs composition might be responsible for the lower polymerization activity seen at early stages of development.

Microtubule assembly from hypothyroid preparations (Fig.2) was also tested (19,20): at day 15 postnatal the polymerization activity was similar to that measured at earlier stages (3-5 days) with the euthyroid preparations. Analysis of the MAPs present at day 15 postnatal also showed a higher proportion of immature Tau than in the control of the same age (20). This suggested that the transition between juvenile and mature Tau is delayed in hypothyroidism, a conclusion which might be sufficient to explain the lower polymerization activity produced by thyroid hormone deficiency.

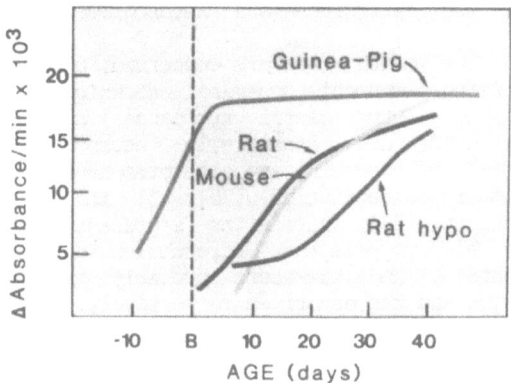

Figure 2. Changes in the rate of in vitro microtubule assembly during brain development. These rates were measured for the guinea-pig a species which has a mature brain at birth and the rat and mouse which develop their brain postnatally. The figure also shows that the changes in the rates of microtubule assembly is delayed for the hypothyroid rat brain.

THE POLYMERIZATION ACTIVITY OF JUVENILE Tau AND MAP2

"In vitro" reconstitution experiments showed that pure tubulin polymerizes very efficiently "in vitro" in the presence of either Tau or MAP2 (21). The structure of the microtubules obtained in such conditions is very similar and apparently identical to that seen in the intact cell. Juvenile Tau and MAP2 also produce microtubules when incubated with pure tubulin; however the rates of assembly and overall polymerisation were lower than those observed with the adult corresponding MAPs (14). This suggests that the equilibrium of tubulin assembly is different depending on which MAP is used i.e. juvenile or adult.

The mechanism of action of MAPs on tubulin assembly is still poorly understood. One possibility is that the different MAPs modify the conformation of this protein in a way which facilitates its interaction with other tubulin molecules; another possibility is that tubulin contains the information required for self-polymerization, the role of MAPs being to stabilize the microtubule lattice thus displacing the equilibrium towards assembly.

Although it has been shown that injecting adult Tau into a cell results in the stabilization of the microtubules (22), none of these mechanisms is entirely proved so far. However if one remembers that neurite outgrowth is highly dependent on the efficiency of microtubule assembly it is clear that both mechanisms would account for the reduced rate of neurite outgrowth seen at early developmental stages and in the brain of the young hypothyroid animals.

REGIONAL AND CELLULAR DISTRIBUTION OF THE MAJOR MAPs

Although the presence of up to 35 MAPs in "in vitro" assembled microtubules has been reported (10) several years ago, the precise significance of such a heterogeneity is still uncompletely understood. The function of some of the MAPs which are not promotors of tubulin assembly has been recently elucidated: for instance some minor MAPs (23) are responsible for the movement of vesicles along the axonal microtubules i.e. the retrograde and anterograde axoplasmic transport (24).

As far as the major MAPs are concerned, a very important finding was achieved by using immunohistochemical techniques applied to whole brain sections. These experimental approaches allowed to show that: 1) MAP2 and Tau proteins are essentially neuronal markers; however immunologically related proteins are also present in the astroglial cells but at a much lower concentration (25); 2) in the neurons MAP2 is essentially dendritic (26), whereas Tau is present in the different types of axons (27,28). This suggests that microtubules differing in the type of MAP they contain, and therefore probably in their properties, are present in the axons and the dendrites respectively.

Recently, developmental studies were undertaken (29) to know how juvenile and adult Tau are expressed in the different types of axons which begin to be formed in the cerebellum of the rat during the first postnatal weeks, i.e. from birth until adulthood. We found that immature Tau is expressed in the growing axons: its presence was observed, at early stages, only in those axons that are in vicinity of their future postsynaptic counterparts. For instance, few days after birth in the cerebellum the climbing fibers clearly express immature Tau; at this stage the climbing fibers make transitory contacts with the cell body of the Purkinje cells whereas, at later stages they synapse with the dendrites of these macroneurons. The parallel fibers, i.e. the axons of the granule cells, which beg' to grow few days after birth, express the Tau antigen only several days later (approximately at day 10) and only in the regions where their postsynaptic counterpart, the dendrites of the Purkinje cells are already developped and stained by an anti-MAP2 antibody. In other words the presence of Tau in the axons and of MAP2 in the dendrites seem to be synchronized and expressed only when the two types of neuronal processes begin to synapse.

A similar conclusion applies to the cerebellum of a hypothyroid rat (Fig.3) but at day 15 postnatal the maturation process is delayed and several abnormalities are observed (Brion et al., unpublished results). For instance the expression of Tau in the parallel fibers resembles at day 15 that observed at day 10 in the euthyroid control. Major abnormalities were also seen at the level of the Purkinje cells. Labeling with an anti-MAP2 antibody showed that: 1) as previously described by

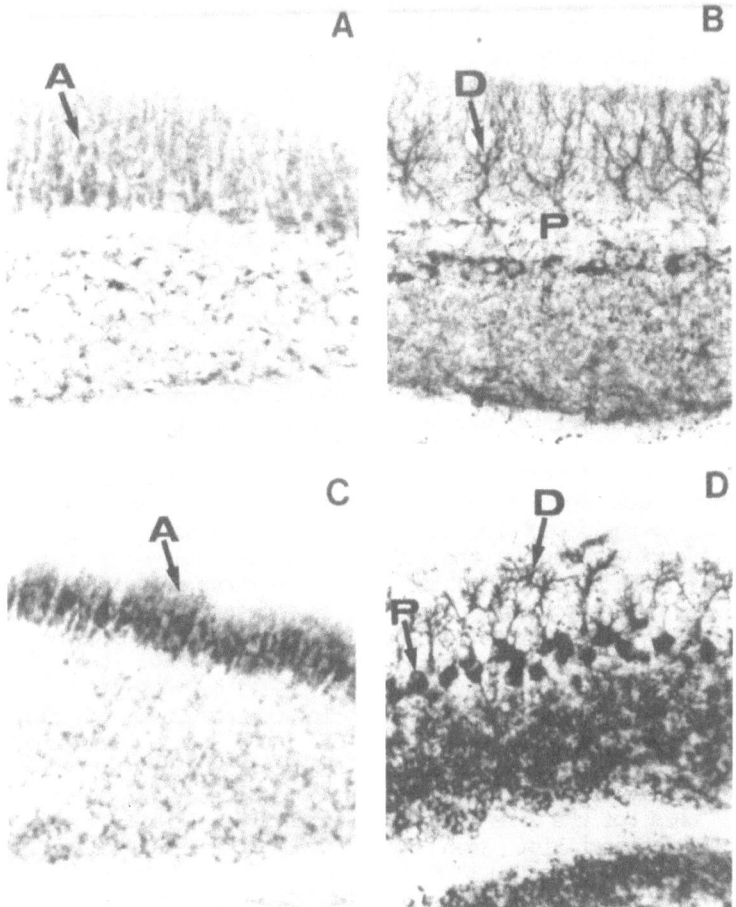

Figure 3. Immunohistochemical staining of 14 day old euthyroid (A,B) and hypothyroid (C,D) rat cerebellum with anti-Tau (A,C) and anti-MAP2 (B,D) antibodies. Purkinje cell body (P) and their dendritic tree (D). Parallel fiber axons (A).

Legrand (2) the dendritic tree is markedly reduced in size with a long primary dendrite and a defective arborization; 2) two or three layers of Purkinje cells are piled up in several regions of the cerebellum; 3) some Purkinje cells seem to develop their dendritic tree downwards i.e. towards the granular layer: 4) the cell body of all these macroneurons is heavily stained by the anti-MAP$_2$ antibody i.e. a situation which is observed in the euthyroid control at earlier stages (5-8 days postnatal).

EXPRESSION OF Tau mRNA DURING BRAIN DEVELOPMENT

A Tau cDNA probe, prepared from an immature brain cDNA library, hybridizes with Tau mRNAs of 6 kb encoding both for juvenile and mature Taus (29). This probe was therefore used (30) to quantitate both immature and adult Tau mRNAs during brain development. Northern and dot blot analysis showed that the abundance of Tau mRNA doubles from a late fetal stage (-4 days) until birth, remains constant until day 6 postnatal and

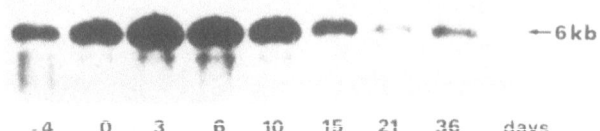

←6kb

-4   0   3   6   10   15   21   36   days

Figure 4. Developmental changes of mouse brain Tau mRNA concentration (from birth to 36 days postnatal).

then decreases progressively until day 21; from this stage until 36 days postnatal the Tau mRNA content is approximately 10 fold lower than at day 6 (Fig.4). The concentration of the tubulin mRNA, which also markedly decreases during brain development (31) closely follows that of Tau mRNAs. This suggests that the transcription and/or the stability of the messages for Tau and tubulin are coordinated. In addition, the very marked decrease in concentration of both mRNAs seen after the first postnatal week contrasts with the relative stability in abundance of the corresponding proteins; the concentration of tubulin decreases only two fold between day 10 postnatal and adulthood (32) whereas that of juvenile + adult Taus does not change as dramatically as their mRNAs during development (unpublished results). One may therefore speculate that the rate of transcription and/or the stability of the messages for Tau and tubulin are negatively regulated after the first postnatal week whereas the stability of the corresponding proteins increases during development. One possibility is that Tau and tubulin are stabilized both because they are assembled into the microtubules and because they closely interact with other components of the cytoskeleton (33-37).

Recently experiments were performed to know whether hypothyroidism has some effect on the expression of Tau mRNAs. Surprisingly, at different stages of brain development, the concentration of Tau mRNAs was found to be higher in the hypothyroid preparations compared to the euthyroid ones of the same age. This suggests that the biphasic evolution of Tau mRNA concentration seen during euthyroid brain development is delayed in hypothyroidism.

DISCUSSION AND CONCLUSIONS

The conclusions suggested by the kinetic data obtained few years ago were that: 1) the microtubule polymerization activity increases during brain development. This evolution seems to be related to the differential expression of the microtubule-associated proteins, such as Tau proteins, differing in size and in polymerization activity; 2) thyroid hormones increase the polymerization activity probably because they accelerate the transition from the immature to the more active mature Tau forms. However, little was known on the mechanism that generates Tau heterogeneity at different stages of brain development, and on the site of action of thyroid hormones.

Two immature Tau mRNAs have been recently cloned (38). Their aminoacid sequence is identical with the exception that one of the two clones contains an additional stretch of 23 residues at the carboxyl terminal. Actually, we have recently isolated (unpublished results) three juvenile Tau species, differing slightly in molecular weight, which give very similar peptide maps upon partial proteolysis; in these conditions a large "core" peptide of the same size is produced with the three juvenile Tau forms. This "core peptide" reacts with an adult anti-Tau monoclonal antibody. The same antibody also reveals a "core" peptide of approximately the same size produced by partial hydrolysis of the adult Tau forms. This suggests a similar structure and some homology between the different juvenile and adult Tau forms, at least for the sequences which contain the common epitope. One should expect, however, to find a lower degree of homology between the juvenile and adult species since their peptide maps as shown by silver staining are different (unpublished results). The sequences of the different juvenile and adult Tau proteins, which will probably be available soon, will probably provide an answer to this problem.

Finally, and since only one Tau gene has been detected (29,39) it might be that the juvenile and adult Tau proteins are generated by a splicing mechanism. If this possibility is confirmed an important question will be to know how such a mechanism is regulated to account for the sharp differential expression between juvenile and mature Tau which is observed dur' g brain development. Such an information is probably also essential to understand how thyroid hormones modify the expression of these proteins.

Another important question is related to the functional significance of the presence, both at immature and mature stages, of several Tau and MAP2 proteins differing in size; as a general rule the smaller species are present at immature stages, for instance the "small" MAP2 of 65 kilodaltons and the "small" Taus of 48 kilodaltons. These "small" species probably contain the sites of interaction with both tubulin and the calmodulin-Ca^{2+} complex (40) which are also present in the "big" species (41). It has been shown (42,43) that the long MAP2 molecule can be cut in two large fragments, one which contains the tubulin binding site, and another one which is probably responsible for connecting the microtubules to the microfilaments and the neurofilaments. It might be therefore that the juvenile "small" species do not contain the sites allowing the interaction of the microtubules with other components of the cytoskeleton. In other words the Tau and MAP_2 genes would contain several domains coding for the sites responsible respectively for different binding activities (tubulin, calmodulin, microfilaments, neurofilaments, etc); a developmentally regulated splic' g mechanism would therefore generate a variety of molecules differing by the domains they contain. This would confer lability and therefore plasticity to the microtubules present in the growing neurites. In contrast, microtubule stability would increase during brain development both because adult MAPs stabilize the polymer lattice and because they allow their interaction with other components of the cytoskeleton. It remains to be determined whether the effect of thyroid hormones on these processes is direct i.e. if it takes place directly at the level of the differential expression of Tau proteins. The immunohistochemical data also suggest that thyroid hormones might regulate other aspects of neuronal differentiation such as the number of layers of the Purkinje cells, their orientation, the mechanism of transport of the neuritic components duri g outgrowth, etc.

REFERENCES

1. J.T. Eayrs. Influence of the thyroid on the central nervous system. Br. Med. Bull 16: 122 (1960).
2. J. Legrand. Analyse de l'effet morphogénétique des hormones thyroïdiennes sur le cervelet du jeune rat. Arch. Anat. Microsc. Morphol. Exp. 56: 205 (1967).
3. J. Nunez. Thyroid hormones in: Handbook of Neurochemistry 8: 1. A. Lajta Ed., Plenum Publishing Company New York (1985).
4. K.M. Yamada, B.S. Spooner and M.K. Wessels. Axon growth: role of microfilaments and microtubules. Proc. Natl. Acad. Sci. USA 66: 1206 (1970).

5. N.W. Seeds, A.G. Gilman, T. Amano and M.W. Nirenberg. Regulation of axon formation by clonal lines of neuronal tumor. Proc. Natl. Acad. Sci. USA 66: 160 (1970).

6. M.P. Daniels. Colchicine inhibition of nerve fiber formation in vitro. J. Cell Biol. 53: 164 (1972).

7. R.C. Weizenberg. Microtubule formation in vitro in solutions containing low calcium concentrations. Science 177: 1104 (1972).

8. M.L. Shelanski, F. Gaskin and R. C. Cantor. Microtubule assembly in the absence of added nucleotides. Proc. Natl. Acad. Sci. USA 70: 765 (1973).

9. P. Dustin. In: "Microtubules". Springer Verlag Berlin (1984).

10. D.B. Murphy, K.A. Johnson and G.G. Borisy. Role of tubulin-associated proteins in microtubule nucleation and elongation. J. Mol. Biol. 117: 33 (1977).

11. R.D. Sloboda, S.A. Rudolph, J.L. Rosenbaum and P. Greengard. Cyclic AMP-dependent endogenous phosphorylation of a microtubule-associated protein. Proc. Natl. Acad. Sci. USA 72: 177 (1975).

12. D.W. Cleveland, S.Y. Hwo and M.W. Kirschner. Purification of Tau, a microtubule-associated protein that induces assembly of microtubules from purified tubulin. J. Mol. Biol. 116: 207 (1977).

13. A. Mareck, A. Fellous, J. Francon and J. Nunez. Changes in composition and activity of microtubule-associated proteins during brain development. Nature 284: 353 (1980).

14. J. Francon, A.M. Lennon, A. Fellous, A. Mareck, M. Pierre and J. Nunez. Heterogeneity of microtubule-associated proteins and brain development. Eur. J. Biochem. 129: 465 (1982).

15. D. Couchie and J. Nunez. Immunological characterization of microtubule-associated proteins specific for the immature brain. FEBS Lett. 188: 331 (1985).

16. B. Riederer and A. Matus. Differential expression of distinct microtubule-associated proteins during brain development. Proc. Natl. Acad. Sci. USA 82: 6006 (1985).

17. L.I. Binder, A. Frankfurter, H. Kim, A. Caceres, M.R. Payne and L.I. Rebhun. Heterogeneity of microtubule-associated protein 2 dur g rat brain development. Proc. Natl. Acad. Sci. USA 81: 5613 (1984).

18. J. Francon, A. Fellous, A.M. Lennon and J. Nunez. Requirement for "factors" for tubulin assembly during brain development. Eur. J. Biochem. 85: 43 (1978).

19. A. Fellous, A.M. Lennon, J. Francon and J. Nunez. Thyroid hormones and neurotubule assembly in vitro during brain development. Eur. J. Biochem. 101: 365 (1979)

20. J. Nunez. Microtubules and brain development: the effects of thyroid hormones. Neurochem. Int. 7: 959 (1985).

21. A. Fellous, J. Francon, A.M. Lennon and J. Nunez. Microtubule assembly in vitro. Purification of assembly promoting factors. Eur. J. Biochem. 78: 167 (1977).

22. D.G. Drubin and M.W. Kirschner. Tau protein function in living cells. J. Cell Biol. 103: 2739 (1986).

23. R.D. Vale, B.J. Schnapp, T.S. Reese and M.P. Sheetz. Organelle, bead and microtubule translocations promoted by soluble factors from the squid giant axon. Cell 40: 559 (1985).

24. R.D. Allen, D.G. Weiss, J.H. Hayden, D.T. Brown, H. Fujiwake and M. Simpson. Gliding movement of and bidirectional transport along single native microtubules from squid axoplasm: evidence for an active role of microtubules in cytoplasmic transport. J. Cell Biol. 100: 1736 (1985).

25. D. Couchie, C. Fages, A.M. Bridoux, B. Rolland, M. Tardy and J. Nunez. Microtubule-associated proteins and in vitro astrocyte differentiation. J. Cell Biol. 101: 2095 (1985).

26. R. Bernhardt and A. Matus. Initial phase of dendrite growth: evidence for the involvment of high molecular weight microtubule-associated proteins (HMWP) before the appearance of tubulin. J. Cell Biol. 92: 598 (1982).

27. L.I. Binder, A. Frankfurter and L.I. Rebhun: The distribution of Tau in the mammalian central nervous system. J. Cell Biol. 101: 1371 (1985).

28. J.P. Brion, J. Guilleminot, D. Couchie, J. Flament-Durand and J. Nunez. Both adult (52-70 kDa) and juvenile (48 kDa) Tau microtubule associated proteins are axon-specific in the developing and adult rat cerebellum. Neuroscience (1988) in press.

29. D.G. Drubin, D. Caput and M.W. Kirschner. Studies on the expression of the microtubule-associated protein Tau during mouse brain development with newly isolated complementary DNA probes. J. Cell Biol. 98: 1090 (1984).

30. D. Couchie, C. Charrière-Bertrand and J. Nunez. Expression of the mRNA for Tau proteins during brain development and in cultured neurons and astroglial cells. J. Neurochem. (1988) in press.

31. J.F. Bond and S.R. Farmer. Regulation of tubulin and actin mRNA production in rat brain: expression of a new beta-tubulin mRNA with development. Mol. Cell. Biol. 3: 1333 (1983).

32. A.M. Lennon, J. Francon, A. Fellous and J. Nunez. Rat, mouse and guinea-pig brain development and microtubule assembly. J. Neurochem. 35: 804 (1980).

33. E. Nishida, T. Kuwaki and H. Sakai. Phosphorylation of microtubule associated proteins (MAPs) and pH of the medium control interaction between MAPs and actin filaments. J. Biochem. Tokyo 90: 575 (1981).

34. S.C. Selden and T.D. Pollard. Phosphorylation of microtubule-associated proteins regulates their interaction with actin filaments. J. Biol. Chem. 258: 7064 (1983).

35. R. F. Sattilaro. Interaction of microtubule-associated protein 2 with actin filaments. Biochemistry 25: 2003 (1986).

36. J.F. Leterrier., R.K.H. Liem and M.L. Shelanski. Interactions between neurofilaments and microtubule-associated proteins: a possible mechanism for intraorganellar bridging. J. Cell Biol. 95: 982 (1982).

37. R. Heimann, M.L. Shelanski and R.K.H. Liem. Microtubule-associated proteins bind specifically to the 70 kDa neurofilament protein. J. Biol. Chem. 260: 12160 (1985).

38. G. Lee, N. Cowan and M.W. Kirschner. The primary structure and heterogeneity of Tau protein from mouse brain. Science 239: 285 (1988).

39. R.L. Neve, P. Harris, K.S. Kosik, D.M. Kurnit and T.A. Donlon. Identification of cDNA clones for the human microtubule-associated protein Tau and chromosomal localization of the genes for Tau and microtubule-associated protein 2. Mol. Brain Res. 1: 271 (1986).

40. C. Erneux, H. Passareiro and J. Nunez. Interaction between calmodulin and microtubule-associated proteins prepared at different stages of brain development. FEBS Lett. 172: 315 (1984).

41. Y.C. Lee and J. Wolff. Calmodulin binds to both microtubule-associated protein 2 and Tau proteins. J. Biol. Chem. 259: 1226 (1984).

42. R.B. Vallee, M.J. Dibartolomeis and W.E. Theurkauf. A protein kinase bound to the projection portion of MAP_2 (microtubule-associated protein 2) J. Cell Biol. 90: 568 (1981).

43. R.A. Gottlieb and D.B. Murphy. Analysis of the microtubule-binding domain of MAP_2. J. Cell Biol. 101: 1782 (1985).

ACKNOWLEDGMENTS

We are indebted to Dr M.W. Kirschner for the Tau cDNA probe and to Dr U. Littauer for the tubulin cDNA probe. This work was partially supported by a grant from l'Association pour la Recherche sur le Cancer.

Figure 2 was reproduced from J. Nunez, Neurochem Int. 7: 959 (1985) and Figure 4 from Couchie et al., J. Neurochem. in press.

We also thank Mrs Cook (ULB Brussels) and G. Tournier for their technical assistance and N. Scharapan for the preparation of the manuscript.

INVESTIGATION OF MYELINOGENESIS **IN VITRO** : TRANSIENT EXPRESSION OF 3,5,3'-TRIIODOTHYRONINE NUCLEAR RECEPTORS IN SECONDARY CULTURES OF PURE RAT OLIGODENDROCYTES

L.L. Sarliève, F. Besnard, and G. Labourdette
Centre de Neurochimie du CNRS et U44 de l'INSERM
5, rue Blaise Pascal
67084 Strasbourg, Cedex, France

B. Yusta, A. Pascual, and A. Aranda
Unidad de Endocrinologia Experimental
Instituto de Investigaciones Biomedicas
Consejo Superior de Investigaciones Cientificas and Facultad
de Medicina, Universidad Autonoma de Madrid, 28029 Madrid, Spain

M. Luo, J. Puymirat, and J.H. Dussalut
Unité de Recherche en Ontogénèse et Génétique Moléculaire
Centre Hospitalier de l'Université Laval, Sainte, Foy
Quebéc, Canada G1V 4G2

INTRODUCTION

The glial compartment of the central nervous system (CNS) consists of a number of different cell types which interact closely together. It has been recognized for many years that most interfascicular oligodendrocytes of the white matter undergo massive membrane synthesis at myelination, leading to the formation and growth of myelin sheaths around axons in the CNS. Such oligodendrocytes probably subsequently maintain the integrity of myelin throughout life (for a review, see refs. 1,2). On the other hand, "critical periods" of accelerated development characterized by increased sensitivity and influenced by external environmental factors have been identified during brain development (3). For example, in the rat and mouse, the "critical period" during which hormones influence brain development is associated, among other events, with rapid myelinogenesis occuring in both species between the 10th and 30th day after birth (4). During this period, striking morphological and biochemical changes have been described. The biochemical parameters which best seem to correlate with these temporal changes are the enzymes and compounds most closely associated with myelination. Cerebrosides, galactosyl glycerol lipids, sulfatides, sulfogalactosyl glycerol lipids, and the enzymes catalyzing their synthesis, the myelin basic protein, myelin proteolipid protein (PLP) or a synthetic polypeptide composed of the C-terminal amino acids of the PLP sequence, Wolfgram protein, 2',3'-cyclic-nucleotide phosphohydrolase (CNP) and pH 7.2 cholesterol ester hydrolase are very useful molecular markers for myelination (for an extensive review, see refs. 5,6).

Experimental and clinical data have indicated that thyroid hormones (TH) have their most critical influence on the brain during late fetal and early postnatal periods (7). Thus, hypothyroidism during this so-called "critical period", unless treated in early infancy, leads to permanent mental retardation and behavioral abnormalities (8). These deficits are thought to be partly associated with defects in neuronal growth, synaptogenesis (9) and myelination (10-14), resulting from TH deficiency during the perinatal period (for a review, see also Nunez et al., present book). In contrast, in the hyperthyroid state, myelin synthesis commences and terminates earlier (15).

The mature CNS appears to be less vulnerable to altered thyroid status, though there are some reports that describe significant alterations in neurotransmitter levels (16-18) or β-adrenoceptor concentration (19) and in RNA polymerase I and other enzyme activities (20,21). Nevertheless, in the adult animal, dependency on thyroid hormones is still controversial (22).

The identification of the regulators of myelination and their mechanism of action at the molecular level have only partially been elucidated by studies on the intact animal. Although **in vivo** studies (10-15) have implicated thyroid hormones as potentially important regulators, these studies using whole animals were unable to demonstrate whether thyroxine (T4) acts directly or indirectly on the myelin-producing cells. Towards this aim we have previously reported that the activity of enzymes involved in the production of myelin galactolipids was decreased in the brain of the Snell dwarf mouse suffering from a retarded myelination due, at least in part, to altered thyroid function. One of the enzymes, the 3'-phosphoadenosine-5'-phosphosulfate cerebroside sulfotransferase (PAPS : CST, EC 2.8.2.11) found in oligodendroglia (23) catalyzes the last steps of myelin sulfatide biosynthesis. We have demonstrated (24) that the treatment of dwarf mice with T4 during the early postnatal period restored the level of CST to normal.

Nevertheless, manipulation of one hormone **in vivo** invariably affects the availability and concentration of many other hormones. Therefore, cells grown in culture offer the possibility of examining the direct interaction between hormones such as L-3,5,3'-triiodothyronine (T3) or other hormones and a myelin-producing cell such as the oligodendrocyte, without significant interference from other hormones or factors. We have previously described a monolayer primary culture system of cells dissociated from cerebral hemispheres of 14-day-old mouse embryos, which present successive distinct periods of cell proliferation and/or maturation. These periods are characterized essentially as neuronal (neuron-enriched) from 1 to 12 days **in vitro** (DIV), and glial (glial-enriched) between 12 and 60 DIV (25,26). Furthermore, myelin-related membranes are produced in this culture system (26,27).

However, how the interaction of T3 with neural cells is transformed into a final biological response is not fully understood. We have therefore studied the mechanisms of T3 action on the synthesis of CST and have shown that the increase of CST activity may be attributed to enzyme induction. Moreover T3 could act transcriptionally (23).

Since the first step in the action of T3 is the interaction of the hormone with its receptor (28), we have also studied the concentration and equilibrium dissociation constants of T3 nuclear binding sites in cells after various periods of culture. Our data demonstrate that T3 receptors are localized predominantly in neuronal nuclei (29). Moreover, specific T3 binding sites were also observed when a great enrichment of glial cells (70 % of astrocytes and 25 % of oligodendrocytes) was attained (around 20 DIV) in this culture system (26).

However it is not possible to conclude from these results whether both types of glial cells contain the nuclear T3 receptors (NT3R). In addition non-specific T3 binding sites either in nuclei of cultured rat astrocytes (30) or oligodendrocyte nuclei prepared from adult rat cerebral cortex with discontinuous sucrose density gradient ultra-centrifugation (8,31) have been reported. Therefore, in a collaborative effort between three different laboratories the aim of the present investigation was to look for NT3R in almost pure cultured rat oligodendrocytes (OL) using binding assays for a quantitative evaluation and double indirect immuno-fluorescence with a monoclonal antibody against NT3R and a polyclonal antiserum against bovine galactocerebroside (GC), to distinguish which type of OL (morphologically immature or mature or both) express NT3R.

CULTURE CONDITIONS AND IMMUNOCYTOCHEMISTRY

Cell cultures

Pure cultured astrocytes were prepared by a modification of the method of Booher and Sensenbrenner (32). Cerebral hemispheres from newborn Wistar rats were mechanically dissociated in Waymouth's medium supplemented with sodium pyruvate (110 mg/ml), antibiotics and 10 % heat-inactivated fetal calf serum. Cells from one brain were seeded in 6 Petri dishes (100 mm diameter) and the cultures were maintained in the serum containing medium for 21 days and then switched for 3 days either to a medium containing thyroid hormone depleted serum (33) or to a serum-free chemically defined medium. This consisted of Waymouth's medium supplemented with 5 µg/ml insulin and 0.5 µg/ml fatty acid-free bovine serum albumin and antibiotics. Pure cultured astrocytes were also prepared from 14-day-old embryonic chick brain as described above (32).

To obtain a pure culture of oligodendrocytes we first prepared a primary culture (34) enriched in these cells. The preparation was similar to that for astrocytes described above with two differences : 10 % calf serum was used instead of fetal calf serum, and cells were seeded at a twice higher density. After 20 to 30 days these cultures contain up to 50 % oligodendrocytes located on the top of the astrocyte layer. These cells were dislodged mechanically by a modification of the method of McCarthy and De Vellis (35) as described by Besnard et al. (36) and they were seeded on poly-L-lysine coated Petri dishes (60 mm diameter) in the same culture medium. After 3 days the secondary cultures were switched to the chemically-defined serum-free medium described for astrocytes and cultured for 2 additional days before the binding experiments. For the immunocytochemical localization of NT3R, after 1 day the secondary cultures of oligodendrocytes were switched to a chemically-defined serum-free medium consisting of Dulbecco's modified Eagle's medium to which were added 4.5 g of glucose per l, 5 µg of insulin per ml, 10 µg of transferrin per ml, 8 nM selenium (as $Na_2SeO_3 5H_2O$), 15 mM Hepes, 1.2 g of $NaHCO_3$ per l and antibiotics. The cultures were used after 4 days in this medium.

Mixed neuronal-glial cultures were also prepared according to the method of Yavin and Menkes (37). The cerebral hemispheres of 13-to 15-day-old Wistar rat embryos were dissociated mechanically and cultured in Dulbecco's modified Eagle's medium containing 20 % heat-inactivated fetal calf serum and antibiotics in poly-L-lysine precoated Petri dishes (100 mm diameter). Such a system presents successive and distinct periods of cell proliferation and/or maturation. These periods are characterized essentially as neuronal during the first two weeks in culture and as glial thereafter (26). Therefore the cells were used either when a high enrichment of neurons was attained (7 days) or in the virtual absence of neuronal cell population (33 days). In these older cultures the majority of cells are astrocytic in nature and the astroglial gliofilament marker, glial fibrillary acidic protein (GFAP) (38,39) increased continuously, reaching by 38 days of culture an 18-fold higher level than the concentration in adult forebrain (40). In order to ensure a complete cellular depletion of thyroid hormones, the cells were cultured for the last 3 days in a serum-free growth medium (29) for the neuron-enriched cultures (7 days), or in the medium containing thyroid hormone depleted serum (33) for the astrocyte-enriched cultures (33 days).

All cultures were maintained at 37°C in humid atmosphere of 95 % air-5 % CO_2, and their development was followed by phase-contrast microscopy.

Immunocytochemical procedures

Polyclonal antisera against bovine glial fibrillary acidic protein (GFAP) purchased from Dakopatts (Sebia, Issy-les-Moulineaux, France) were used at a 1/200 dilution. Polyclonal antisera against galactocerebrosides (GC) prepared as described by Bologa et al. (41) were used at 1/50 to 1/100 dilutions and monoclonal antibody to myelin basic protein (mMBP) from Hybritech Inc. (San Diego, CA, USA) was diluted (1/200). When surface antigens (GC) were studied, either unfixed cells or cultures fixed with paraformaldehyde (4 % in PBS, 15 min at 4°C) were used. For the internal antigens MBP and GFAP the cells were fixed with paraformaldehyde as described for GC antigen and with methanol (10 min at -20°C), respectively. After fixation the cultures were washed with PBS, incubated with antisera for 1 h at room temperature and then washed again with PBS.

When the cells were immunostained first with GC rabbit antiserum or mMBP (ascitic fluid), the second incubation was performed with a fluorescein isothiocyanate (FITC) labelled goat anti-rabbit immunoglobulin G (IgG, 1/100 Biosys SA, Compiègne, France) serum for 1 h at room temperature. If GFAP rabbit antiserum was used in the first step, the second antibody was a goat anti-rabbit IgG serum, conjugated with peroxidase (Biosys, SA). After incubation, the preparations were washed 3 times with PBS and treated with a mixture containing 4-chloro-1-naphtol (0.018 %, v/v) and H_2O_2 (0.002 %, v/v) in water for 20 min. The preparations were then washed with PBS and mounted in buffered glycerol.

To localize NT3R immunocytochemically, the oligodendrocytes grown on glass coverslips were washed twice with PBS and fixed with 3 % v/v paraformaldehyde in PBS pH 7.4 for 10 min at room temperature. They were then incubated with 2 % BSA in PBS for 60 min at room temperature. For double immunostaining the cells were incubated with 2B3-NTR mAb prepared as described by Luo et al. (42) (1:200 dilution in PBS) and antigalactocerebroside serum (1:100 dilution in PBS) for 10 min at room temperature followed by incubation for 18 h at 4°C. After rinsing three times for 5 min in PBS the glass coverslips were incubated with FITC labelled goat anti-mouse IgG serum and rhodamine labelled goat anti-rabbit IgG serum (1:100) for one hour at room temperature and rinsed three times in PBS. Then, the slides were mounted in kaiser's glycerol gelatin (Merck), sealed with eukitt (0. Kindler) and examined by phase-contrast and ultraviolet light (UV) optics in a Zeiss MC 63 stereomicroscope.

The percentage of cells stained with cell-type specific markers using the indirect immunocytochemical technique was determined by counting cells in 3-4 separate fields from triplicate cultures, with a minimum of 400 cells counted. Values (%) given are means of two different experiments. GFAP-containing cells were counted after nuclear staining with 1 % toluidine blue for 60 s.

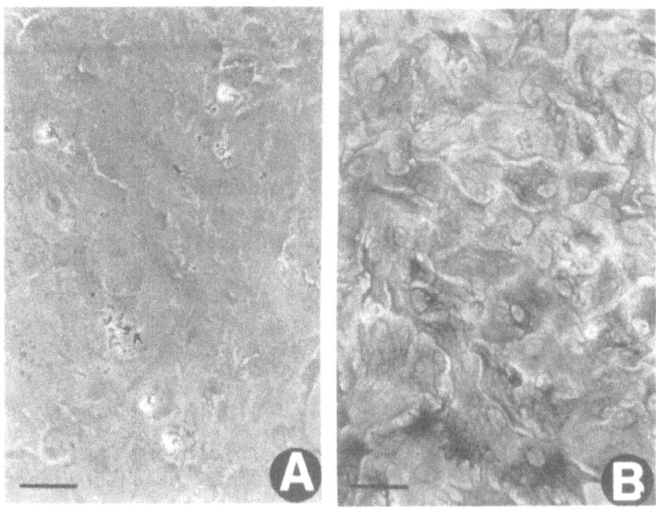

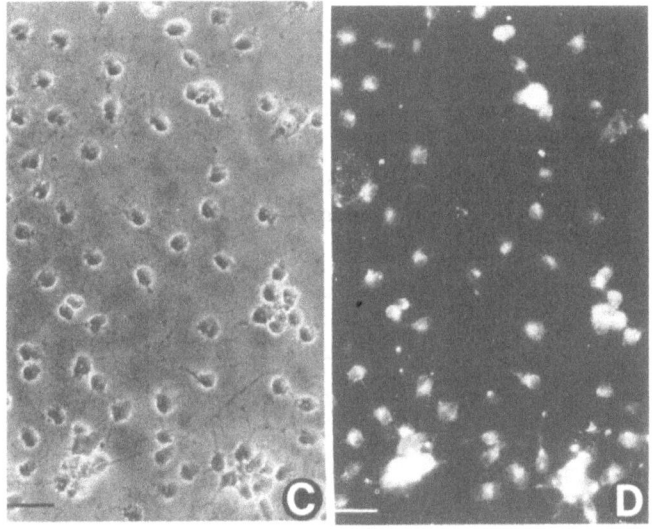

Fig. 1. Immunostaining of glial cells with cell-type specific markers.
The upper panels show rat pure astrocytes after 24 days in
culture. (A), phase-contrast; (B), same Petri dish
immunostained with anti-glial fibrillary acidic protein (GFAP)
serum + goat anti-rabbit immunoglobulin G (IgG) serum,
conjugated with peroxidase. Bar corresponds to 100 µm. The
lower panels represent rat pure oligodendrocytes after 30 days
in culture (25 days in primary culture + 5 days in secondary
culture); (C), phase-contrast; (D), same field immunostained
with antigalactocerebroside (GC) serum + fluorescein
isothiocyanate (FITC)-labelled goat anti-rabbit IgG serum. Bar
corresponds to 40 µm.

IDENTIFICATION OF CULTURED CELLS BY IMMUNOCYTOCHEMISTRY (INDIRECT IMMUNO-STAINING)

Astrocytes

Dissociated brain cells from newborn rat cultured for 24 days on a plastic surface developed into a pure glial cell population (Fig. 1A). In agreement with previous observations (43) the neuronal cells degenerated during the first 2 days and some oligodendrocytes survived for about a week. On the other hand, the flat polygonal shaped cells multiplied actively and formed a monolayer within 2 weeks. The cells contained in these astrocyte primary cultures were identified by their positive immunoreaction with anti-GFAP antiserum (38, 39). As shown in Fig. 1B, 95 % of the cells shown in Fig. 1A were GFAP-positive. The remaining 5 % consists mainly of ameboid-microglial or macrophage-like cells (44,45).

Oligodendrocytes

At 5 days in secondary culture, the oligodendrocytes with network forming processes (Fig. 1C) were immunocytochemically identified using antibodies against galactosylceramides (galactocerebrosides GC) (46) and myelin basic protein (MBP) (47). The staining of cultures with the anti-GC and monoclonal anti-MBP revealed that more than 90 % and 76 % of the cells were GC-positive (Fig. 1D) and MBP-positive (not shown), respectively, while only 2-5 % were GFAP-positive. The remaining 5-8 % consist mainly of ameboid-microglial cells (45).

Moreover, by light microscopy, as shown in Fig. 2, two sub-populations of oligodendrocytes were observed at 5 days in secondary culture : 1) cells of medium size (6 µm), round and rather dark with several primary processes, representing 77 % of the total GC-positive cells, were considered as morphologically immature and 2) larger cells (12-15 µm) irregular in shape with an extensive and elaborate network of thinner lateral processes. These oligodendrocytes account for 23 % of the total cells expressing GC and were regarded as morphologically mature.

QUANTITATIVE EVALUATION BY BINDING ASSAYS AND CHARACTERISTICS OF NT3R IN VARIOUS TYPES OF CULTURED BRAIN CELLS

Characteristics of nuclear T3 binding in brain cell cultures

Figure 3 shows typical Scatchard plots of L-[^{125}I]T3 nuclear binding to intact rat oligodendrocytes (R-oligo) and astrocytes (R-astro). Binding obtained in chick astrocytes (Ch-astro) and in mixed neuronal-glial cell populations (R-MNG) at 7 and 33 days in culture, i.e., when these cultures were either enriched in neurons (R-MNG-7) or astrocytes (R-MNG-33), respectively, is also illustrated. Only a single class of high affinity-low capacity receptors with an equilibrium dissociation constant (Kd) which was similar for the various types of cells (i.e., R-oligo: 0.13 nM, R-astro: 0.11 nM, R-MNG-7: 0.13 nM and R-MNG-33: 0.11 nM), except for the Ch-astro (0.30 nM), was observed.

The oligodendroglial nuclear T3 receptor had a maximal binding capacity almost as high as that found in the neuron-enriched cultures (56 and 65 fmol/100 µg DNA, respectively). On the other hand, rat astrocytes had a receptor concentration 2-3 fold lower than oligodendrocytes. This was observed both in pure rat cultures (R-astro) and in the mixed cultures in the astroglial period (R-MNG-33), and was also confirmed in the chick cells (Ch-astro) where the maximal binding capacity was the lowest. It is interesting to note that after oligodendrocytes were removed from the top of the primary cultures from which they were prepared, the maximal binding

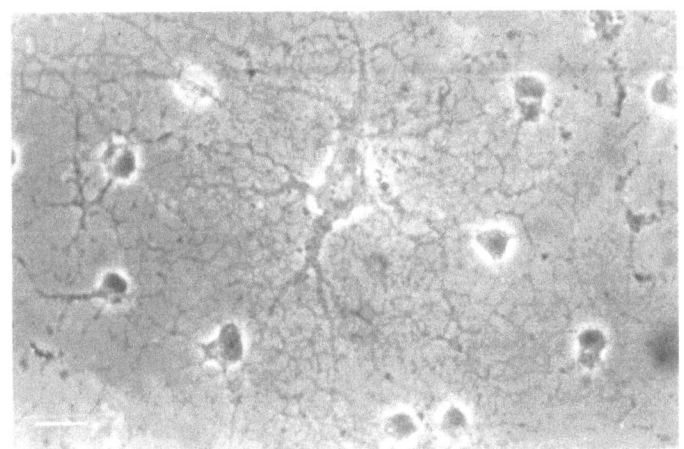

Fig. 2. Phase-contrast of rat pure oligodendrocytes after 25 days
in culture (20 days in primary culture + 5 days in secondary culture). Note in the center of the micrograph a
morphologically mature oligodendrocyte (OL) with 6 main
processes and the profuse network of many small lateral
processes. This cell is surrounded by several morphologically immature OL. Bar corresponds to 11 μm.

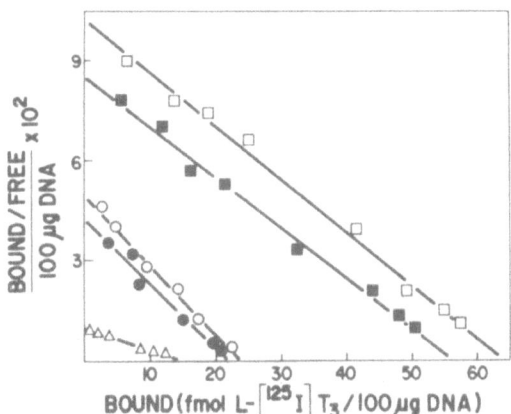

Fig. 3. Scatchard analysis of L-[^{125}]T3 binding to nuclei in intact
cells. T3 binding was determined as described by Samuels
and Tsai (70) in : □,rat mixed neuronal glial cultures after
7 days in culture (R-MNG-7); ■, rat oligodendroglial cultures after 30 days in culture (R-oligo); o,rat astroglial
cultures after 24 days in culture (R-astro); ●,rat mixed
neuronal-glial cultures after 33 days in culture (R-MNG-
33); and △,chicken astroglial cultures after 13 days in
culture (Ch-astro). Each experimental point is the mean of
triplicate cultures. Individual experimental values were
within 10 % of each other (from Yusta et al., Endocrinology 122:in press (1988) (71). With permission).

119

capacity of the remaining cells was approximately 2-fold lower than in pure secondary oligodendrocyte cultures.

When studied in isolated nuclei, receptor concentration was again about three times higher in oligodendrocytes than in astrocytes, whilst the Kd were similar in both types of cells (Fig. 4). The affinity obtained using isolated nuclei was somewhat lower than that found in the intact cells (Kd = 0.31-0.32 nM and Kd = 0.11-0.13 nM, respectively), in agreement with our previous observations using glioma and neuroblastoma cell lines (48).

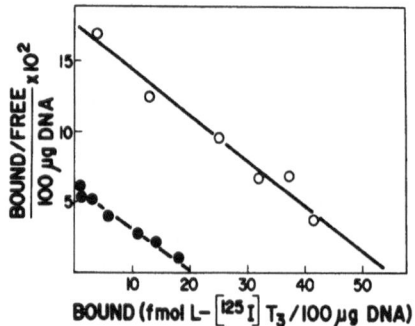

Fig. 4. Scatchard analysis of L-[^{125}I]T3 binding to isolated nuclei. Binding was determined as described by Samuels and Tsai (70) in rat oligodendroglial (O) and astroglial (●) cells after 30 and 24 days in culture, respectively (from Yusta et al., Endocrinology 122:in press (1988) (71). With permission).

Binding affinity of T3 analogs

Table 1 shows the relative affinity of T3 analogs for the nuclear binding site of oligodendrocytes and astrocytes. This affinity was calculated by determining the ratio of the concentration of T3 required for a 50 % decrease in tracer binding to the concentration of each analog required for a 50 % decrease under identical conditions.

Taking the effect of T3 as 1.0 for oligodendrocytes, tri-iodothyroacetic acid (TRIAC) was at least as effective as T3 which had 10-fold greater potency than T4. Tetraiodothyroacetic acid (TETRAC) had the lowest potency (6 % of T3). Similar results were obtained with the astrocyte receptors, where TRIAC and T3 were the more effective compounds, whereas the other analogs tested were less potent in the order : TRIAC = T3 > T4 > TETRAC.

Extraction and sedimentation of thyroid hormone receptors

Nuclear receptors solubilized from nuclei of pure cultures of rat oligodendroglial and astroglial cells were sedimented in isokinetic sucrose gradients. As illustrated in Fig. 5 the L-[^{125}I]T3 appeared mainly

Table 1. Relative Binding Affinities of Iodothyronines
for the Nuclear T3 Receptor.

Analog	Relative Affinity	
	Oligodendroglial Nuclei	Astroglial Nuclei
L–T3	1.0	1.0
L–TRIAC	1.1	1.2
L–T4	0.1	0.2
L–TETRAC	0.06	0.1

Binding affinity was determined in isolated nuclei from
rat oligodendrocytes and astrocytes after 30 and 24 days
in culture, respectively. Relative affinity is defined
as the ratio of the concentration of L–T3 required for
a 50 % decrease in tracer binding to the concentration
of analog required for the same decrease. Results are
means of triplicate data with less than 5–10% variation.

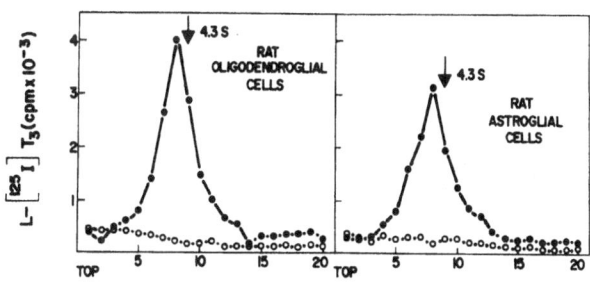

FRACTION NUMBER

Fig. 5. Isokinetic binding sedimentation of T3 receptors from
oligodendrocytes and astrocytes. The receptors were
extracted from nuclei of rat oligodendroglial and
astroglial cells after 30 and 24 days in culture,
respectively, with 0.4 M KCl and labelled **in vitro**
with L–[^{125}I]T3 in the absence (●—●) or presence
(○—○) of an excess of non–radioactive T3. The
arrows represent the position of hemoglobin which
has a sedimentation coefficient 4.3 S and was used
as an internal marker (from Yusta et al., Endocrinology 122:in press (1988) (71). With permission).

associated with a peak with a sedimentation coefficient of approximately
3.8 S as already observed for other neural cells (30,48). As can also be
observed in Fig. 4, this peak is inhibited in the presence of an excess of
unlabelled T3.

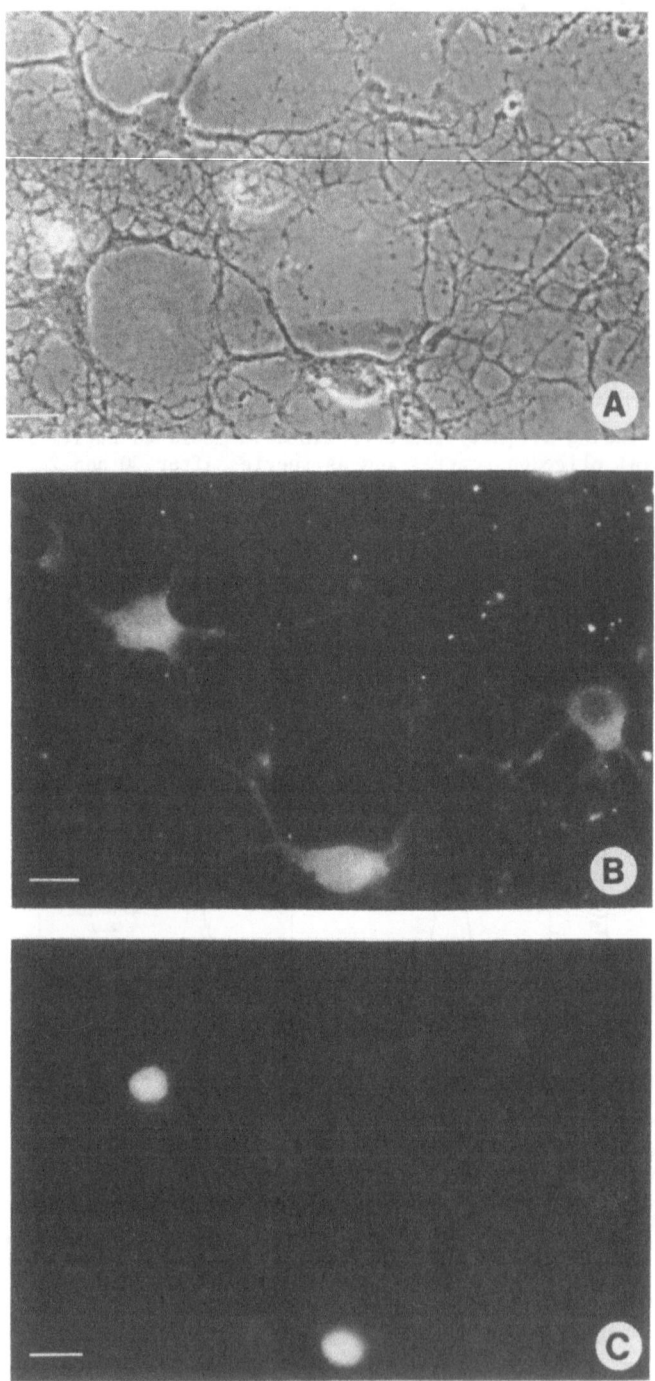

Fig. 6. Rat pure oligodendrocytes after 25 days in culture.
Micrographs of the same field : (A), phase-contrast;
(B), immunofluorescence double staining with cell type
specific marker galactocerebroside (GC) and (C), with
a monoclonal antibody against NT3R. Bar corresponds
to 7 μm.

Table 2. Expression of Nuclear T3 Receptors (NT3R) in two
Subpopulations of Oligodendrocytes : Effect of T3.

Experiment[a]	Galactocerebrosides (GC)-and (NT3R)-[b] Positive Oligodendrocytes		
	$\dfrac{GC^+ \; NT3R}{GC+}$	$\dfrac{GC^+ \; NT3R^+}{MI \; GC+}$	$\dfrac{GC^+ \; NT3R^+}{MM \; GC+}$
Control	70 % ± 3.5	77 % ± 4.5	44 % ± 7.5
+T3	65 % ± 4	74 % ± 2	38 % ± 4

[a]The cells were grown for 20 days in primary culture plus
24 h as secondary culture in presence of a serum-containing
medium. Then, they were switched to a chemically-defined
serum-free medium and grown for 4 additionnal days either in
absence (Control) or presence of 10^-M T3 (+T3). Results
(Oligodendrocytes expressing both GC and NT3R antigens) are
in % of GC positive cells.

[b]GC^+ cells were divided into two groups : I) morphologically
immature (MI) and II) morphologically mature (MM).
Values are means ± SD of 24 separate fields randomly chosen
in triplicate Petri dishes from two different experiments.

IMMUNOCYTOCHEMICAL LOCALIZATION OF NUCLEAR TRIIODOTHYRONINE RECEPTORS
(NT3R) IN MORPHOLOGICALLY IMMATURE AND MATURE CULTURED OLIGODENDROCYTES
AND EFFECT OF T3

Immunocytochemical localization of NT3R

Double indirect immunofluorescence studies with the monoclonal
antibody against NT3R and the polyclonal antiserum against galactocerebroside are shown in Fig. 6. Quantitative analysis demonstrates that 70 % of
the GC-positive cells were also labelled with NT3R mAb (Table 2). However,
while GC was expressed mostly on the surface of cultured oligodendrocytes
or along their branched processes (Fig. 6B), only nuclei were intensely
stained with NT3R mAb (Fig. 6C).

Furthermore, as shown in Table 2, 77 % of morphologically immature
oligodendrocytes stained with the anti-GC serum also expressed NT3R
antigen. In contrast, only 44 % of morphologically mature oligodendrocytes
were NT3R positive.

Effect of T3 on cultured oligodendrocytes expressing NT3R

Since a reduction in the number of receptors after the exposure of a
target cell to the homologous hormone (down-regulation) is a very common
finding we studied the effect of adding T3 (10^-M) tó the chemically
defined medium on the expression of NT3R in pure cultured
oligodendrocytes. The results, shown in Table 2, revealed no significant
effect of T3 either on the number of NT3R-positive and morphologically
immature oligodendrocytes or on the number of NT3R-positive and
morphologically mature oligodendrocytes.

CONCLUSIONS

Nuclear triiodothyronine receptor binding study

Although the presence of thyroid hormone receptors in the neuronal nuclei is well documented, there is still controversy concerning the presence of specific T3 binding sites in nuclei prepared from different glial cell populations. Recent data obtained in our and other laboratories have clearly demonstrated the presence of receptors in a glial cell line (48) and in primary cultures of mouse (29,49), rat (50) and chick (29) glial cells composed mainly of astrocytes. By contrast, it has been reported that oligodendroglial nuclei isolated from various brain regions do not contain the receptor (8,31). Since it is generally accepted that the presence of receptors in a given cell is a prerequisite for the cell to be responsive to thyroid hormones, the reported effects of T3 on myelination have been postulated to be mediated through neurons (30,31). Therefore, as a result of the recent developments in the cultivation of pure oligodendrocytes which are capable of proliferating (36) expressing specific oligodendrocyte and myelin markers (46,51), and producing myelin-related membranes (6,26,27), we have been able to re-examine the presence of nuclear receptors either in intact cells or in isolated nuclei from rat oligodendroglial cells. Our oligodendrocyte cultures were more than 90 % pure, as shown by immunostaining with anti-GC, an oligodendroglial marker (46). This method permits unambiguous cell identification, in contrast with the methods used in previous reports on thyroid hormone binding in brain, where the nuclei are isolated by gradient centrifugation and identified only on the basis of nuclear morphology (20,31,52-55).

Present results demonstrate the presence of nuclear high-affinity T3 receptors in pure secondary cultures of oligodendrocytes. To our knowledge, the present report is the first to demonstrate unequivocally that this cell type expresses the nuclear receptor, and that therefore the effects of thyroid hormone on myelination could be directly exerted on the oligodendrocyte.

The properties of the oligodendrocyte receptor were closely similar to those found in pure cultures of astrocytes or in other systems (mixed neuronal-glial cultures) studied in parallel. The affinity of T3 for the receptor, the relative affinity for various hormone analogs, and the iso-kinetic sucrose gradients agree with those previously reported using nuclei from brain cultures (30), including transformed brain cells (48) and other non-neural cells (56,57). The tentative identification of the thyroid hormone receptor with the c-erb A gene product, together with the existence of several c-erb A genes (58,59) suggested the existence of a family of closely related thyroid hormone receptors which could be expressed in a tissue-specific fashion. Recently, the identification of a novel c-erb A gene product which is expressed at high levels in the rat brain has been reported (60). It is interesting to point out that this protein binds T3 and TRIAC with similar potency (60), as also observed by us in this study, whereas in other systems (56,57,59) TRIAC is significantly more potent than T3.

Our data show that the maximal binding capacity of rat oligodendrocytes (50-60 fmol/100 µg DNA) was similar to that obtained under the same experimental conditions in mixed neuronal-glial cultures at 7 days : i.e., when a high enrichment of neurons was attained. Furthermore, this value is also similar to that found in cultured neurons from various species (29,30,50) or in neurons obtained from different brain regions (49).

124

On the other hand, the receptor level in oligodendrocytes was always 2-4 fold higher than that found in astrocytes, in agreement with the data obtained in cultured astrocytes by us and other groups (29,30,61). Therefore, one cannot exclude the possibility that the highest receptor levels obtained in glial cells which have also been reported (49,62) could be due to the presence of oligodendrocytes in these cultures in addition to astrocytes.

Another interesting feature is that the maximal binding capacity we found in rat mixed neuronal-glial cultures at 33 days : i.e., when the majority of the cells are astrocytes (22 fmol/100 µg DNA) was nearly identical to that obtained in pure rat astrocytes primary cultures (24 fmol/ 100 µg DNA). These results could indicate that the presence of neurons during the first 2 weeks in mixed neuronal-glial cultures were unable to influence the pattern of receptor development in cultured glial cells.

Double immunostaining study

In these studies we have further shown by double indirect immunofluorescence with anti-galactocerebrosides (an early OL surface marker) and with a monoclonal antibody (mNT3R) detecting the NT3R (the predominant 57 KD protein) in nuclear extracts from liver, brain, kidney and spleen (42), that the nucleus of about 80 % of morphologically immature OL were stained by mNT3R. Binding studies and immunocytochemical localization of NT3R were performed in parallel on oligodendrocytes of the same age in culture (25-30 days), corresponding in the rat, to the period of most rapid myelination. This age overlaps with the so-called "critical period" during which hypothyroidism leads to defects in myelination (10-15). Indeed, numerous **in vitro** studies from various laboratories on the effect of TH (see Sarlièvre (6), Ferret–Sena et al. (23) for references and refs. 63-68) and the demonstration in the present study of a high level of NT3R in pure oligodendrocytes suggest that T3 has a direct action through NT3R on CNS myelin-producing cells. It is important to note that GC-positive rat Schwann cells (the peripheral nervous system myelin-producing cells) in culture, also express NT3R (Luo and Barakat, unpublished results).

The developing brain is generally more responsive to TH than the mature brain. However numerous NT3R, were recently demonstrated in neuronal nuclei prepared from the cerebral cortex of the mature rat brain (20), which could explain some reports describing significant alterations in neurotransmitter levels (16-18) or enzyme activities (20,21), in the adult rat brain. Conversely, a reduction of the cell type containing the receptor (only 44 % of mature oligodendrocytes also express NT3R), may be responsible for the age related appearance and disappearance of TH sensitivity of several metabolites or enzymes characteristic of the oligodendrocytes (29,68,69). Thus, the more mature are the oligodendrocytes in culture, the less numerous are the myelin-producing cells expressing NT3R. These results accord with earlier reports (8,31) claiming an absence of such receptors in oligodendrocyte nuclei prepared from adult rat brain.

Finally, our data suggest that TH and their receptors in oligodendrocytes, act as signals required to synchronize the sequential expression of the genetic program of this cell type.

ACKNOWLEDGEMENTS

We wish to thank Dr. O.K. Langley for linguistic criticism of the manuscript, Mrs. C. Thomassin-Orphanides for typing the manuscript and "La

Fondation pour la Recherche Médicale" (Paris) for finantial support of Dr. M. Luo during his stay in Strasbourg.

This study was supported in part by grants from "Comision Asesora de Investigacion Cientifica y Tecnica" and "Fondo de Investigaciones Sanitarias de la Seguridad Social" (Spain).

REFERENCES

1. M.G. Rumsby, Oligodendrocyte- myelin sheath interrelationships, in: "Search for the Cause of Multiple Sclerosis and other Chronic Diseases of the Central Nervous System", A. Boese, ed., Verlag Chemie, Weinheim-Deerfield Beach, Florida-Basel, pp. 50-63 (1980).
2. C.S. Raine, Morphology of myelin and myelination, in: "Myelin", 2nd Edition, P. Morell, ed., Plenum Press, New York, pp. 1-50 (1984).
3. A.N. Davison and J. Dobbing, The developing brain, in: "Applied Neurochemistry, A.N. Davison and J. Dobbing, eds., Contemporary Neurology Series, F.A. Davis Company, Philadelphia, pp. 253-286 (1968).
4. J.A. Benjamins and G.M. McKhann, Development, regeneration and aging of the brain, in: "Basic Neurochemistry" 3rd Edition, G.J. Siegel, R.W. Albers, B.W. Agranoff and R. Katzmann, eds., Little, Brown and Company, Boston, pp. 445-469 (1976).
5. W.T. Norton and W. Cammer, Isolation and characterization of myelin, in: "Myelin", 2nd Edition, P. Morell, ed., Plenum Press, New York, pp. 147-195 (1984).
6. L.L. Sarliève, Myelinogenesis in primary cultures, in: "A Multidisciplinary Approach to Myelin Diseases", G. Serlupi-Crescenzi, ed., Proceedings of a NATO ARW, Plenum Press, New York, pp. 171-191 (1988).
7. J. Legrand, Hormones thyroïdiennes et maturation du système nerveux, J. Physiol. Paris 78:603 (1982-1983).
8. J.H. Dussault and J. Ruel, Thyroid hormones and brain development, Ann. Rev. Physiol. 49:321 (1987).
9. J. Nunez, Microtubules and brain development : the effects of thyroid hormones, Neurochem. Int. 7:959 (1985).
10. R. Balazs, B.W.L. Brooksbank, A.N. Davison, J.T. Eayrs and D.A. Wilson, The effect of neonatal thyroidectomy on myelination in the rat brain, Brain Res. 15:219 (1969).
11. R. Balasz, B.W.L. Brooksbank, A.J. Patel, A.L. Johnson and D.A. Wilson, Incorporation of [^{35}S]sulfate into brain constituents during development and the effects of thyroid hormone on myelination, Brain Res. 30:273 (1971).
12. J.M. Matthieu, P.J. Reier and J.A. Sawchak, Proteins of rat brain myelin in neonatal hypothyroidism, Brain Res. 84:443 (1975).
13. T.J. Flynn, D.S. Deshmukh and R.A. Pieringer, Effects of altered thyroid function on galactosyl diacylglycerol metabolism in myelinating rat brain, J. Biol. Chem. 252:5864 (1977).
14. J. Clos, J. Legrand, N. Limozin, C. Dalmasso and G. Laurent, Effects of abnormal thyroid state and undernutrition on carbonic anhydrase and oligodendroglia development in the rat cerebellum, Dev. Neurosci. 5:243 (1982).
15. S.N. Walters and P. Morell, Effects of altered thyroid states on myelinogenesis, J. Neurochem. 36:1792 (1981).
16. J.H. Jacoby, G. Mueller and R.J. Wurtman, Thyroid state and brain monoamine metabolism, Endocrinology 97:1332 (1975).
17. J.M. Ito, T. Valcana et P.S. Timiras, Effect of hypo- and hyperthyroidism on regional monoamine metabolism in the adult rat brain, Neuroendocrinology 24:55 (1977).

18. P. Savard, Y. Mérand, T. Di Paolo et A. Dupont, Effects of thyroid state on serotonin, 5-hydroxyindoleacetic acid and substance P contents in discrete brain nuclei of adult rats, Neuroscience 10:1399 (1983).

19. G. Gross, O.E. Brodde et H.J. Schumann, Decreased number of β-adreno-ceptors in cerebral cortex of hypothyroid rats, Eur. J. Pharmacol. 61:191 (1980).

20. T. Yokota, H. Nakamura, T. Akamizu, T. Mori and H. Imura, Thyroid hormone receptors in neuronal and glial nuclei from mature rat brain, Endocrinology 118:1770 (1986).

21. A. Dembri, O. Michel, M. Belkhiria and R. Michel, Effects of short and long-term thyroidectomy on mitochondrial and nuclear activity in adult rat brain (abstract n° 139), Annu. Meet. Eur. Thyroid Ass. Madrid 1983, Anns Endocr. 44:78A (1983).

22. Z. Gottesfeld, C.J. Garcia and R.B. Chronister, Perinatal, not adult, hypothyroidism suppresses dopaminergic axon sprouting in the deafferented olfactory tubercle of adult rat, J. Neurosci. Res. 18:568 (1987).

23. V. Ferret-Sena, A. Sena, G. Rebel, A. Pascual, L. Freysz, G. Vincendon and L.L. Sarliève, Nuclear triiodothyronine receptors and mechanisms of triiodothyronine and insulin action on the synthesis of cerebroside sulfotransferase by cultures of cells dissociated from brain of embryonic mice, in: "NATO ASI Series : Enzymes of Lipid Metabolism II", L. Freysz, H. Dreyfus, R. Massarelli and S. Gatt, eds., Plenum Press, New York, pp. 597-613 (1986).

24. L.L. Sarliève, R. Bouchon, C. Koehl and N.M. Neskovic, Cerebroside and sulfatide biosynthesis in the brain of Snell Dwarf mouse : effects of thyroxine and growth hormone in the early postnatal period, J. Neurochem. 40:1058 (1983).

25. L.L. Sarliève, J.P. Delaunoy, A. Dierich, A. Ebel, M. Fabre, P. Mandel, G. Rebel, G. Vincendon, M. Wintzerith and A.N.K. Yusufi, Investigations on myelination in vitro. III. Ultrastructural, biochem-ical, and immunohistochemical studies in cultures of dissociated brain cells from embryonic mice, J. Neurosci. Res. 6:659 (1981).

26. M. Fabre, O.K. Langley, L. Bologa, J.P. Delaunoy, A. Lowenthal, V. Ferret-Sena, G. Vincendon and L.L. Sarliève, Cellular development and myelin production in primary cultures of embryonic mouse brain, Dev. Neurosci. 7:323 (1985).

27. L.L. Sarliève, M. Fabre, J. Susz and J.M. Matthieu, Investigations of myelination in vitro : IV. "Myelin-like" or premyelin structures in cultures of dissociated brain cells from 14-15 day-old embryonic mice, J. Neurosci. Res. 10:191 (1983).

28. P. De Nayer, Thyroid hormone action at the cellular level, Hormone Res. 26:48 (1987).

29. A. Pascual, A. Aranda, V. Ferret-Sena, M.M. Gabellec, G. Rebel and L.L. Sarliève, Triiodothyronine receptors in developing mouse neuronal and glial cell cultures and in chick-cultured neurones and astrocytes, Dev. Neurosci. 8:89 (1986).

30. J.M. Kolodny, J.L. Leonard, P.R. Larsen and J.E. Silva, Studies of nuclear 3,5,3'-triiodothyronine binding in primary cultures of rat brain, Endocrinology 117:1848 (1985).

31. J. Ruel, R. Faure and J.H. Dussault, Regional distribution of nuclear T3 receptors in rat brain and evidence for preferential localization in neurons, J. Endocrinol. Invest. 8:343 (1985).

32. J. Booher and M. Sensenbrenner, Growth and cultivation of dissociated neurons and glial cells from embryonic chick, rat and human brain in flask cultures, Neurobiology 2:97 (1972).

33. H.H. Samuels, F. Stanley and J. Casanova, Depletion of L-3,5,3'-triiodothyronine and L-thyroxine in euthyroid calf serum for use in cell culture studies of the actions of thyroid hormone, Endocrinology 105:80 (1979).

34. G. Labourdette, G. Roussel and J.L. Nussbaum, Oligodendroglia content of glial cell primary cultures, from newborn rat brain hemispheres, depends on the initial plating density, Neurosci. Lett. 18:203 (1980).

35. K.D. McCarthy and J. De Vellis, Preparation of separate astroglial and oligodendroglial cell cultures from rat cerebral tissue, J. Cell Biol. 85:890 (1980).

36. F. Besnard, M. Sensenbrenner and G. Labourdette, Culture of oligo-dendrocytes from brain of newborn rat, in: "A Dissection and Tissue Culture Manual of the Nervous System", A. Shahar, J. de Vellis, A. Vernadakis and B. Haber, eds., AR Liss Inc., New York, in press (1988).

37. E. Yavin and J.H. Menkes, The culture of dissociated cells from rat cerebral cortex, J. Cell Biol. 57:232 (1973).

38. A. Bignami, L.F. Eng, D. Dahl and C.T. Uyeda, Localization of the glial fibrillary acidic protein in astrocytes by immunofluorescence, Brain Res. 43:429 (1972).

39. E. Bock, M. Moller, C. Nissen and M. Sensenbrenner, Glial fibrillary acidic protein in primary astroglial cell cultures derived from newborn rat brain, FEBS Lett. 83:207 (1977).

40. E. Bock, Z. Yavin, O.S. Jorgensen and E. Yavin, Nervous system-specific proteins in developing rat cerebral cells in culture, J. Neurochem. 35:1297 (1980).

41. S.L. Bologa, H.P. Siegrist, A. Z'graggen, K. Hofmann, U. Weismann, D. Dahl and N. Herschkowitz, Expression of antigenic markers during the development of oligodendrocytes in mouse brain cell cultures, Brain Res. 210:217 (1981).

42. M. Luo, R. Faure, J. Ruel and J.H. Dussault, A monoclonal antibody to the rat nuclear T3 receptor : production and characterization, Endocrinology 124:in press (1988).

43. M. Sensenbrenner, G. Labourdette, J.P. Delaunoy, B. Pettmann, G. Devilliers, G. Moonen and E. Bock, Morphological and biochemical differentiation of glial cells in primary culture, in: "Tissue Culture in Neurobiology", E. Giacobini, A. Vernadakis and A. Shahar, eds., Raven Press, New York, pp. 385-395 (1980).

44. B.H.J. Juurlink and L. Hertz, Plasticity of astrocytes in primary cultures : an experimental tool and a reason for methodological caution, Dev. Neurosci. 7:263 (1985).

45. D. Giulian, Ameboid microglia as effectors of inflammation in the central nervous system, J. Neurosci. Res. 18:155 (1987).

46. M.C. Raff, R. Mirsky, K.L. Fields, R.P. Lisak, S.H. Dorfman, D.H. Silberberg, N.A. Gregson, S. Leibowitz and M.C. Kennedy, Galacto-cerebroside is a specific cell-surface antigen marker for oligo-dendrocytes in culture, Nature 274:813 (1978).

47. N.H. Sternberger, Y. Itoyama, M.W. Kies and H. DeF. Webster, Myelin basic protein demonstrated immunocytochemically in oligodendroglia prior to myelin sheath formation, Proc. Natl. Acad. Sci. USA 75:2521 (1978).

48. J. Ortiz-Caro, B. Yusta, F. Montiel, A. Villa, A. Aranda and A. Pascual, Identification and characterization of nuclear T3 receptors in cells of glial and neuronal origin, Endocrinology 119:2163 (1986).

49. J. Puymirat and A. Faivre-Bauman, Evolution of triiodothyronine nuclear binding sites in hypothalamic serum-free cultures : evidence for their presence in neurons and astrocytes, Neurosci. Lett. 68:299 (1986).

50. M. Luo, R. Faure and J.H. Dussault, Ontogenesis of nuclear T3 receptors in primary cultured astrocytes and neurons, Brain Res. 381:275 (1986).

51. A. Espinosa de los Monteros, G. Roussel and J.L. Nussbaum, A procedure for long-term culture of oligodendrocytes, Dev. Brain Res. 24:117 (1986).

52. M.A. Haidar, S. Dube and P.K. Sarkar, Thyroid hormone receptors of developing chick brain are predominantly in the neurons, Biochem. Biophys. Res. Commun. 112:221 (1983).

53. J.M. Kolodny, P.R. Larsen and J.E. Silva, In vitro 3,5,3'-triiodothyronine binding to rat cerebrocortical neuronal and glial nuclei suggests the presence of binding sites unavailable in vivo, Endocrinology 116:2019 (1985).

54. D. Gullo, A.K. Sinha, R. Woods, K. Pervin and R.P. Ekins, Triiodothyronine binding in adult rat brain : compartmentation of receptor, populations in purified neuronal and glial nuclei, Endocrinology 120:325 (1987).

55. D. Gullo, A.K. Sinha, A. Bashir, M. Hubank and R.P. Ekins, Differences in nuclear triiodothyronine binding in rat brain cells suggest phylogenetic specialization of neuronal functions, Endocrinology 120:2398 (1987).

56. D. Koerner, H.L. Schwartz, M.I. Surks and J.H. Oppenheimer, Binding of selected iodothyronine analogues to receptor sites of isolated rat hepatic nuclei : high correlation between structural requirements for nuclear binding and biological activity, J. Biol. Chem. 250:6417 (1975).

57. H.H. Samuels, F. Stanley and J. Casanova, Relationship of receptor affinity to the modulation of thyroid hormone nuclear receptor levels and growth hormone synthesis by L-triiodothyronine and iodothyronine analogs in cultured GH1 cells, J. Clin. Invest. 63:1229 (1979).

58. J. Sap, A. Munoz, K. Damm, Y. Goldberg, J. Ghysdael, A. Leutz, H. Beug and B. Vennstrom, The c-erb-A protein is a high-affinity receptor for thyroid hormone. Nature 324:635 (1986).

59. C. Weinberger, C.C. Thompson, E.S. Ong, R. Lebo, D.J. Gruol and R.M. Evans, The c-erb-A gene encodes a thyroid hormone receptor, Nature 324:641 (1986).

60. C.C. Thompson, C. Weinberger, R. Lebo and R.M. Evans, Identification of a novel thyroid hormone receptor expressed in the mammalian central nervous system, Science 237:1610 (1987).

61. F. Courtin, F. Chantoux and J. Francon, Thyroid hormone metabolism by glial cells in primary culture, Mol. Cell Endocrinol., 48:167 (1986).

62. G. Shanker, N.R. Bhat and R.A. Pieringer, Investigations on myelination in vitro: thyroid hormone receptors in cultures of cells dissociated from embryonic mouse brain, Bioscience Rep. 1:289 (1981).

63. N.R. Bhat, L.L. Sarliève, G. Subba Rao and R.A. Pieringer, Investigations on myelination in vitro. Regulation by thyroid hormone in cultures of dissociated brain cells from embryonic mice, J. Biol. Chem. 254:9342 (1979).

64. G. Shanker, A.T. Campagnoni and R.A. Pieringer, Investigations on myelinogenesis in vitro: developmental expression of myelin basic protein mRNA and its regulation by thyroid hormone in primary cerebral cell cultures from embryonic mice. J. Neurosci. Res. 17:220 (1987).

65. G. Shanker and R.A. Pieringer, Investigations on myelinogenesis in vitro: II. The occurence and regulation of protein kinases by thyroid hormone in primary cultures of cells dissociated from embryonic mouse brain, Bioscience Rep. 7:159 (1987).

66. F. Montiel, L. Sarliève, A. Pascual and A. Aranda, Multihormonal control of proliferation and cytosolic glycerol phosphate dehydrogenase, lactate dehydrogenase and malic enzyme in glial cells in culture, Neurochem. Int. 9:247 (1986).

67. G. Almazan, P. Honegger and J.M. Matthieu, Triiodothyronine stimulation of oligodendroglial differentiation and myelination. A developmental study, Dev. Neurosci. 7:45 (1985).

68. J.W. Koper, R.C. Hoeben, F.M.H. Hochstenbach, L.M.G. van Golde and M. Lopes-Cardozo, Effects of triiodothyronine on the synthesis of sulfo-lipids by oligodendrocyte-enriched glial cultures, Biochim. Biophys. Acta 887:327 (1986).

69. G. Shanker, S.G. Amur and R.A. Pieringer, Investigations on myelinogenesis **in vitro**: a study of the critical period at which thyroid hormone exerts its maximum regulatory effect on the developmental expression of two myelin associated markers in cultured brain cells from embryonic mice, Neurochem. Res. 10:617 (1985).

70. H.H. Samuels and J.R. Tsai, Thyroid hormone action in cell culture: demonstration of nuclear receptors in intact cells and isolated nuclei, Proc. Natl. Acad. Sci. USA 70:3488 (1973).

71. B. Yusta, F. Besnard, J. Ortiz-Caro, A. Pascual, A. Aranda and L. Sarlière, Evidence for the presence of nuclear 3,5,3'-triiodothyronine receptors in secondary cultures of pure rat oligodendrocytes, Endocrinology 122:in press (1988).

DEVELOPMENTAL AND THYROID HORMONE REGULATION OF

Na,K-ATPase ISOFORMS IN THE BRAIN

Alicia McDonough and Cheryl Schmitt

Department of Physiology and Biophysics
University of Southern California
School of Medicine
Los Angeles, California, 90033

INTRODUCTION

The sodium pump (Na,K-ATPase) is a plasma membrane bound enzyme found in all animal cells and subserves many specialized functions in nerve cells. It maintains transmembrane cationic gradients by pumping three sodiums out of the cell for two potassiums into the cell fueled by ATP hydrolysis (1). The resultant electrochemical gradients are crucial for normal neuronal activity, and drive the transport of nutrients and reuptake of neurotransmitters into the nerve cells (1, 2).

The sodium pump is composed of two subunits: alpha and beta. Three isoforms of the alpha catalytic subunit have been identified: alpha (alpha 1), alpha+ (alpha 2), and alpha 3 (3). Only one beta glycoprotein subunit mRNA has been identified (4). The brain expresses all three alpha isoforms. Alpha (alpha 1) is found in all tissues examined. Alpha+, is also found in eye, heart and skeletal muscle. Although these two isoforms have nearly identical molecular weights (1), alpha+ has a slower mobility in SDS-PAGE (5). No peptide form has been identified for alpha 3 but its identity is inferred from isolation of a unique cDNA with sequence homology to the other two isoforms (3).

During development there is a 10-fold change in Na,K-ATPase activity in rat brain (6) accompanied by the onset of electrical activity and change in ionic composition in the brain (7,8). These changes start a few days before birth and approach adult levels by 30 days of age. In this chapter we review our studies examining how the abundance of the alpha and alpha+ isoforms of Na,K-ATPase change during this period (9).

Thyroid hormone regulates a significant portion of the increase in Na,K-ATPase activity during development (6). However, in the adult brain Na,K-ATPase synthesis is not

regulated by T3 even though T3 regulates Na,K-ATPase
synthesis in many adult tissues (e.g. kidney, heart, liver
and skeletal muscle) (10, 11). These findings support the
concept of the existence of a critical period for the
action of T3 on Na,K-ATPase synthesis in developing brain.
In this chapter we review our studies examining the onset
of this critical period for T3 induction of the alpha and
alpha+ isoforms (12).

DEVELOPMENTAL INCREASES IN Na,K-ATPase ACTIVITY, AND ALPHA
AND ALPHA+ ABUNDANCE IN RAT BRAIN

In order to assess the developmental increases in
Na,K-ATPase activity and abundance, membrane fractions were
prepared from fetal and neonatal animals as previously
described (9). Na,K-ATPase activity was measured
enzymatically as the ouabain inhibitable fraction of the
umoles of inorganic phosphate liberated per mg protein per
hour. Figure 1 summarizes the results obtained from three
litters. Sodium pump activity increased 10 fold between
15.5 days gestation and 20 days of age, which is similar to
that previously reported (6). Whether the increased
activity is a consequence of an increased enzymatic
turnover or increased sodium pump abundance cannot be
concluded from this data. We next tested the hypothesis
that the increase resulted from an increase in abundance of
alpha and/or alpha+ forms of Na,K-ATPase.

Abundance of the Na,K-ATPase isoforms was studied by
immunoblotting (Western blot). In short, this is
accomplished by resolving the brain membrane fractions by
SDS-PAGE, blotting the gel onto diazotized paper, and
incubating the paper with anti-Na,K-ATPase antisera then
iodinated Protein A followed by autoradiography, as
previously described (9). Abundance of the isoforms was
quantitated by scanning densitometry. Resolution of the
alpha and alpha+ isoforms was found to be optimal when gels
were made and run in unpurified SDS at pH 9.0. The Western

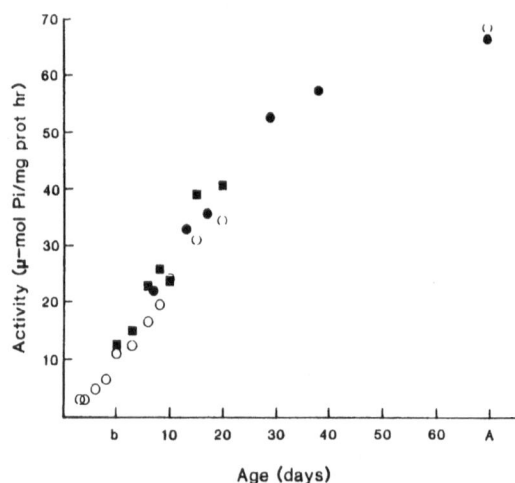

FIGURE 1. Na,K-ATPase activity during neonatal
development. Assayed as described (9). The symbols
represent separate litters.

blot signals were found to be linear in the range of 10-40 ug protein per lane. Developmental experiments were conducted by resolving 30 ug protein per sample.

All of the antisera were generated against Na,K-ATPase purified from the kidney, which is alpha form. The antisera differed in their reactivity to alpha vs. alpha+. Thus, although we cannot assess the magnitude of alpha relative to alpha+, we can assess the fold change in alpha and alpha+ during development. This is illustrated in Figure 2 where a blot was incubated sequentially with an antiserum that had a higher affinity for alpha+, then an antiserum that had a higher affinity for alpha. A second litter is shown that was incubated with a mixture of the two antisera.

The changes in alpha isoform abundance of the two litters shown in Figure 2 and an additional litter were quantitated by scanning densitometry. We assume that the subunit itself, and thus the affinity of the antiserum for the subunit, does not change during development. As shown in Figure 3, the abundance of both alpha and alpha+ increased 10 fold from 18.5 days gestation to 20 days of age. Alpha+ increases earlier in development than alpha. Thus, the 10 fold increase in Na,K-ATPase activity during this developmental period can be accounted for by a similar increase in the abundance of both the alpha and alpha+

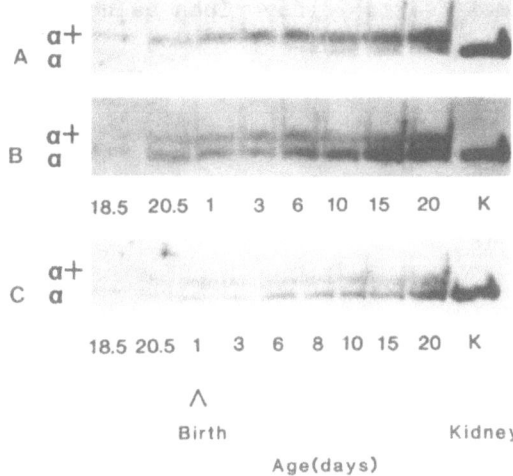

FIGURE 2. Immunodetection of Na,K-ATPase alpha and alpha+ subunits in developing rat brain. 30 ug of crude membrane protein from littermates of different days during development were prepared and separated by 7.5% SDS-PAGE, blotted onto diazophenylthioether paper, and probed with antisera. A: anti-guinea pig Na,K-ATPase antiserum diluted 1:100. B: the paper from A was reincubated with anti-rat Na,K-ATPase antiserum diluted 1:400. C: a mixture of both antisera was used to probe the samples of a second litter. Numbers represent days of gestation (18.5, 20.5) or days of age. K, 2 ug purified kidney Na,K-ATPase.

isoforms of the pump. This conclusion is verified
quantitatively in Figure 4 where the relative change in
Na,K-ATPase activity is plotted against the relative change
in abundance of the alpha and alpha+ isoforms.

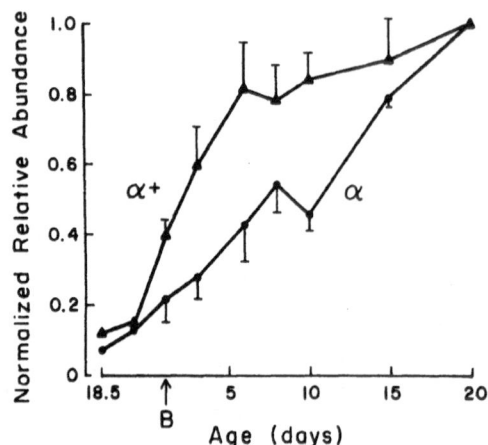

FIGURE 3. Developmental changes in abundance of alpha and
alpha+. Abundance of alpha and alpha+ during development
calculated from Western blots of samples obtained from
three litters. All alpha+ normalized to 20 day alpha+ and
all alpha normalized to 20 day alpha values. Mean +/-
standard error of the mean plotted.

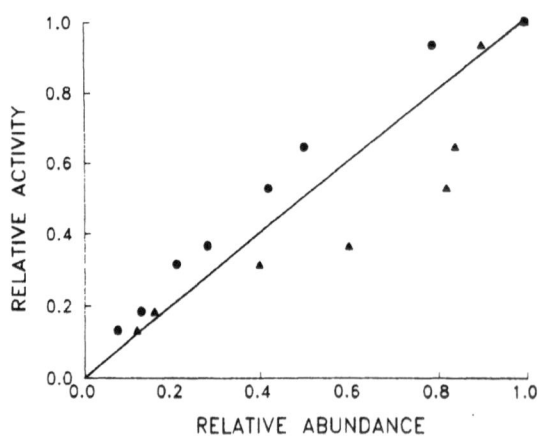

FIGURE 4. Na,K-ATPase activity vs. alpha isoform abundance
during development. Na,K-ATPase activity and alpha
(circles) and alpha+ (triangles) abundance were expressed
relative to the 20 day values. A reference line with a
slope of 1 is indicated.

THYROID HORMONE REGULATION OF Na,K-ATPase ACTIVITY AND
ALPHA ISOFORM ABUNDANCE DURING DEVELOPMENT

The presence of isoforms of Na,K-ATPase suggests the
possibility of differential regulation of the isoforms.
Since thyroid hormone has been shown to regulate a
component of the increase in Na,K-ATPase activity during
development (6), we examined T3 regulation of Na,K-ATPase
activity and the abundance of the alpha isoforms during
this period.

To determine the effect of T3 on Na,K-ATPase during
development, pregnant rats were rendered hypothyroid with a
low-iodine diet, perchlorate in the drinking water and a
subcutaneous implant of propylthiouracil (9). Following
birth, the mothers remained on this regime to assure the
suckling pups remained hypothyroid. Hypothyroid status was
verified by radioimmunoassay of T3. Na,K-ATPase activity
was determined in control and hypothyroid neonates as
summarized in Figure 5A. The activities of the hypothyroid
and euthyroid groups diverge after 14 days, and are
statistically different by 15 days and after where the
hypothyroid values are 15-20% less than the euthyroid
values.

Resolution and detection of the alpha isoforms allowed
us to determine whether thyroid hormone status regulated
the abundance of one or both isoforms. The changes in
alpha and alpha+ abundances were quantified by scanning
densitometry, normalized to euthyroid values and are
presented in Figure 5B. The line at 1.0 and shaded area
indicate the mean and standard error of euthyroid abundance
of alpha and alpha+ during this period. Abundance of both
isozymes becomes significantly different from euthyroids by
14 days and thereafter indicating that thyroid status
affects both isoforms of Na,K-ATPase. The lower abundance
of alpha isoforms confirms the lower enzymatic activity.
We conclude that brain Na,K-ATPase responds to thyroid
hormone status at around 14 days of age. The onset of
thyroid hormone responsivity might be earlier if the very
low (rather than zero) levels of T3 detected in some of the
neonates was sufficient to induce Na,K-ATPase.

DISCUSSION
In this study we have shown that thyroid hormone status
causes specific alterations in Na,K-ATPase enzyme abundance
and activity during development when normalized to a
constant amount of membrane protein suggesting specific
effects, direct or indirect, on Na,K-ATPase synthesis. The
increase in abundance could also reflect T3 mediated
decrease in degradation rate but this is unlikely given the
large increase in abundance.

Na,K-ATPase activity and abundance is the same in the
brains of hypothyroid and euthyroid neonates before 15 days
of age. Since we have shown that Na,K-ATPase abundance is
regulated by ionic substrates as well as hormones (14), the
increased sodium influx seen in brain cells after birth
might stimulate upregulation of Na,K-ATPase synthesis

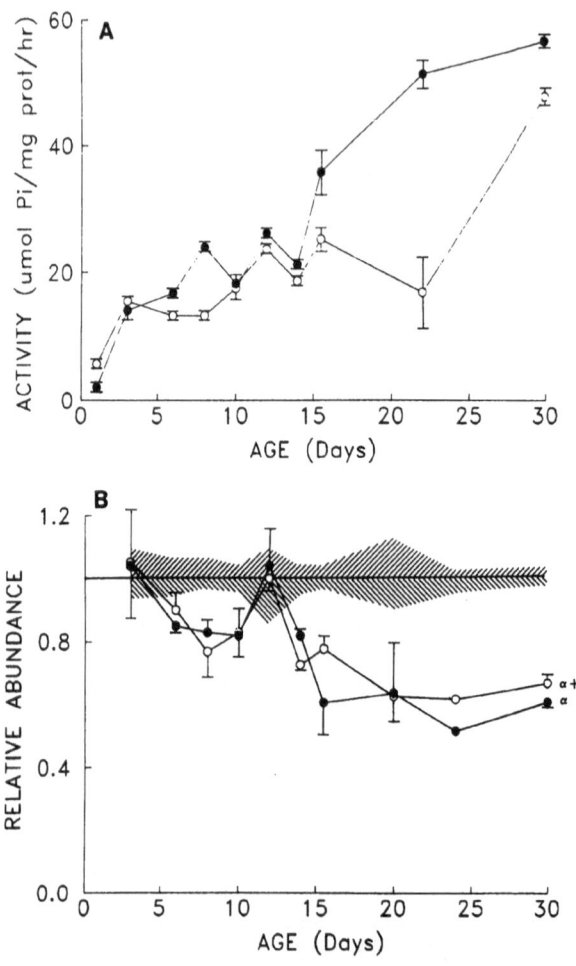

FIGURE 5A. Na,K-ATPase activity in hypothyroid and
euthyroid brain. Na,K-ATPase activity was assayed in brain
microsomes at the ages indicated. Each point is the mean
+/- standard error. Closed circles - euthyroid, open
circles - hypothyroid.

FIGURE 5B. Abundance of alpha and alpha+ isozymes in
hypothyroid and euthyroid brain. Relative abundance of
isoforms in neonatal brain calculated by scanning
immunoblots of samples obtained from three hypothyroid and
five euthyroid litters. Abundance expressed as
hypothyroid/paired mean euthyroid values. Each point is
the mean +/- the standard error. Solid line at 1.0
indicates the normalized mean euthyroid values. The shaded
area above the line represents the standard error of the
normalized euthyroid data for alpha+, while the shaded area
below is the standard error of alpha. Closed circles -
alpha, open circles - alpha+.

irrespective of thyroid status. Subsequent differentiation processes beyond 15 days, including those seen in this study, may be T3 dependent. The possibility remains that even during the critical period T3 regulates Na,K-ATPase abundance indirectly by upregulating the fluxes of ionic substrates (15).

In conclusion, we suggest the following scheme to integrate how changes in thyroid hormone receptors, T3 concentration and morphological alterations in developing brain might account for changes in Na,K-ATPase activity and abundance during development. Brain T3 receptors and plasma T3 concentration increase rapidly after birth approaching adult concentrations between 11 and 21 days. This period corresponds to one in which decisive maturational effects are evident in brain, including rapid myelinogenesis, and proliferation of axons and dendrites. During this period T3 may regulate Na,K-ATPase isozymes directly or, alternately, may regulate synthesis indirectly by increasing the fluxes of ionic substrates (e.g. sodium channels). Since a novel T3 receptor expressed in brain has been recently identified (16), we suggest that after 22 neonatal days the thyroid regulation is uncoupled, perhaps through a switch in expression of thyroid hormone receptors to a form that does not bind to Na,K-ATPase or the intermediary regulator.

REFERENCES

1. Cantley, L.C. Structure and mechanism of (Na,K)-ATPase. Curr. Top. Bioenerg., 11:201-237, 1981.

2. Iverson, L.L. and J.S. Kelly. Uptake and metabolism of aminobutyric acid by neurones and glial cells. Biochem. Pharmacol., 24:933-938, 1975.

3. Shull, G.E., J. Greeb, and J.B. Lingrel. Molecular cloning of three distinct forms of the Na,K-ATPase alpha subunit from rat brain. Biochem. 25:8125-8132, 1986.

4. Young, R.M., G.E. Shull and J.B. Lingrel. Multiple mRNAs from rat kidney and brain encode a single Na,K-ATPase beta subunit protein. J. Biol. Chem. 262:4905-4910, 1987.

5. Sweadner, K.J. Two molecular forms of Na,K-ATPase in brain J. Biol. Chem 254:6060-6067, 1979.

6. Valcana, T. and P.S. Timiras. Effect of hypothyroidism on ionic metabolism and Na-K activated ATP phosphohydrolase activity in the developing rat brain. J. Neurochem.16: 935-943, 1969.

7. Abdel-Latif, A.A., J. Brody, and H. Ramahi. Studies on sodium-potassium adenosine triphosphatase of the nerve endings and appearance of electrical activity in developing brain. J. Neurochem. 14:1133-1141, 1967.

8. Vernadakis,A. and Woodbury, D.M. Electrolyte and amino acid changes in rat brain during maturation. Am. J. Physiol. 203:748-752, 1962.

9. Schmitt, C.A. and A.A. McDonough. Developmental and thyroid hormone regulation of two molecular forms of Na,K-ATPase in brain. J. Biol. Chem. 261:10439-10444, 1986.

10. Ismail Beigi, F. and I.S. Edelman. The mechanism of the calorigenic action of thyroid hormone. J. Gen. Physiol. 52:710-722, 1971.

11. Lin, M.H.and T. Akera. Increased Na,K-ATPase concentrations in various tissues of rat caused by thyroid hormone treatment. J. Biol. Chem. 253:723-726.

12. Schmitt, C.A. and A.A. McDonough. Thyroid hormone induces Na,K-ATPase during a critical period of development in neonatal rat brain. Fed. Proc. 46:361, 1987.

13. McDonough, A.A., T.A. Brown, B. Horowitz, R. Chiu, J. Schlotterbeck, J. Bowen, and C. Schmitt. Thyroid hormone coordinately regulates Na,K-ATPase alpha and beta subunit mRNA levels in kidney. Am. J. Physiol. 254:C323-C329, 1988.

14. Bowen, J.W. and A.McDonough. Pretranslational regulation of Na,K-ATPase in cultured canine kidney cells by low K^+. 252:C179-C189, 1987.

15. Atterwill, C.K., D.J. Atkinson, I. Bermudez. Effect of thyroid hormone and serum on the development of Na,K-ATPase and associated ion fluxes in cultures from rat brain. Neurosci. 14:361-373, 1985.

16. Thompson,C.C., C. Weinberger, R. Lebo, and R.M. Evans. Identification of a novel thyroid hormone receptor expressed in the mammalian central nervous system. Science 237:1610-1614, 1987.

THYROID HORMONES AND NEUROTRANSMITTER SYSTEMS: AN INTERACTION WITH NERVE GROWTH FACTOR IN THE REGULATION OF CHOLINERGIC NEURONES

Ambrish J. Patel, Anthony Hunt and Jozsef Kiss

MRC Collaborative Center and Div. of Neurophysiology
and Neuropharmacology
National Institute for Medical Research
Mill Hill, London NW7 1AA, United Kingdom

INTRODUCTION

Congenital hypothyroidism is well established as a cause of mental subnormality in man. However, there are only a few neuropathological studies on brains of patients with an early abnormal thyroid state, and none of these studies gives information about the underlying biological mechanisms.[1,2] For this, one has to turn to animal models. In the human foetus, thyroid function commences towards the end of the first trimester of pregnancy, whereas in rat pups it begins near term. This permits examination of events occurring in rat brain, after manipulation of thyroid hormone levels, from the early postnatal period. In the present chapter, we will consider primarily the findings of our laboratory; the references to other studies can be found in our recent papers and reviews.[3-11]

Our earlier studies were mainly focused on the effect of thyroid hormone on certain morphological and biochemical aspects of brain development in rats. Briefly, it was found that thyroid hormone primarily affects cell acquisition in those parts of the brain where postnatal neurogenesis is significant, namely the cerebellum, dentate gyrus and olfactory bulbs.[12,13] It would appear that the role of thyroid hormone in the formation and maintenance of nerve cells is related to changes in either cell migration or maturation, rather than to alterations in the replication of germinal cells.[9,14-16] Severe retardation in the maturation of Purkinje cells in the cerebellum, of cortical neurones in the forebrain and of both pyramidal and granule cells in the hippocampus is found during the period of thyroid deficiency.[10,11] But only in the cerebellum does the restricted availability of postsynaptic contact sites, due to hypoplasia of the Purkinje cells, result in a transient increase in the degeneration of differentiating granule cells.[9,14,16] Studies of the development of the hippocampal structure indicate that thyroid hormone is important in the establishment of the CA1 to CA4 gradient of pyramidal cell differentiation and in the development of the spatio-temporal relationship between the pyramidal and granule cells of the hippocampus.[10] Thyroid hormone also has a marked influence on the biochemical markers of neuronal differentiation which, in general, is retarded in thyroid deficiency and advanced in hyperthyroidism.[4,17-20] On detailed examination, it would appear that alterations in thyroid status lead to unsynchronized shifts in the development of the different transmitter and peptidergic systems, and in the specific cell surface molecules that are believed to play an important role in cell-to-cell interaction and

139

recognition. Distortions in the development of neuronal circuits are also observed in electron-microscopic and neurophysiological studies.[21-22] These findings have led us to propose that such distortions, rather than synchronized shifts in normal developmental relationships, may be instrumental in causing the lasting functional impairments that are often seen in neuropathological conditions associated with alterations in thyroid status during early life.

Earlier studies on hypothyroid animals were mainly concentrated on the postnatal period, and little is known about the effects of hormonal rehabilitation after weaning. In addition, major advances in our knowledge of cell-specific markers in the central nervous system,[23] and about tissue culture methods,[24] now makes it possible to ask penetrating questions about the direct effect of the thyroid hormone and its possible interaction with centrally produced humoural factors.

REVERSIBILITY OF THE CELL-TYPE SPECIFIC EFFECTS OF THYROID HORMONE DEFICIENCY DURING EARLY LIFE

We have followed the development of neurotransmitter systems in terms of the activities of choline acetyltransferase (ChAT), glutamate decarboxylase (GAD) and glutaminase, enzymes critical for the synthesis of the neurotransmitters acetylcholine, γ-aminobutyric acid (GABA) and glutamate, respectively (Table 1).[5,6] Thyroid deficiency is found to result in a marked retardation in the developmental changes of the activities of ChAT and GAD. The effects are not uniform, both their severity and duration varying considerably from one brain region to another (Table 1). In spite of continued thyroid hormone deficiency, ChAT and GAD activities in the hippocampus and GAD activity in the cerebellum are restored to normal by 25 days. These results are consistent with earlier findings showing a marked retardation in the development of choline uptake and density of muscarinic cholinergic and GABA-ergic receptors in the brain of hypothyroid rats.[4,17,25] None of these studies has examined the reversibility of the effects of neonatal hypothyroidism. We have found that the alterations in ChAT activity in the cerebral cortex and cerebellum, and in GAD activity in the basal forebrain and cerebral cortex, are restored to normal levels at 102 days of rehabilitation. Only in the basal forebrain is ChAT activity not increased during rehabilitation, remaining persistently reduced in comparison with controls throughout the experimental period studied (Table 1). The effect on subcortical cholinergic neurones is specific to hypothyroidism, as a similar reduction was not observed in rats rehabilitated after neonatal undernourishment, which is believed to accompany thyroid deficiency during early life.[5,26] The selective persistent reduction in ChAT activity in the basal forebrain of the rat after thyroid deficiency during early life may be due to one or more of the following: cell death, cell atrophy, or reduction in ChAT synthesis without apparent morphological alterations in cholinergic cells. Our preliminary findings using ChAT immunohistochemistry[27] indicate that, in comparison with controls, the intensity of the immunoreaction product is much lower in the ChAT-positive neurones of the basal forebrain of rehabilitated neonatally hypothyroid rats (Fig. 1).

In contrast to the developmental patterns of the activities of ChAT and GAD, that of glutaminase specific activity in all brain regions of hypothyroid rats is very similar to that for control animals (Table 1).[6] Recently, Balazs et al.[28] studied the effect of thyroid hormone on the maturation of cultured cerebellar granule cells, which are believed to be glutamatergic. A number of indices were used, including the developmental changes affecting the neuronal cell adhesion molecule (N-CAM or D2 protein), muscarinic receptor binding, voltage (veratridine)-sensitive sodium uptake and Na^+,K^+-ATPase enzyme activity, all of which in the cerebellum in vivo

Table 1. The activity of enzymes, as markers of nerve cell types, in different regions of rat brain during neonatal thyroid deficiency and following long-term rehabilitation[a]

Marker enzyme	Age (days)	Basal forebrain	Cerebral cortex	Hippocampus	Cerebellum
ChAT	5	85*	96	98	98
(cholinergic)	15	67*	65*	60*	117*
	25	75*	78*	101	113*
	130	76*	100	93	104
GAD	5	85*	106	83*	97
(GABA-ergic)	15	87*	90*	91	82*
	25	88*	90*	97	94
	130	101	99	95	101
Glutaminase	5	93	96	103	95
(glutamatergic)	15	94	106	105	100
	25	96	89	99	92
	130	90	95	105	106

[a]Young rats were made thyroid-deficient by fostering at birth to mothers that had been given 50 mg of propylthiouracil daily by stomach tube for 2 days before they gave birth and then throughout the lactation period. On day 28, the hypothyroid rats were rehabilitated by weaning the young on a normal diet. Four control and 4 experimental rats were studied at 5, 15, 25 and 130 days after birth. Whole homogenates from different brain regions were used for the estimation of choline acetyltransferase (ChAT), glutamate decarboxylase (GAD) and glutaminase activities. Enzyme activity (per mg protein) is expressed as a percentage of the respective control values. Significant differences between the mean values of control and experimental groups are indicated by asterisks: $P<0.05$. The data are taken from Patel et al.[5,6]

are influenced by thyroid hormone.[17,19,29] In cultured granule cells none of these parameters was significantly altered by thyroid hormone.[28] Thus the effect of thyroid hormone on neuronal maturation appears to be cell-type specific. For example, glutamatergic cells do not appear to be a target of thyroid hormone action, whereas cholinergic cells are more severely affected than the others.

INTERACTION BETWEEN THE EFFECTS OF THYROID HORMONE AND NERVE GROWTH FACTOR (NGF) ON CHOLINERGIC NEURONES

In order to obtain more precise information about the persistent effect of thyroid deficiency on subcortical cholinergic nerve cells without the complications of secondary reactions in in vivo studies, conditions have been established for the culturing in chemically defined medium of neurones enriched in cholinergic cells from the septal-diagonal band region of the embryonic rat brain.[8] At 10 days in vitro, the cultures contain more than 96% nerve cells of which about 18% are cholinergic neurones, while the proportion of astrocytes was less than 2%. Recently, it has been suggested

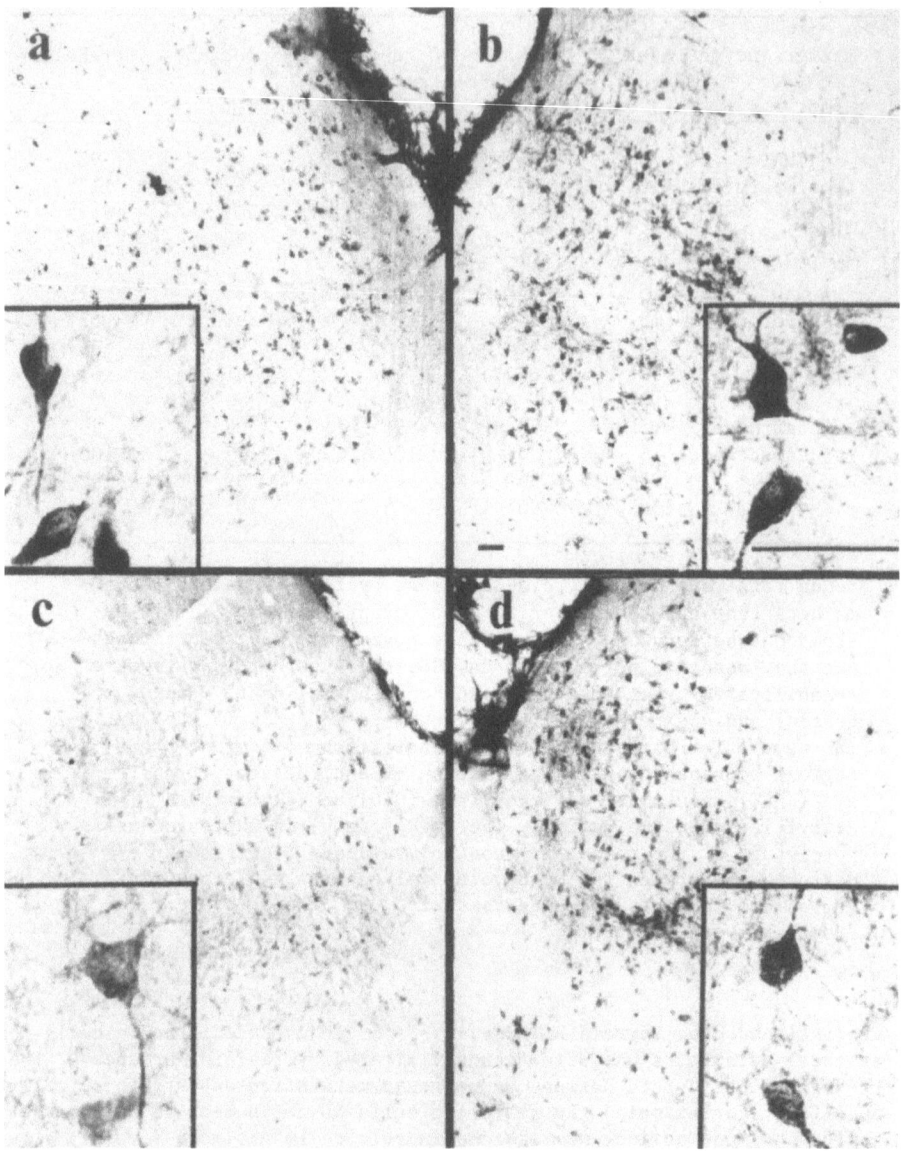

Figure 1. Immunohistological demonstration of comparative topography
and intensity of immunoreaction products of ChAT- and
NGF-R-positive neurones in the basal forebrain. Control
rats: (a) ChAT-positive and (b) NGF-R-positive cells.
Rehabilitated neonatally hypothyroid rats: (c) ChAT-
positive and (d) NGF-R-positive cells. Two immediately
adjacent 80 μm thick Vibratome cut coronal sections are
processed for ChAT or NGF-R immunohistochemically as
described by Kiss et al.[27] The sections are mounted in
mirror position. In the insets, cells are shown at high
magnification to demonstrate the intensity of the immunoreaction product more clearly. Bars = 50 μm.

that NGF may be involved in the development of central cholinergic neurones, and that NGF-like immunoreactive substances are detectable in the brain.[30,31] Therefore, we have examined the possibility that the effect of thyroid hormone may be mediated through processes evoked by NGF.[8,32-34] During development, the specific activity of ChAT increases markedly in cholinergic cultures. The addition of thyroid hormone or NGF to the culture medium increases the amount of ChAT activity in a dose-dependent manner (Fig. 2). The elevation of ChAT activity is due to an increase in the amount of enzyme per cholinergic cell, since neither treatment with NGF nor with thyroid hormone has a significant effect on the number of cells in the cultures, including the cholinergic neurones.[8] When cultures are supplemented with both agents at maximal effective concentrations, the stimulation in ChAT activity is much greater than the sum of the individual effects (Fig. 2). The interaction between the effects of thyroid hormone and NGF in cholinergic cells is not an in vitro artefact, but can be reproduced in vivo in the basal forebrain of the developing rat.[7] Treatment with thyroxine from birth to 10 days increases ChAT activity by about 20% in the basal forebrain of normal rats (Table 2). Similarly, intraventricular administration of low doses of NGF for 10 postnatal days results in about a 40% increase in ChAT activity.[35,36] However, when the young rats are given both thyroxine and NGF, the elevation in ChAT activity is not only significantly greater than after treatment with either thyroid hormone or NGF alone, but also the stimulation in enzyme activity is much higher (about 80%) than the additive increase after separate treatment with each of these humoural agents (Table 2). Although, the synergistic effect in vivo of these compounds (Table 2) is less marked than in vitro (Fig. 2). These observations show

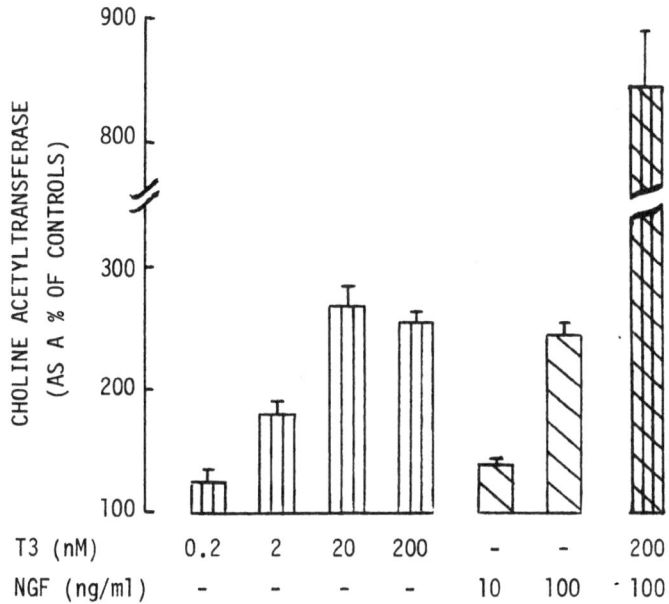

Figure 2. Interaction between the effects of thyroid hormone and NGF on ChAT activity. Neuronal cultures enriched in cholinergic cells were grown for 10 days in either normal chemically defined medium or the same containing different concentrations of either 3,3',5-triiodo-L-thyronine (T3) or NGF or both agents. Results are expressed as a percentage of control ChAT activity. Each column is the mean value with S.E.M. indicated by the bar. The data are taken from Hayashi and Patel.[8]

Table 2. Effect of thyroid hormone (T4) and NGF on the activities
of ChAT, GAD, glutaminase, GS and CNPase, as markers of
neural cell types, in the basal forebrain of 10–day–old
normal and hypothyroid (Tx) rats[a]

Experimental		ChAT	GAD	Glutaminase	GS	CNPase
I	T4	120*	88	86	98	123*
II	NGF	139*	87	83	109	92
III	T4 + NGF	176*	90	90	98	120*
IV	Tx	92	84*	104	101	87
V	Tx + T4	127*§	109§	107	95	119*§
VI	Tx + NGF	107	96	98	107	90
VII	Tx + T4 + NGF	172*§	116§	116	114	131*§

[a]A group of normal or Tx rats received a daily subcutaneous
injection of L–thyroxine (0.3 μg/g body weight) from birth
(I or V), or intraventricular injections of NGF (2 μg) at
postnatal days 1, 3, 5, 7 and 9 (II or VI), or both
L–thyroxine and NGF (III or VII). The rats were killed at
postnatal day 10. Whole homogenates of the basal part of
the forebrains were used for the estimation of choline
acetyltransferase (ChAT), glutamate decarboxylase (GAD),
glutaminase, glutamine synthetase (GS) and 2',3'-cyclic
nucleotide-3'-phosphohydrolase (CNPase) activities.
Enzyme activity (per mg protein) is expressed as a
percentage of the respective control values. Significant
differences (P<0.05) between the mean values of control
and experimental groups are indicated by *, and between
Tx and Tx given humoral agent(s) are shown by §. The
data are taken from Patel et al.[7]

that subcortical cholinergic neurones are subject to regulation by an
interaction between thyroid hormone and local humoural factors such as
NGF.[7,8]

There are a number of possibilities that could account for the interaction between thyroid hormone and NGF. For example, treatments with
insulin and insulin–like growth factor II have been found to increase
specific and saturable NGF binding sites in cultured human neuroblastoma
cells.[37] It is possible that analogous mechanisms between thyroid hormone
and NGF may operate in subcortical cholinergic cells. Another possibility
could be that if thyroid hormone controls the formation of ChAT or of
proteins essential for cholinergic cell maturation at a pretranslational
level[38] and NGF regulates the de novo synthesis of these crucial proteins
at transcriptional level,[30] then a combination of these effects could
synergistically potentiate ChAT activity.

IS THYROID HORMONE REQUIRED FOR THE NGF EFFECT ON THE SUBCORTICAL NEURONES
IN VIVO?

Young rats made hypothyroid at birth are intraventricularly given the
minimum amount of NGF that produces a significant increase in basal forebrain ChAT activity in normal animals.[7,36] In contrast to normal young,
treatment with low doses of NGF has no significant effect on the activity
of ChAT in thyroid deficient rats (Table 2).[7,32] However, when hypothyroid

rats are treated with both NGF and thyroxine, ChAT activity in the basal forebrain is significantly higher than in the rats given thyroxine alone. Furthermore, the enzyme activity value in the hypothyroid rats given both thyroxine and NGF reaches the same level found in normal animals after treatment with both humoural agents (Table 2). The results indicate that low doses of NGF fail to increase significantly ChAT activity in thyroid deficient rats although the stimulation of enzyme activity induced by thyroxine is potentiated by NGF in these animals. The in vivo concentration of NGF in brain is very low (less than 2 ng/g tissue) and this could emphasize further the importance of thyroid hormone for the in situ action of local humoural factors like NGF on cholinergic nerve cells.

CELLULAR SPECIFICITY FOR THE EFFECTS OF THYROID HORMONE AND NGF IN THE BRAIN

In the peripheral nervous system, the administration of NGF elevates the activities of tyrosine hydroxylase and dopamine-β-hydroxylase, the marker enzymes for catecholaminergic neurones, and enhances sympathetic neuronal survival and advances growth of nerve fibres.[30,31] Both neuronal and non-neuronal cells have been shown to express NGF receptors (R) in the periphery.[39,40] The initial studies in the brain have also indicated very wide regional distribution for NGF and NGF-R (high affinity saturable binding sites) and a possible role in a number of biochemical reactions.[31] However, our recent studies, on the effects of NGF on neuronal cell types (in terms of marker enzymes)[23] and on the immunocytochemical localization of NGF-R containing cells (using a monoclonal antibody derived by Chandler et al.[41]), now show that the effect of NGF is selective for subcortical cholinergic neurones.[7,27] NGF, when administered intraventricularly, has no effect on the activities of GAD (i.e. GABA-ergic cells), glutaminase (i.e. glutamatergic cells), glutamine synthetase (i.e. astrocytes), or of 2',3'-cyclic nucleotide-3'-phosphohydrolase (CNPase; i.e. oligodendroglial cells) in the brain of either normal or neonatally hypothyroid rats (Table 2). Also, in contrast to its effect on ChAT activity, NGF did not potentiate the increase in CNPase and GAD activities induced by thyroxine in hypothyroid rats (Table 2). Similarly, administration of NGF does not affect monoaminergic cell markers in the brain.[35]

The findings on the normal morphology and distribution of NGF-R- and ChAT-containing cells of the adult rat forebrain[27] are in good agreement with the biochemical observations.[7,35,36] Unlike in the peripheral nervous system, only neurones showed immunoreactivity to NGF-R in the brain. Both the NGF-R and ChAT immunoreactive cells appear to form a continuous anteroposterior band, which includes the olfactory tubercle, the medial septal nucleus, the vertical and horizontal limbs of the diagonal band and the basal nucleus.[27] In each subdivision of the basal forebrain, the topographic organization, the localization, the intensity of the immunoreaction and the total number of NGF-R- and of ChAT-immunoreactive neurones are strikingly similar, indicating that nearly all NGF-R containing cells are cholinergic neurones (Fig. 1). On the other hand, in the basal forebrain of rehabilitated neonatally hypothyroid rats the intensity of the immunoreaction product in NGF-R-positive cells (which is comparable to that seen in NGF-R- and ChAT-positive cells in controls) is greater than that of the ChAT-containing neurones (Fig. 1). As a result, the number of ChAT positive cells appeared to be lower than the NGF-R-positive cells after thyroid deficiency during early life (Fig. 1). In cells other than the basal forebrain neurones, including those of the cerebral cortex, the hippocampus and the cerebellum, the reaction to the antibody was very diffuse and faint, and thus proved to be very difficult to dissociate with reasonable certainty from background non-specific staining. These biochemical and morphological findings indicate that the effect of NGF is specific to cholinergic cells, whereas thyroid hormone also affects the development of GABA-ergic neurones, astroglial cells and oligodendrocytes.

OTHER FUNCTIONAL IMPLICATIONS

The present findings on congenital hypothyroidism may also be relevant to senescence and many developmental and degenerative disorders of the brain. Abnormalities in the formation of acetylcholine have been associated with age-dependent atrophy of the basal forebrain cholinergic neurones, and as in Alzheimer-type dementia in man, these changes in the cholinergic system appear to contribute to age-related cognitive impairments in rodents.[42,43] Senescence reduces the conversion of thyroxine to triiodothyronine and the level of thyroid hormone receptors, and these alterations may be responsible for the suboptimal thyroid state with aging.[44,45] Both in developing and adult mice, administration of thyroid hormone is found to increase the NGF level in the brain.[46] Also, a continuous intracerebral infusion of NGF over a period of 4 to 5 weeks can partly reverse the atrophy of the cholinergic cell body and improve retention of a spatial memory task in behaviourally impaired aged rats.[43] It is believed, therefore, that such studies of underlying mechanisms of the biological damage to cholinergic cells, and of the operation of the memory and other cognitive processes in neonatally hypothyroid animals, may serve as useful models to test cholinergic drugs and trophic factors, which could be of therapeutic utility in dementias.

SUMMARY

The central nervous system consists of a complex array of many different neuronal and neuroglial cell types whose precise interrelationships determine its functioning. Alterations in the thyroid status affects both the formation and differentiation of brain cells, resulting in a distortion, rather than synchronized shifts, in the organization of neuronal interconnections. Studies on the differentiation of nerve cell types, characterized by different neurotransmitters, have revealed that the effect of thyroid hormone on nerve cell maturation is cell-type specific; cholinergic cells are more severely affected than the others. The retardation in the development of choline acetyltransferase (ChAT) activity observed during the period of thyroid deficiency is completely reversible after long-term rehabilitation in all brain regions except the basal forebrain, suggesting selective sensitivity of subcortical cholinergic neurones to thyroid hormone.

To obtain more precise information, without the complexity of secondary reactions in in vivo studies, conditions have been established for growing (in a chemically defined medium) relatively pure neuronal cultures enriched in cholinergic cells from the septal-diagonal band region of 17-day-old embryonic rat brain. Exposure of the cultures to thyroid hormone and nerve growth factor (NGF) for 10 days enhances the expression of ChAT activity in a dose-dependent manner. The effects of thyroid hormone and NGF are more than additive at optimal concentrations. The elevation of ChAT is due to an increase in the enzyme activity per cholinergic cell, since neither thyroid hormone nor NGF has a significant effect on the number of cells, including the cholinergic neurones. The in vitro effects on ChAT activity can be reproduced in vivo in the basal forebrain of 10-day-old rats after neonatal treatment with NGF (intraventricularly) and thyroid hormone (subcutaneously). However, in hypothyroid rats significant effects of NGF at low doses are not detectable, although the stimulation of ChAT activity induced by thyroid hormone treatment is potentiated by NGF. The effect of NGF is specific to cholinergic cells, while thyroid hormone also affects the development of GABA-ergic neurones, astrocytes and oligodendroglial cells. The findings on the normal morphology and distribution of NGF-receptor (R)- and ChAT-containing cells of the adult rat forebrain are in good agreement with these biochemical observations. Unlike the peripheral nervous system, only neurones show immunoreactivity to NGF-R antibody in the brain. In each subdivision of the basal forebrain, the topographic organization, the localization, the intensity of the immunoreaction and the total cell number of NGF-R- and ChAT-containing neurones is strikingly similar, indicating that nearly all NGF-R-immunoreactive cells are cholinergic neurones. The observations show that subcortical cholinergic neurones are subject to regulation by an interaction between thyroid hormone and local humoural factors such as NGF.

ACKNOWLEDGEMENTS

We would like to acknowledge the collaboration of Dr. M. Hayashi and are grateful to Mr. J. McGovern for his expert help in part of these studies. We also thank Rita Nani for typing the manuscript. This paper is dedicated to Dr. Robert Balazs on his sixty-fifth birthday.

REFERENCES

1. G. D. Grave, "Thyroid Hormones and Brain Development", Raven, New York (1977).
2. R. MacFaul, S. Dorner, E. M. Brett, and D. B. Grant, Neurological abnormalities in patients treated for hypothyroidism from early life. Arch. Dis. Childhood 53:611-619 (1978).
3. A. J. Patel and R. Balazs, Hormones and cell proliferation in the rat brain, in: "Progress in Psychoneuroendocrinology", F. Brambilla, G. Racagni, and D. de Wied, eds., pp. 621-632, Elsevier, Amsterdam (1980).
4. A. J. Patel, R. Balazs, R. M. Smith, A. E. Kingsbury, and A. Hunt, Thyroid hormone and brain development, in: "Multidisciplinary Approach to Brain Development", C. di Benedetta, R. Balazs, G. Gombos, and G. Porcellati, eds., pp. 261-277, Elsevier, Amsterdam (1980).
5. A. J. Patel, M. Hayashi, and A. Hunt, Selective persistent reduction in choline acetyltransferase activity in basal forebrain of the rat after thyroid deficiency during early life, Brain Res. 422:182-185 (1987).
6. A. J. Patel, A. Hunt, and M. Hayashi, Effect of thyroid deficiency on the regional development of glutaminase, glutamatergic neuron marker, in the rat brain, Int. J. Dev. Neurosci. 5:295-303 (1987).
7. A. J. Patel, M. Hayashi, and A. Hunt, Role of thyroid hormone and nerve growth factor in the development of choline acetyltransferase and other cell-specific marker enzymes in the basal forebrain of the rat, J. Neurochem. 50:803-811 (1988).
8. M. Hayashi and A. J. Patel, An interaction between thyroid hormone and nerve growth factor in the regulation of choline acetyltransferase activity in neuronal cultures, derived from the septal-diagonal band region of the embryonic rat brain, Dev. Brain Res. 36:109-120 (1987).
9. A. Rami, A. Rabie, and A. J. Patel, Thyroid hormone and development of the rat hippocampus: cell acquisition in the dentate gyrus, Neuroscience 19:1207-1216 (1986).
10. A. Rami, A. J. Patel, and A. Rabie, Thyroid hormone and development of the rat hippocampus: morphological alterations in granule and pyramidal cells, Neuroscience 19:1217-1226 (1986).
11. J. Legrand, Effects of thyroid hormones on central nervous system development, in: "Neurobehavioural Teratology", J. Yanai, ed., pp. 331-363, Elsevier, Amsterdam (1984).
12. A. J. Patel, A. Rabie, P. D. Lewis, and R. Balazs, Effect of thyroid

deficiency on postnatal cell formation in the rat brain: a biochemical investigation, Brain Res. 104:33-48 (1976).

13. A. Rabie, A. J. Patel, M. C. Clavel, and J. Legrand, Effect of thyroid deficiency on the growth of the hippocampus in the rat. A combined biochemical and morphological study, Dev. Neurosci. 2:183-194 (1979).

14. P. D. Lewis, A. J. Patel, A. L. Johnson, and R. Balazs, Effect of thyroid deficiency on cell acquisition in the postnatal rat brain: a quantitative histological study, Brain Res. 104:49-62 (1976).

15. A. J. Patel, P. D. Lewis, R. Balazs, P. Bailey, and M. Lai, Effects of thyroxine on postnatal cell acquisition in the rat brain, Brain Res. 172:57-72 (1979).

16. A. J. Patel and A. Rabie, Thyroid deficiency and cell death in the rat cerebellum during development, Neuropath. Appl. Neurobiol. 6:45-49 (1980).

17. A. J. Patel, R. M. Smith, A. E. Kingsbury, A. Hunt, and R. Balazs, Effects of thyroid state on brain development: muscarinic acetyl-choline and GABA receptors, Brain Res. 198:389-402 (1980).

18. R. M. Smith, A. J. Patel, A. E. Kingsbury, A. Hunt, and R. Balazs, Effects of thyroid state on brain development: β-adrenergic receptors and 5'-nucleotidase activity, Brain Res. 198:375-387 (1980).

19. A. J. Patel, A. Hunt, and E. Meier, Effects of undernutrition and thyroid state on the ontogenetic changes of D1, D2 and D3 brain-specific proteins in rat cerebellum, J. Neurochem. 44:1581-1587 (1985).

20. P. L. Woodhams, J. McGovern, G. P. McGregor, D. J. O'Shaughnessey, M. A. Ghatei, M. A. Blank, T. H. Adrian, Y. Lee, J. M. Polak, S. R. Bloom, and R. Balazs, Effects of changes in neonatal thyroid status in the development of neuropeptide systems in the rat brain, Int. J. Dev. Neurosci. 1:155-164 (1983).

21. F. Hajos, A. J. Patel, and R. Balazs, Effect of thyroid deficiency on the synaptic organization of the rat cerebellar cortex, Brain Res. 50:387-401 (1973).

22. F. Crepel, Excitatory and inhibitory processes acting upon cerebellar Purkinje cells during maturation in the rat: influence of hypo-thyroidism, Exp. Brain Res. 20:403-420 (1974).

23. O. K. Langley, M. S. Ghandour, and G. Gombos, Immunohistochemistry of cell markers in the central nervous system, in: "Handbook of Neurochemistry", Vol. 7, A. Lajtha, ed., pp. 545-611, Plenum, New York (1984).

24. J. E. Bottenstein and G. Sato, "Cell Culture in the Neurosciences", Plenum, New York (1985).

25. R. N. Kalaria and A. N. Prince, Effects of thyroid deficiency in the development of cholinergic, GABA, dopaminergic and glutamate neuron markers and DNA concentrations in the rat corpus striatum, Int. J. Dev. Neurosci. 3:655-666 (1985).

26. A. J. Patel, M. del Vecchio, and D. J. Atkinson, Effect of under-nutrition on the regional development of transmitter enzymes: glutamate decarboxylase and choline acetyltransferase, Dev. Neurosci. 1:41-53 (1978).

27. J. Kiss, J. McGovern, and A. J. Patel, Immunohistochemical localization of cells containing nerve growth factor receptors in the different regions of the adult rat forebrain, Neuroscience: in press (1988).

28. R. Balazs, V. Gallo, C. K. Atterwill, A. E. Kingsbury, and O. S. Jørgensen, Does thyroid hormone influence the maturation of cerebellar granule neurones? Biomed. Biochim. Acta 44:1469-1482 (1985).

29. C. K. Atterwill, D. J. Atkinson, I. Bermudez, and R. Balazs, Effect of thyroid hormone and serum on the development of Na^+,K^+-adenosine triphosphatase and associated ion fluxes in cultures from rat brain, Neuroscience 14:361-373 (1985).

30. H. Thoenen and D. Edgar, Neurotrophic factors, _Science_, 229:238-242 (1985).
31. S. R. Whittlemore and A. Seiger, The expression, localization and functional significance of β-nerve growth factor in the central nervous system, _Brain Res. Rev._ 12:439-464 (1987).
32. P. Honegger, Nerve growth factor-sensitive brain neurons in culture, _Monogr. Neural Sci._ 9:36-42 (1983).
33. F. Hefti, J. Hartikka, F. Eckenstein, H. Gnahn, R. Heumann, and M. Schwab, Nerve growth factor increases choline acetyltransferase but not survival or fiber outgrowth of cultured fetal septal cholinergic neurons, _Neuroscience_ 14:55-68 (1985).
34. F. Hefti, J. Hartikka, and M. B. Bolger, Effect of thyroid hormone analogs on the activity of choline acetyltransferase in cultures of dissociated septal cells, _Brain Res._ 375:413-416 (1986).
35. H. Gnahn, F. Hefti, R. Heumann, M. E. Schwab, and H. Thoenen, NGF-mediated increase of choline acetyltransferase (ChAT) in the neo-natal rat forebrain: evidence for a physiological role of NGF in the brain? _Dev. Brain Res._ 9:45-52 (1983).
36. W. C. Mobley, J. L. Rutkowski, G. I. Tennekoon, J. Gemski, K. Buchanan, and M. V. Johnston, Nerve growth factor increases choline acetyltransferase activity in developing basal forebrain neurons, _Mol. Brain Res._ 1:53-62 (1986).
37. E. Recio-Pinto, F. F. Lang, and D. N. Ishii, Insulin and insulin-like growth factor II permit nerve growth factor binding and the neurite formation response in cultured human neuroblastoma cells, _Proc. Natl. Acad. Sci. U.S.A._ 81:2562-2566 (1984).
38. S. Seelig, C. Liaw, H. C. Towle, and J. H. Oppenheimer, Thyroid hormone attenuates and augments hepatic gene expression at a pretranslational level, _Proc. Natl. Acad. Sci. U.S.A._ 78:4733-4737 (1981).
39. H. Rohrer and I. Sommer, Simultaneous expression of neuronal and glial properties by chick ciliary ganglion cells during development, _J. Neurosci._ 3:1683-1693 (1983).
40. A. Zimmermann and A. Sutter, β-Nerve growth factor (βNGF) receptors on glial cells. Cell-cell interaction between neurons and Schwann cells in cultures of chick sensory ganglia, _EMBO J._ 2:879-885 (1983).
41. C. E. Chandler, L. M. Parsons, M. Hosang, and E. M. Shooter, A mono-clonal antibody modulates the interaction of nerve growth factor with PC12 cells, _J. Biol. Chem._ 259:6882-6889 (1984).
42. M. W. Decker, The effects of aging on hippocampal and cortical projections of the forebrain cholinergic system, _Brain Res. Rev._ 12:423-438 (1987).
43. W. Fisher, K. Wictorin, A. Bjorklund, L. R. Williams, S. Varon, and F. H. Gage, Amelioration of cholinergic neuron atrophy and spatial memory impairment in aged rats by nerve growth factors, _Nature_ 329:65-68 (1987).
44. K. R. Latham and Y. C. L. Tseng, Nuclear thyroid hormone activities and serum iodothyronine levels in aging rats, _Age_ 2:48-54 (1985).
45. M. Margarity, T. Valcana, and P. S. Timiras, Thyroxine deiodination, cytoplasmic distribution and nuclear binding of thyroxine and triiodothyronine in liver and brain of young and aged rats, _Mech. Aging Dev._ 29:181-189 (1985).
46. P. Walker, M. L. Weil, M. E. Weichsel, Jr., and D. A. Fisher, Nerve growth factor, _in_: "Fetal Brain Disorders - Recent Approaches to the Problem of Mental Deficiency", B. S. Hetzel and R. M. Smith, eds., pp. 187-203, Elsevier, Amsterdam (1981).

ONTOGENY OF THYROID HORMONE-PROCESSING SYSTEMS IN RAT BRAIN

Mary B. Dratman, Floy L. Crutchfield and Janice T. Gordon

Department of Medicine, VA Medical Center and The Medical
College of Pennsylvania, Philadelphia, PA 19104

INTRODUCTION

Recent morphologic, biochemical and functional evidence supports a
direct role for thyroid hormones in adult brain (1-4). Nevertheless, the
concept that adult brain is unresponsive to thyroid hormones continues, at
least in some quarters, to prevail. By contrast, a role for the hormone
during brain development has, for some time, been assumed, even though observations suggesting a cause and effect relationship between triiodothyronine (T3) in the developing brain and a T3-dependent response have been
presented only recently. Evidence that T3 nuclear receptors are homologous to the products of the c-erb A protooncogene family provides a compelling rationale for involvement of these receptors in early events associated with blast cell replication and specification. This rationale is
now coupled with evidence that a high degree of T3 nuclear receptor occupancy coincides in time with the period of active neurogenesis in the fetal
lamb (5). However, there is as yet no demonstrated link between T3 nuclear
receptor complex formation and the somewhat later effects of the hormone
on growth of nerve cell processes, synaptogenesis and myelin formation,
and, as yet, no evidence for participation of the T3 nuclear receptor in
adult brain activities.

ONTOGENETIC FEATURES OF NEUROTRANSMITTER ACTIONS IN BRAIN

Investigations of Jean Lauder have shown that many substances which
regulate certain critical phases of neurogenesis and nerve cell specialization were originally identified and are better known for their important
neuroregulatory or direct neurotransmitter roles in the differentiated
brain (6). In this context, Lauder cites evidence for early and important
growth promoting activities of biogenic amines, acetyl choline, GABA, and
substance P, expressed well before the apparatus of neurotransmission has
been set in place. As outlined in Table I, mitogenesis and differentiation of neuroblasts and morphogenetic cell movements are induced by dopamine, for example, during this early, pre-neurotransmission phase of brain
development. Although receptors for those early activities are unknown,
they are likely to be different from those mediating the information-trans-
mitting, synaptic activities of dopamine in the adult brain, and may well
be localized in the cell nucleus. During (and presumably after) the construction of the apparatus of neurotransmission, dopamine appears to act
as a "trophic" agent inducing nerve cell specialization such as neurite
elongation, myelinogenesis and synaptogenesis. Whether the receptors for

TABLE I   ROLE OF NEUROACTIVE AGENTS IN THE BRAIN

DEVELOPING BRAIN		MATURE BRAIN
Apparatus for Neurotransmission		
not yet in place	forming or in place	in place
mitogenic and differentiating effects on neuroblasts;	pre and transsynaptic actions as trophic agents inducing further developmental specializations (neurite elongation; myelinogenesis; synaptogenesis, etc.)	transsynaptic actions as neuromodulators neuroregulators or neurotrans- mitters
control of morphog- etic cell movements		

these functions are the same as those involved in dopaminergic actions as classically understood, remains to be seen. Finally, when the apparatus for neurotransmission is fully in place, transsynaptic actions of dopamine are in the forefront, and are brought about through the formation of com- plexes with well-characterized pre- and post-synaptic receptors. Thus, the same molecular specie may have markedly different neural activities, possibly mediated through different receptors, during different phases of brain ontogenesis.

DO THYROID HORMONES INFLUENCE GROWTH AND SYNAPTIC ACTIVITY IN BRAIN?

Our laboratory has been interested in the succession of early growth-promoting and subsequent synaptic activities of known neurotransmitters because a similar model might serve to link the early neurogenetic effects of thyroid hormones in the growing brain with their later behavioral and autonomic nervous system effects in adults. Evidence derived from study-ing ontogenetic features of thyroid hormone-dependent brain development in the rat supports this possibility. Thus, the actions of T3 on neuro-blast proliferation and differentiation (which are probably mediated through nuclear receptors) occur before the development of a synaptosomal apparatus which (as will be seen) is later active in processing the hor-mone. Further along in the course of development, effects of the hormone on behavior and autonomic nervous system functions emerge, which may rea-sonably be linked to neuromodulatory, synaptosomally-based actions of the hormone in the central and peripheral adrenergic nervous system. This succession of thyroid hormone-dependent events is not unlike the onto-genetic succession of events characteristic of the actions of many known neurotransmitters, as outlined in Table I.

Fate of the Hormone in Brain

In the early 70's, when different approaches to studying these pro-blems experimentally were considered, the distribution of thyroid hormones in the brain had not yet been systematically examined. Since that time a number of studies have addressed the issue of regional, cellular and sub-cellular localization and metabolism of thyroid hormones in brain (7-9).

As a result we now have available considerable information regarding the
fate of the hormone in the adult CNS. Among the more striking observations
in the area are those demonstrating marked stability of thyroid hormone
levels in both adult and developing rat brain (2,10). This stability is
even maintained in the face of long standing extremes of hormone availability. Coordinated adjustments in the activity of brain and liver help
to maintain those stable conditions, suggesting that brain iodothyronine
homeostasis is important for the function of the organism as a whole.

Subcellular Distribution and Metabolism

Thyroxine (T4) enters the extracellular fluid space through blood brain
barrier (11) and the choroid plexus (12) delivery systems. The potentially important role of the latter system has become a matter of considerable
interest, following the demonstration of Dickson et al that T4 binding prealbumin (transthyretin) is strongly localized and independently controlled
within choroid plexus cells (12). T4 (and any T3 entering the brain) is
transported from the extracellular fluid space into groups of selected
nerve somata and their axonal and dendritic terminals (synaptosomes). Moreover, with time after i.v. administration of a T4 or a T3 pulse, there is
progressive accumulation of iodothyronines in synaptosomal particles
(Fig.1).

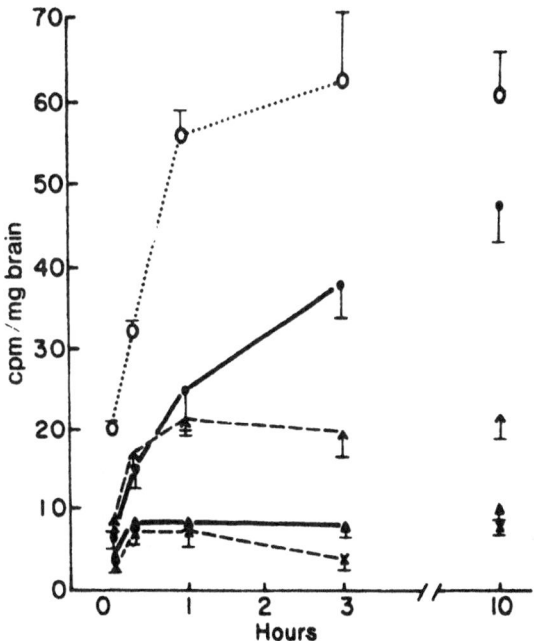

Fig. 1.   RADIOACTIVITY IN SUBCELLULAR PARTICLES OF S_1 FRACTION
OF RAT BRAIN AFTER I.V. ^{125}I-T3

Data points show mean ± SEM; o---o: synaptosomes;
●---●: myelin; x---x: mitochondria; other symbols
refer to intervening (unidentified) particles in
the gradient. Abscissa shows time after i.v.
hormone.

In brain, T4 is converted to T3 at a low K_m and faster fractional rate than has been demonstrated in liver and other somatic tissues (2,13). As compared with other brain fractions, the rate of T4 conversion to T3 is highest in synaptosomes (Fig. 2).

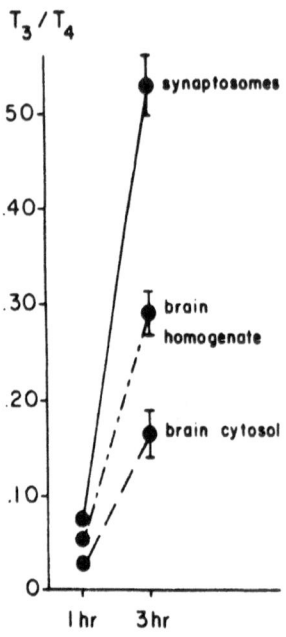

Fig. 2.   FRACTIONAL CONVERSION OF T3 TO T3 *IN VIVO* IN BRAIN CYTOSOL AND SYNAPTOSOMES

Rats were given an i.v. pulse of ^{125}I-T4; brain homogenates and subcellular fractions of brain were extracted by methods of Gordon, et al (19); iodothyronines in the extracts were separated by HPLC. Data points show mean T3/T4 ± SD at 1 and 3 hours after isotope administration.

A high affinity mechanism for binding T3 to selected synaptosomal membranes has been described by Mashio and colleagues (14,15). These binding sites may help to maintain the strong concentration gradient from cytosol to synaptosomes noted after labeled T4 and T3 administration (Fig. 3), and may implement actions of T3 within or across nerve endings. The affinity and capacity of T3 synaptosomal binding sites are similar to those of T3 nuclear receptors (14); the ratio of labeled T3 found in these organelles after ^{125}I-T3 administration is also similar through the 1 hour time interval, although concentrations are relatively higher in synaptosomes at later times (Table II). In addition to T3 directly taken up or formed from synaptosomally-concentrated T4, the evidence thus far indicates that T3, directly taken up or formed from T4 within the nerve cell body, is axonally transported to synaptic elements in terminal fields (Figs. 2 and 4), thereby continuously adding to the intrasynaptosomal T3 pool. This may explain why enriched synaptosomal elements derived from brain homogenates show a T3 concentration gradient from cytosol to synaptosomes through at least 24 hours after i.v. ^{125}I-T4 administration.

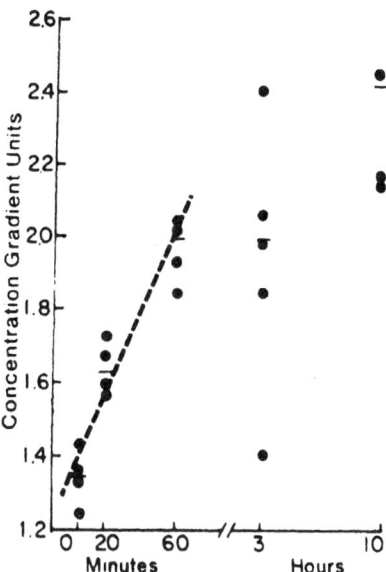

Fig. 3. RATIOS OF LABELED HORMONE IN SYNAPTOSOMES RELATIVE
TO BRAIN CYTOSOL AFTER I.V. ^{125}I-T_3.

The concentration gradient from cytosol to synaptosomes is linear through the first hour and shows
a progressive, non-linear increase thereafter.
Each data point represents the ratio in an individual animal.

TABLE II Ratio of synaptosomal to nuclear ^{125}I-T_3

hours after pulse	ratio per brain
1	1.00
3	1.26
24	1.41

*Hormone was administered i.v.; synaptosomes were
separated by gradient centrifugation of S_1 derived from labeled brain homogenates (9); nuclei
were recovered from the low speed homogenate
pellet after passage through Ficoll gradients.
T_3 was isolated as described in legend, Fig. 2.

155

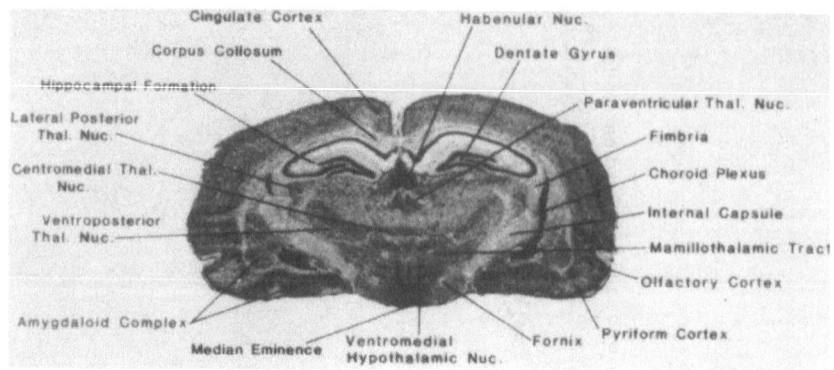

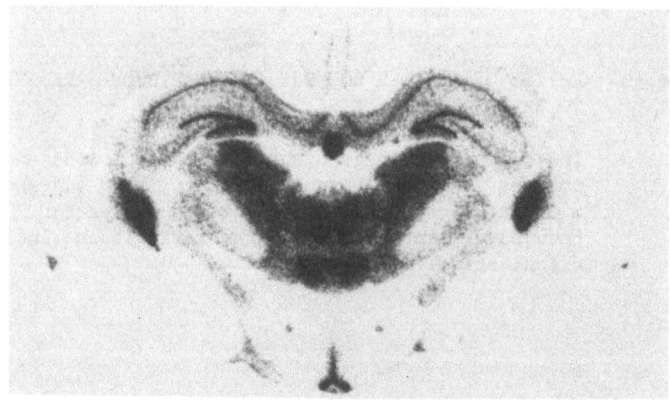

Fig. 4. AUTORADIOGRAPHIC EVIDENCE FOR AXONAL TRANSPORT
AND TERMINAL FIELD LOCALIZATION OF I.V. ^{125}I-T3.

Top panel: Histologic section prepared from
forebrain region of rat brain at the level of
the mid hypothalamus.

Lower panel: Film autoradiogram prepared from
same region of brain 48 hours after hormone
injection. Note persistence of choroid plexus
labeling, unlabeled fimbria but labeling in
internal capsule and fornix indicating selectivity of white matter label and axonal transport. Thalamic nuclei (important terminal
field relays) are seen to be differentially
labeled, with particularly strong concentration
of isotope in the paraventricular thalamic
nuc.

On the other hand, a morphologically distinct type of nerve ending, the cerebellar mossy fiber terminal, exhibits its highest ratio to cytosolic T_3 concentration 3 hours after i.v. ^{125}I-T_3 and undergoes a progressive net fall thereafter (Fig. 5). It is therefore likely that rates of labeled hormone turnover in synaptosomes differ from one neural network to another, due, for example, to differences in endogenous hormone stores or local differences in impulse traffic, to name only the more obvious possibilities. Whether such factors might contribute to the observed T_3 processing differences between cerebellar mossy fiber synaptosomes and mixed, conventional synaptosomal populations, can be tested experimentally.

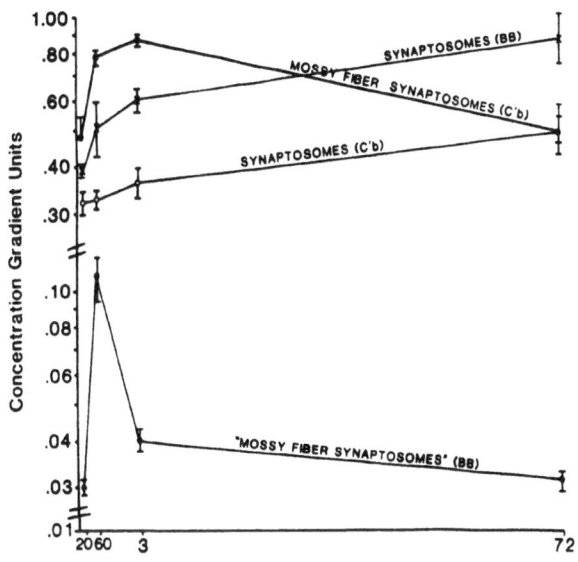

Fig. 5. RATIOS OF LABELLED T3 CONCENTRATIONS IN CYTOSOL AND SYNAPTOSOMES IN DIFFERENT SYNAPTOSOMAL POPULATIONS

Conventional synaptosomes and mossy fiber terminals were isolated on density gradients (9,21) and T_3 concentrations in the different fractions were measured as described in Fig. 2. Data points are mean concentration gradient units derived from two 2-pool brain homogenates; error bars show range of differences in the replicate observations. Abscissa: time after i.v. ^{125}I-T_3; 20 and 60 = minutes; 3 and 72 = hours.

Autoradiographic Studies

The selectivity of hormone localization within the brain is hardly appreciated if the data are derived from regional dissections, but is clearly demonstrated in thaw-mount autoradiograms prepared after i.v. administration of high SA ^{125}I-iodothyronine (7). As an example, Fig. 6 shows time-dependent changes in labeling of cerebellum in film autoradiograms prepared from coronal sections of labeled rat brain taken at intervals through 48 hours after hormone administration. Although most attributes of labeling rates and patterns are unique for every region and subregion of brain, certain of the observations can be generalized, viz: Except for a time lag, T4 is distributed like T3 unless conversion of T4 to T3 is impeded (in which case the labeling patterns are unresolved); At

1 hour, there are high concentrations of label in choroid plexus and diffuse distribution in the extracellular fluid space; Within 3 hours for T3
and 10 hours for T4 there is progressive (saturable) resolution within
nerve cell bodies of selected brain nuclei; By 24-48 hours, label at original sites of concentration in grey matter shows marked diminuition or
disappearance while label in selected fiber tracts appears and becomes
progressively more resolved; By 48-72 hours, the label is noted in terminal fields relevant to originally labeled brain nuclei; In certain relay
stations or regions with reverberating circuits (e.g. cerebellum), occupancy eventually diminishes but the pattern of distribution is maintained
for the entire period of observation.

Functional effects. Although the distribution and binding of iodothyronines in brain have been reasonably well-characterized, only a few
studies have provided experimental evidence of a functional role for the
hormone in adult brain. The work of Ruiz-Marcos and colleagues (16) is
of particular interest. Their investigations have shown that long-standing adult-onset hypothyroidism causes reversible loss of dendritic spines
of pyramidal cells in visual and auditory cortex (3). The deficit is most
prominent in the terminal dendritic tree. Although the authors undoubtedly make different interpretations of their data, Ruiz-Marcos' observations
might conceivably reflect loss of "trophic" effects normally exerted by
thyroid hormone-processing axon terminals on pyramidal cell dendrites.
Light microscopic autoradiograms show that neuronal cell bodies in layers
2, 3 and 4 in the auditory and visual cortex are heavily labeled with
^{125}I-T3 within 3 hours after an i.v. hormone pulse. Further evidence suggests that the cellular label is translocated into axons (1). The axon

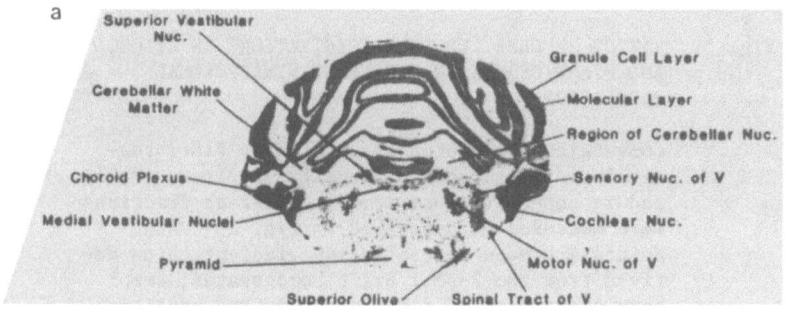

Fig. 6.   FILM AUTORADIOGRAMS FROM MID-CEREBELLAR REGION OF
          RAT BRAIN STEM.

          (a) above: Histologic (coronal) section of cere-
          bellum at level of cochlear nn.
          (b) facing page: Autoradiograms of cerebellum prepar-
          ed 1, 3, 10, 24 and 48 hours after i.v. ^{125}I-T4 (left)
          or ^{125}I-T3 (right). Note intense labeling of choroid
          plexus in all T4-labeled sections, diffuse labeling
          after both isotopes at 1 hour, time lag for T4 as com-
          pared with T3 but eventual resolution within the granule
          cell layer in each instance. At early times, pyramids
          are unlabeled but are clearly distinguishable from back-
          ground at 48 hours. For T3 but not yet for T4, labeling
          of spinal tract of V can be distinguished. Additional
          details are found in reference 1.

terminals of these cells synapse with dendrites of pyramidal cells in the deeper cortical layers and could serve as a source of thyroid hormone-mediated trophic effects. This interpretation is relevant to evidence derived from serial film autoradiography (unpublished observations).

Another experimental result compatible with direct effects of the hormone on nerve cell activity comes from studies showing earlier heart rate stimulating properties of intrathecally as compared with intravenously delivered T3 to hypothyroid rats (Fig. 7). Thus, though sparse, presently available functional results demonstrate that thyroid hormones play a continuing role in adult brain, which may be synaptically mediated.

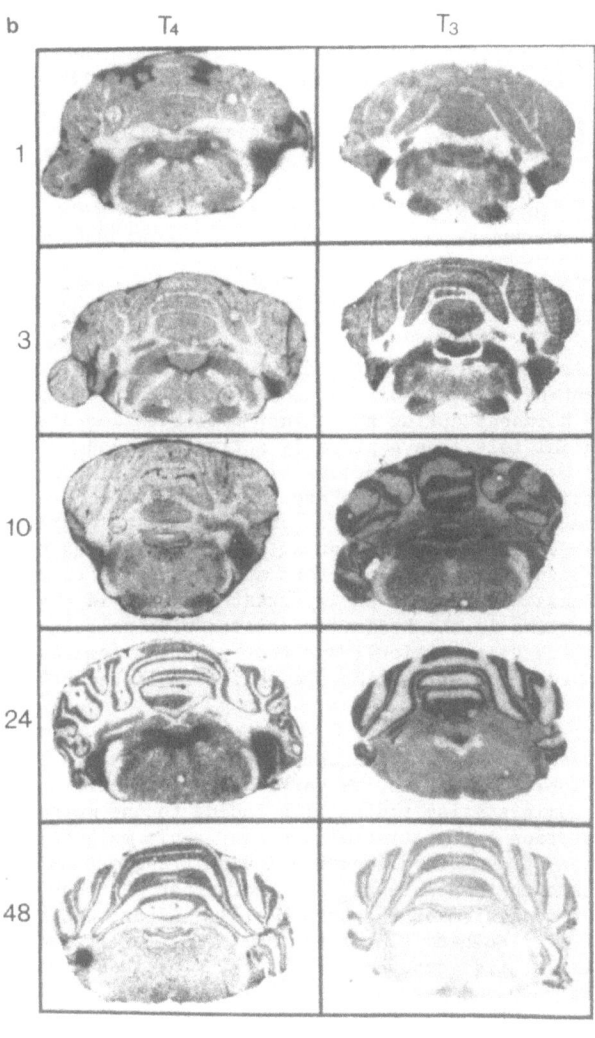

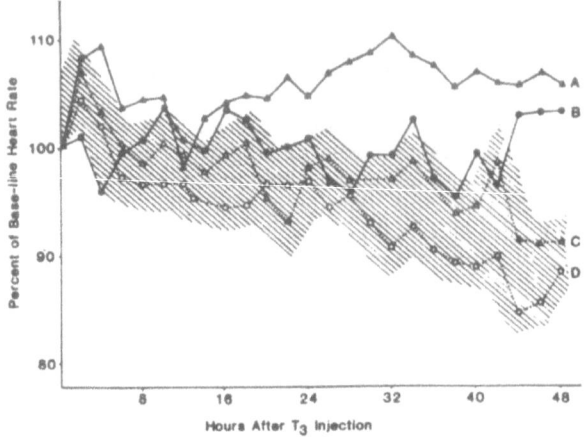

Fig. 7.   EFFECT OF ROUTE OF ADMINISTRATION ON FRACTIONAL
          HEART RATE RESPONSE TO T3

          Pairs of rats were given 1 µg T3/100 g BW or NaOH
          by i.v. or intrathecal injection. Mean fractional
          changes every 2 hours indicated by A: intrathecal
          T3; B: i.v. T3; C: intrathecal NaOH; D: i.v. NaOH.
          I.V. T3 results were not significantly different
          from both NaOH controls; intrathecal T3 (A) pro-
          duced a significant increase in heart rate re-
          sponse as compared with B, C, D combined (p<.02).

EARLY ONTOGENETIC STUDIES

     Serial observations made during the early postnatal phase of rat
brain development help to establish a connecting link between the
role of thyroid hormones during neuroblast proliferation/differentiation
and their potential role as synaptically active compounds. These ontogenetic studies have revealed that certain well-established iodocompound
processing functions of adult brain are not yet active during late fetal
and early postnatal life; that synaptosomal processing characteristic of
the adult animal brain is not present at birth, but begins to develop at
the time that brain nuclear T3 receptor capacity begins to decline; that
the weaning process constitutes an important landmark in thyroid hormone
processing mechanisms in the brain (associated with selective reshuffling
of regional hormone distribution); and that male and female differences
in brain iodothyronine distribution are well-established before sexual
maturation.

     To non-invasively investigate the ontogeny of thyroid hormone pro-
cessing in developing rat brain, we carried out a modified isotope equilibrium study, administering radioactive iodide to the mothers from the
15th day of gestation to the time of weaning. This methodology insured
efficient and non-stressful radioactive iodide transport to the progeny
before birth, through the placenta, and thereafter, through the powerful
iodide transport mechanism of the breast (17). After weaning, labeled
iodide was made available to the offspring through their own drinking
water. As a result, iodocompounds of the rat pup tissues and bodily fluids
became non-invasively and pervasively labeled with ^{125}I-iodine (17,18).
It was therefore possible to study the time course of iodocompound-processing activities in brain through application of straightforeward radio-
chemical methods of analysis.

Using paper and reverse phase HPLC chromatography of whole brain homogenate extracts (19) daily changes in brain iodocompounds were measured during the first week of life. Thereafter, measurements were made at intervals throughout the rest of the period of observation (through 40 days). Fig. 8 shows the important changes in brain iodocompound identity and concentration from early to late phases of the nursing period and compares the results with those in serum and skeletal muscle. As noted, highly significant changes in brain iodide, reverse T3 and iodotyrosine concentrations occur over the course of the nursing period. The high concentrations of iodide and iodotyrosines in brain at birth and during the first week of life may indicate that these compounds play a role in early postnatal brain development. They also suggest that the ability to block the entry of iodide and the ability to dispose of iodide generated in the brain may not yet have developed during the first week of postnatal life. As regards the presence of iodotyrosine, it is possible that ether ring-cleavage is very active during this early period and/or that iodotyrosine deiodinase activity has not yet matured. It is noteworthy that despite the marked reduction in iodide and iodotyrosine concentrations as development proceeds, T3:T4 ratios found in brain homogenates during the entire nursing phase are quite stable, are similar to those found in the post-weaning period, and are in line with results obtained from measurements of endogenous T3:T4 ratios in adult brain.

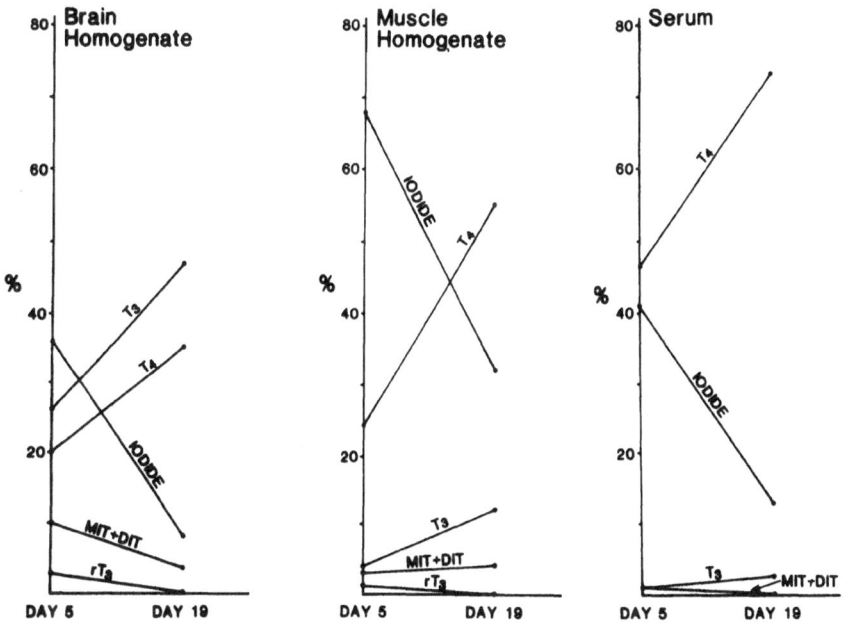

Fig. 8.   IODOCOMPOUNDS IN BRAIN, MUSCLE AND SERUM IN RAT NURSLINGS

Tissues and blood samples were obtained on the 5th and 19th day of life from rat pups pervasively labeled with 125I; brain, muscle and serum iodocompounds were measured by methods described in the text. Changes in iodocompound levels from 5 to 19 days: sig. for T4 in all instances (p<.001); for T3 in brain (p<.001) and muscle (p<.01); for iodotyrosines in brain (p<.01); iodide levels showed a similar, significant rate of decrease in all samples.

Unfortunately, brain tissue labeling produced by the isotope equilibrium methods used in these experiments was not sufficient to produce autoradiographic images. This is attributable to the low concentrations of iodide administered (restricted, in order to maintain general health and active thyroid function) and the limited half-life of the isotope. Therefore, regional distribution was estimated from anatomical dissections of whole brains. On day 1 through 6 of postnatal life, the rat brain is extremely friable and regional landmarks are difficult to distinguish. Nevertheless, it is possible to carry out gross dissections which reveal that regional differences in iodocompound distribution are already established by day 1 of life. These may reflect regional differences in nuclear receptor density, which receptors are reported to be at their height during the first few days of postnatal (rat) life (18).

On and after the 12th day of life, regional dissections can be approached with greater confidence. The results demonstrate that the relationships among the different regions observed during the first week of life are generally maintained during later aspects of the preweaning period. However, weaning induces changes in the absolute and relative patterns of regional labeling within the neuraxis. Thus, the hypothalamus becomes one of the most prominent sites of iodocompound concentration, when previously it was the least. The cerebral cortex, which was the most prominently labeled, now ranks 4th in the list of 5 regions. Whether this shift is related to programmed death of certain iodocompound-concentrating

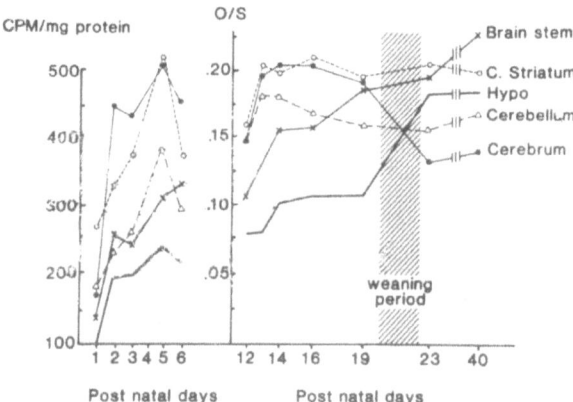

Fig. 9.   REGIONAL DISTRIBUTION OF BRAIN IODOCOMPOUNDS
          DURING DEVELOPMENT

Iodocompound measurements made from postnatal days
1 through 40 showed that, relative to brain stem,
where levels increased progressively during the
period of observation, levels in corpus striatum
were maintained at a generally high level; in cerebellum, at an intermediate level; in hypothalamus,
a slow rise with a marked increase during weaning,
and in cerebrum, initially high, a significant decrease with weaning. Little further regional change
was noted from day 23 to 40, at which time all values
were sig. different from each other (p<.05). O/S =
ratio of brain to serum iodocompounds expressed as
cpm/mg: cpm/µl (22).

eural elements or to other developmental changes is entirely unknown.
owever, to the degree that can be discerned through the medium of region-
l dissections, there is a highly reproducible effect of weaning on re-
ional organization of iodocompound processing systems in the rat brain.

Changes in Subcellular Distribution of Iodocompounds with Development

During the first few days of life, almost all iodocompounds in the
post-nuclear supernatant phase of brain homogenates are found in the soluble fraction, and relatively few are associated with membrane-bound organelles. As development proceeds, however, this relationship is reversed
and soon, labeling of perikaryal particulates predominates, while labeling
of the cytosol is proportionately diminished (Fig. 10a). Among the post-
nuclear particles examined after the 4th day of life, iodocompounds were
always more heavily concentrated in synaptosomes, and, to a significantly
lesser extent, in myelin, whereas in fractions enriched in mitochondria,
the rate of change was much slower and less prominent (Fig. 10b). Since

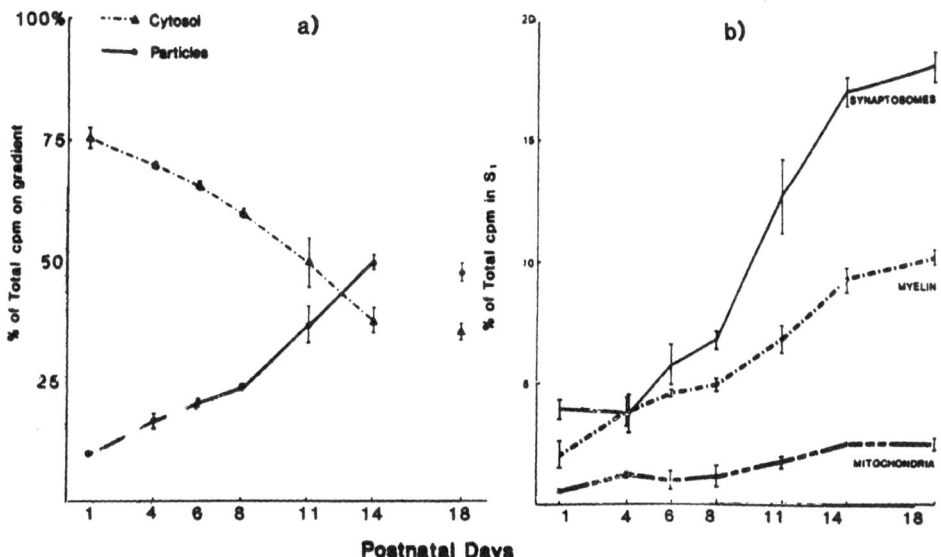

Fig. 10.   IODOCOMPOUNDS IN DEVELOPING RAT BRAIN: LOCALIZATION
IN SUBCELLULAR PARTICLES

Rat brains were obtained at the indicated days of life
(abscissae); after homogenization in .32 M sucrose,
particulate components of the supranuclear fraction
were separated on sucrose density gradients.

(a) Data showing developmental changes in localization of iodocompounds in cytosol and total particles
are expressed as % of total in S_1 ± SD; n = >3 at each
time interval.

(b) Data show mean % of total iodocompounds in synaptosomes, myelin and mitochondria; differences among the
different particles significant (p<.001) both as to
amounts and rates of change.

brain mass increases rapidly during the first 3 weeks of life, we measured the increase in iodocompounds in cytosol and synaptosomes as a function of the increase in protein during the nursing period. The results (Fig. 11a and b) demonstrate that accumulation of T3 in synaptosomes (determined by extraction and chromatography of each brain fraction) is far greater than can be accounted for by the increasing mass of synaptosomal protein (16).

In separate studies, Mashio and colleagues observed progressive increases in high affinity T3 binding sites in synaptosomal membranes derived from the developing rat brain (Fig. 11c). This contrasts with the direction of change in nuclear binding sites, and is compatible with a developmentally-determined shift into a new phase of thyroid hormone action in brain.

To pursue the analogy with known neurotransmitter actions during development: it might be predicted that the formation of a T3-processing system of nerve terminals, now known to occur in the rat from days 4 to 19 of postnatal life, would coincide with T3-dependent changes in nerve cell specialization (neurite elongation, synaptogenesis, myelination). In fact this is the very period of such events. Further investigations should determine whether actions mediated by nuclear receptors, synaptosomal receptors or other, not yet identified, receptors, are responsible.

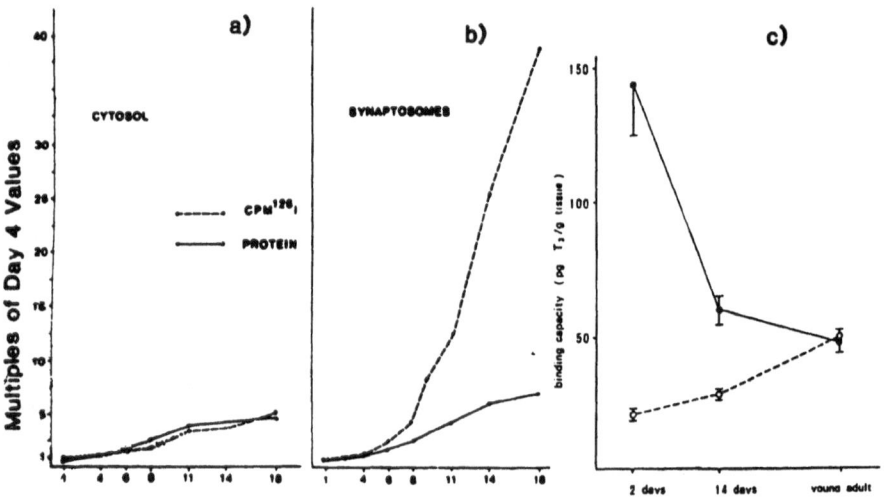

Fig. 11.   DEVELOPMENTAL CHANGES IN IODOCOMPOUNDS AND HIGH AFFINITY T3 BINDING SITES IN SYNAPTOSOMES

(a and b) Rat pup brains were analyzed as described in Legend, Fig. 10. Individual data points are mean levels of iodocompounds in cytosol and synaptosomes per brain, expressed as a multiple of the mean day 4 value. After day 4, ratios of protein to iodocompound content were constant for cytosol but changed significantly for synaptosomes (p<.001); differences between cytosol and synaptosomal iodocompounds/mg protein were also significant (p<.001).

(c) Data of Mashio et al (14,15) showing change in specific binding capacity of T3 nuclear (●——●, and synaptosomal o---o sites during development. Abscissae: postnatal days.

SUMMARY

Ontogenetic features of thyroid hormone processing in brain have been characterized on the basis of new morphologic, biochemical and functional evidence. This evidence raises the possibility that, like other neuroactive substances, thyroxine may participate in developmental events through a series of separate but related mechanisms.

An analogy is drawn between the ontogenesis of substances known to be synaptically active in the differentiated brain (neurotransmitters, neuromodulators, neuroregulators) and selected features of thyroid hormone binding and processing during development.

Many neurotransmitters (for example, dopamine) participate in regulating important stages of brain growth during different phases of development: EARLY: before the appearance of the apparatus of neurotransmission, influencing neuroblast proliferation and differentiation; LATER: during and after the appearance of neurotransmission machinery, influencing nerve cell specializations such as neurite elongation, synaptogenesis, myelination; in DIFFERENTIATED BRAIN, affecting axonal and dendritic plasticity as well as information transfer across synapses.

Early effects of T_3 on neuroblast division and differentiation occur before the neural apparatus for processing thyroid hormones has developed. The period of thyroid hormone-dependent nerve cell specialization occurs later, during the so-called critical period, coinciding with the elaboration of a T_3-processing neural system. In the differentiated brain there is now evidence compatible with a role for T_3 in maintaining neural plasticity and intercellular communication through a transsynaptic mechanism.

REFERENCES

1. M. B. Dratman, F. L. Crutchfield, Y. Futaesaku, M. E. Goldberger, and M. Murray, [125I]Triiodothyronine in the rat brain: evidence for neural localization and axonal transport derived from thaw-mount film autoradiography, Jl. Comp. Neurol. 260: 392 (1987).
2. M. B. Dratman, F. L. Crutchfield, J. T. Gordon, and A. S. Jennings, Iodothyronine homeostasis in rat brain during hypo- and hyperthyroidism, Am. J. Physiol. 245:189E (1983).
3. A. Ruiz-Marcos, F. Sanchez-Toscano, F. Escobar del Rey, and G. Morreale de Escobar, Reversible morphological alterations of cortical neurons in juvenile and adult hypothyroidism in the rat, Brain Res. 3:91 (1980).
4. M. Goldman, M. B. Dratman, F. L. Crutchfield, J. A. Maruniak, A. S. Jennings, and R. D. Gibbons, Heart rate response to triiodothyronine: evidence for a central nervous system site of thyroid hormone action, J. Clin. Invest. 76:1622 (1985).
5. B. Ferreiro, J. Bernal, G. Morreale de Escobar, and B. J. Potter, Preferential saturation of brain 3,5,3'-triiodothyronine receptor during development in fetal lambs, ENDOCRINOLOGY 122: 438 (1988).
6. J. M. Lauder, Hormonal and humoral influences on brain development, Psychoneuroendocrinol. 8:121 (1983).
7. M. B. Dratman, and F. L. Crutchfield, Synaptosomal [125I]triiodothyronine after intravenous [125I]thyroxine, Am. J. Physiol. 4:E638 (1978).
8. M. B. Dratman, Y. Futaesaku, F. L. Crutchfield, N. Berman, B. Payne, M. Sar, and W. E. Stumpf, Iodine 125-labeled triiodothyronine in rat brain: evidence for localization in discrete neural systems, Science 215:309 (1982).

9. M. B. Dratman, F. L. Crutchfield, J. Axelrod, R. W. Colburn, and N. Thoa, Localization of triiodothyronine in nerve ending fractions of rat brain, Proc. Nat. Acad. Sci. 73:941 (1976).

10. M. M. Kaplan, and K. A. Yaskoski, Effects of congenital hypothyroidism and partial and complete food deprivation on phenolic and tyrosyl ring iodothyronine deiodination in rat brain, Endocrinology 110:761 (1982).

11. W. M. Pardridge, Carrier-mediated transport of thyroid hormones through the rat blood-brain barrier: Primary role of albumin-bound hormo ., Endocrinology 105:605 (1979).

12. P. W. Dickson, A. R. Aldred, J. G. T. Menting, P. D. Marley, W. H. Sawyer, and G. Schreiber, Thyroxine transport in choroid plexus, J. Biol. Chem. 262:13907 (1987).

13. M. M. Kaplan, and K. A. Yaskoski, Phenolic and tyrosyl ring deiodination of iodothyronines in rat brain homogenates, J. Clin. Invest. 66:551 (1980).

14. Y. Mashio, M. Inada, K. Tanaka, H. Ishii, K. Naito, M. Nishikawa, and H. Imura, High affinity w,5,3'-L-triiodothyronine binding to synaptosomes in rat cerebral cortex, Endocrinology 110:1257 (1982).

15. Y. Mashio, M. Inada, K. Tanaka, H. Ishii, K. Naito, M. Nishikawa, K. Takahashi, and H. Imura, Synaptosomal T3 binding sites in rat brain: their localization on synaptic membrane and regional distribution, Acta Endocrinol. 104:134 (1983).

16. A. Ruiz-Marcos, F. Sanchez-Toscano, M. J. Obregon, F. Escobar del Rey, and G. Morreale de Escobar, Thyroxine treatment and recovery of hypothyroidism-induced pyramidal cell damage, Brain Res. 239:559 (1982).

17. F. L. Crutchfield, and M. B. Dratman, Growth and development of the neonatal rat: Particular vulnerability of male to disadvantageous conditions during rearing, Biol. of Neonate 38:203 (1980).

18. F. L. Crutchfield, and M. B. Dratman, Early ontogeny of iodocompound-processing neural systems in rat brain, Pediatr. Res. 17:8 (1983).

19. J. T. Gordon, F. L. Crutchfield, A. S. Jennings, and M. B. Dratman, Preparation of lipid-free tissue extracts for chromatographic determination of thyroid hormones and metabolites, Arch. Biochem. Biophys. 216:407 (1982).

20. H. L. Schwartz, Effect of thyroid hormone on growth and development, in: Molecular basis of thyroid hormone action, J. H. Oppenheimer, and H. H. Samuels, eds., Academic Press, New York (1983).

21. M. B. Dratman, and F. L. Crutchfield, Selective localization of triiodothyronine in mossy fiber synaptosomes of rat cerebellum, Abstract, Annual Meeting, The Endocrine Society, 1988.

22. B. J. Winer, Statistical Principles in Experimental Design, McGraw, New York (1962).

DEVELOPMENT OF FETAL THYROID SYSTEM CONTROL

Delbert A. Fisher

Department of Pediatrics, Harbor-UCLA Medical Center

1000 W. Carson St., Torrance, CA 90509

INTRODUCTION

The hypothalamic-pituitary-thyroid system in the fetus is comprised of a complex of hypothalamic centers, anterior pituitary thyrotroph cells, thyroid follicular cells, and peripheral tissues which metabolize and respond to thyroid hormones. Embryogenesis of the hypothalamus and of the pituitary and thyroid glands is largely completed by 12 weeks of gestation in the human fetus. Hypothalamic histogenesis and differentiation and continued growth and functional maturation of the pituitary and thyroid glands proceed into the neonatal period (1). Studies of thyroid system maturation have been conducted in many species, but most detailed data have been developed in the sheep and rat models. The period of thyroid system development in man, a precocial species, extends to one month of postnatal life (some 44 weeks). In the sheep (also a precocial species), comparable thyroid system maturation encompasses 150 days of intrauterine gestation plus two weeks of postnatal life (165 days total). In the rat (an altricial species) thyroid system development requires some 50 days (21 fetal days + 28 postnatal days). Relative thyroid system maturation in these species is quite comparable and thyroid control matures during the latter half of the period of ontogenesis. Thus, the third trimester fetal sheep and neonatal lamb and the neonatal rat have served as useful models for the study of thyroid control maturation. The following discussion of the development of various aspects of control of thyroid hormone production will review data in the three species normalized as relative development time to facilitate species comparisons.

The Pituitary Thyroid-Axis

The central focus of thyroid system control is the pituitary thyrotroph cell. Figure 1 summarizes the major events in thyrotroph cell control of thyrotropin (TSH) synthesis and release. Hypothalamic thyrotropin releasing hormone (TRH) binds to its receptor, stimulating the synthesis and release of TSH. Based on available evidence it appears that TRH-receptor interaction increases intracellular free calcium concentrations both by mobilization from intracellular pools and by increasing transport via cell membrane calcium channels (2). The increased cytoplasmic calcium activates TSH release. Cyclic AMP may play a role: both TRH and cAMP seem capable of increasing calcium inflow into the thyrotroph, as well as mobilizing calcium from intracellular pools. However, cAMP does not appear to be the primary mediator of the TRH effect.

Rather, recent evidence suggests that stimulation of membrane inositol phospholid metabolism with production of inositol triphosphate and diacylglycerol as second messengers may be involved (2). TRH stimulates glycosylation of TSH but does not increase production of the messenger RNA species for the alpha and beta subunits of TSH (3).

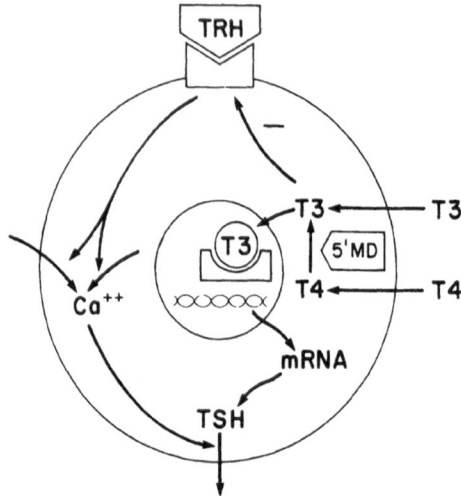

Figure 1. Details of thyrotropin releasing hormone (TRH) and triiodothyronine (T3) regulation of pituitary TSH synthesis and release. See text for details. 5'MD is iodothyronine 5'monodeiodinase, type II.

The second major factor in control of pituitary TSH secretion is thyroid hormone which is known to inhibit TSH release via two major mechanisms, 1) triiodothyronine (T3) binding to the thyrotroph nuclear receptor inhibits TSH mRNA production, both alpha and beta chains, and 2) T3 decreases TRH receptor binding (2,4,5) (Figure 1). The intrapituitary T3 is derived from thyroxine (T4) by monodeiodination as well as by transport from serum. The T4 outer ring monodeiodinase (5'MD) in pituitary tissue is a type II enzyme, as contrasted with type I hepatic 5'MD. The pituitary enzyme differs from hepatic enzyme in having a lower Km, being insensitive to propylthiouracil (PTU) inhibition and being suppressible rather than inducible by thyroid hormone (4,6).

The timing of appearance of the several events in pituitary thyrotroph cell maturation are summarized in Table 1. The less than (<) notation indicates that studies were not conducted at an earlier age.

Studies in the rat have shown the presence of TRH receptors at the time of birth (0.4 of system development)(7). The TSH response to TRH is present at birth in the rat and early in the third trimester in man and sheep (8-11). The pituitary iodothyronine 5'monodeiodinase enzyme for intrapituitary conversion of T4 to T3 is present at birth in the rat and has been measured during the mid-third trimester in the fetal sheep pituitary (12,13). Pituitary nuclear T3 receptors are present in near-adult concentrations in the 5-day neonatal rat and PTU inhibition of pituitary T3 receptor binding was observed at 14 days of postnatal age (about 0.7 of thyroid system maturation) (14). Immunoreac-

Table 1

MATURATION OF PITUITARY THYROTROPH

| | Time of Appearance* | | |
	Man	Sheep	Rat
TRH receptor present			<0.4
TSH response to TRH	<0.6	<0.6	<0.4
T4 5'deiodinase activity present		<0.8	<0.4
T3 receptor present			<0.5
T3 stimulation of T3 receptor binding			<0.7
TSH synthesis	<0.3	<0.3	<0.3
T3 inhibition of TSH synthesis and release	<0.65		<0.5
T3 inhibition of pituitary TRH binding			<0.6
TRH effect on TSH glycosylation			>0.6

*Fractional proportion of thyroid system maturation time. Complete maturation time (=1.0) is 10 months in man, $5\frac{1}{2}$ months in sheep, and 50 days in rats.

tive TSH is present in the pituitary and serum of sheep and human fetuses near the end of the first trimester and is detectable in fetal rats by 18-20 days (1,10,11). The capacity for T3 modulation of TSH synthesis and release is present at 5 days of age in neonatal rats and as early as 26 weeks gestation in the human premature infant (8,10). T3 inhibition of TRH receptor binding has been observed in the 10-day neonatal rat (5). A TRH effect on glycosylation of TSH in rat pituitaries in vitro was not observed at 5 days and by 56 days was easily observed (11). These results indicate that the pituitary cellular mechanisms for control of TSH release are present by 0.4-0.6 of thyroid system maturation. This corresponds to the mid or late second trimester of development in the human fetus.

Thyroid follicular cells are present and can synthesize thyroid hormones by 70-80 days of human gestation (1). However, the capacity of these cells to produce thyroid hormones in response to TSH increases with advancing gestation age. Table 2 summarizes available information regarding the time of appearance of important follicular cell functions during development.

Iodide trapping and T4 synthesis are observed late in the first trimester in sheep and human fetuses (1). Thyroglobulin has been detected as early as 29 days of human fetal age (1). T4 to T3 conversion capacity has been detected at birth in the fetal sheep thyroid (Fisher, unpublished). TSH responsiveness in the developing thyroid appears near the midperiod of thyroid system development, when assessed as the time fetal serum T4 levels begin to increase. In the rat and sheep, serum TSH levels increase before the T4 increase occurs; this is not so clearly evident in the human fetus (1,11). Iodide is known, in large concentrations, to inhibit iodide trapping and to block the effect of TSH on T4 synthesis and release. The adult thyroid has the capacity to defend against excess iodide by decreasing membrane iodide trapping (15). The capacity of the thyroid follicular cell to inhibit iodide transport and defend against iodide-induced hypothyroidism appears as a terminal event in thyroid system ontogenesis in both infant rats and humans (16,17). Studies in rabbits suggest that the failure of fetal thyroid tissue to exhibit autoregulation may relate to reduced iodination of an 8 to 10 kilodalton soluble component of the thyroid gland (15,18).

Table 2

MATURATION EVENTS IN THYROID GLAND FOLLICULAR CELL DEVELOPMENT

	Time of Appearance*		
Event	Man	Sheep	Rat
TSH responsiveness**	0.5	0.3	0.5
Iodide trapping	0.2	0.3	0.3
Thyroglobulin synthesis	0.1		0.3
T4 synthesis	0.2	0.3	0.3
T4 to T3 conversion		<1.0	
Iodide inhibition of iodide trapping	1.0		1.0

*Fractional proportion of thyroid system maturation time. Complete maturation time (=1.0) is 10 months in man, $5\frac{1}{2}$ months in sheep and 50 days in rats.
** Assessed as time serum T4 begins to increase.

The Hypothalamic-Pituitary Axis

One important neuronal TRH control center appears to be the paraventricular nucleus, but TRH is widely distributed in the hypothalamus and highly concentrated in the median eminence (4). One important factor regulating TRH production is environmental temperature. Both peripheral thermal receptors and preoptic neuronal thermal receptors monitor environmental and central body temperature; these receptors modulate preoptic neuronal outflow to the paraventricular nucleus and other TRH synthesizing neurons in the hypothalamus and median eminence which, in turn, modulate TRH secretion (4). Decreasing environmental and/or core body temperature increase TRH output and increase the tonic level of TSH release. Somatostatin (SRIF) and dopamine can inhibit TSH release by actions at the pituitary level, and these inhibitory transmitters contribute to central nervous system modulation of TSH release (4). There is evidence that serotonin may be inhibitory in the adult rat, but this does not seem to be so in other species. Norepinephrine also may be inhibitory. Glucocorticoid can inhibit TSH release at the hypothalamic level, but the mechanism is not known. The exact roles of TRH and non-TRH regulatory factors in TSH control are not clear. Administration of somatostatin antiserum to adult rats increases basal TSH levels and potentiates the TSH response to cold (19). Inhibitory factors probably also play a role in the diurnal variation in TSH secretion, in the inhibitory reactions to stress, in the variation in thyroidal activity in psychosis, etc.

Table 3 summarizes current information regarding the timing of maturation of several major events in hypothalamic control of TSH release.

TRH has been detected in the hypothalamus of all three species near the end of the first trimester of thyroid system ontogenesis. Exogenous T3 has been shown to decrease and thyroidectomy to increase hypothalamic and extraneural tissue TRH concentrations at 130 days of gestation in the sheep (about 0.8 of system development) (20). The increase in hypothalamic TRH concentration begins at 10-12 weeks in humans, at 4-5 days of age in the rat, and at 90-100 days gestation in the sheep (1,21). TSH release becomes responsive to TRH near the mid-period of thyroid system ontogenesis (1,10). Hypothalamic ablation experiments, administration of TRH antiserum, and studies of TRH augmentation of propythiouracil-induced increases in serum TSH in neonatal rats have indicated that the pituitary-TSH release becomes TRH dependent between 5 and 10 days of age (about 0.5 of thyroid system ontogenesis) (22-25). Roughly simultaneous increases in pituitary TSH, serum TSH, and serum T4 levels occur in the human fetus at 18-20 weeks gestation, probably reflecting similar TRH activation of the fetal pituitary thyroidal axis (1,26). In the sheep

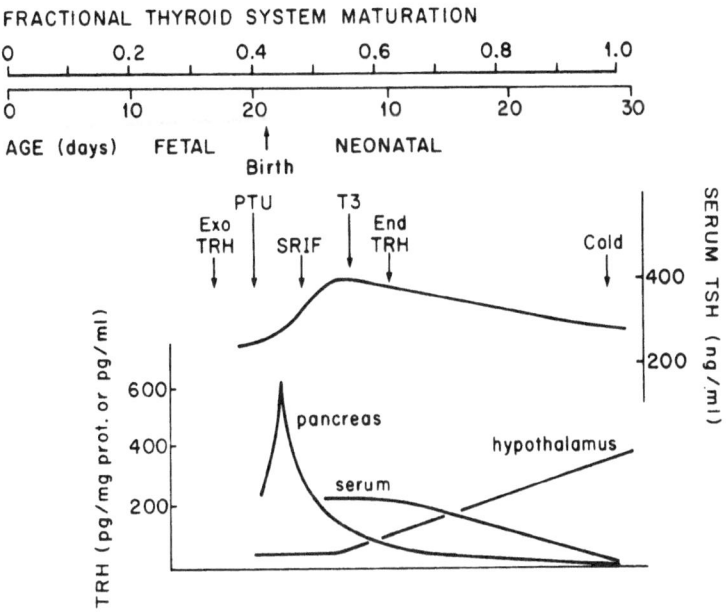

Figure 2. Maturation of TSH secretion in the developing rat. The events depicted are described in more detail in the text. Details of timing are compiled from references 1,10,34.

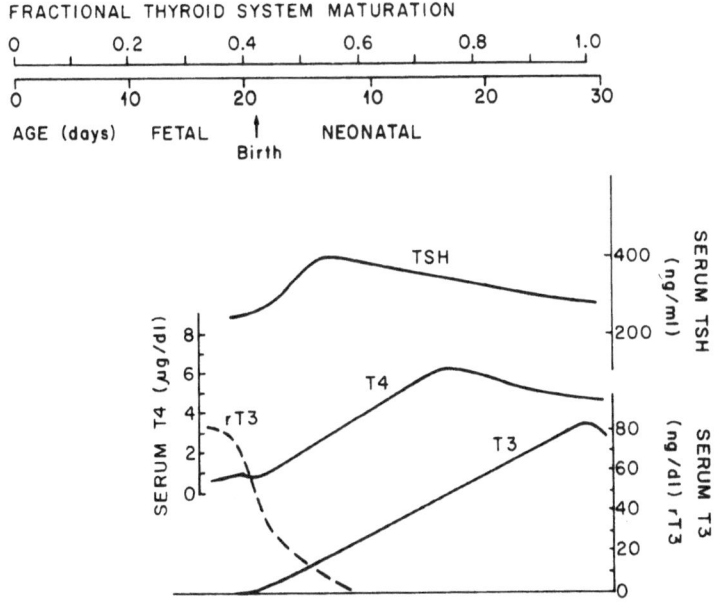

Figure 3. Maturation of thyroid hormone secretion in the developing rat. The events depicted are described in text. Details of timing are compiled from references 1,12,34.

TABLE 3

MATURATION EVENTS IN HYPOTHALAMIC CONTROL OF TSH RELEASE

Event	Time of Appearance*		
	Man	Sheep	Rat
TRH synthesis	<0.3	<0.3	<0.4
TRH stimulation of TSH	0.5	0.5	0.5
T3 inhibition of TRH synthesis		<0.8	
TSH response to cold	<0.6	<0.6	1.0
Somatostatin inhibition of TSH secretion		<1.0	0.4
Dopamine inhibition of TSH secretion	>1.0		>1.0

*Fractional proportion of thyroid system maturation time. Complete maturation time (+1.0) is 10 months in man, 5 months in sheep and 50 days in rats.

the increase in serum T4 also begins at midgestation (1). This "activation" probably reflects maturation of the pituitary-portal vascular system and/or increased TRH secretion since TRH receptors and the postreceptor response to TRH at the thyrotroph level are present prior to this time (Table 1).

The TSH response to cooling has been observed in the sheep fetus at 106 days gestation (0.6 of system ontogenesis) (27). In the human neonate a T4 response to extrauterine cooling has been observed at term (28); moreover, the TSH response of premature infants to extrauterine exposure presumably also reflects a cold response (26). In the neonatal rat a response to cold is not observed until 3-4 weeks of age (29,30).

Somatostatin (SRIF) inhibition of TRH-stimulated TSH secretion has been demonstrated in the neonatal rat using SRIF antiserum as early as 3 days of age (0.4 of thyroid development) (19). However no effect on basal TSH levels was observed during the first 60 days of age (19). Exogenous SRIF inhibits the TSH response to TRH in neonatal lambs (31). Finally, dopamine receptor blockade in neonatal rats had no effect on serum TSH levels during the first 6 weeks of life and no effect at birth in human infants (32,33).

Maturation of Thyroid Control

The timing of maturation events for the developing rat is illustrated in figures 2 and 3. The pituitary-thyroid axis develops and functions autonomously of the hypothalamus during the first half of the period of thyroid system onto-genesis. The system becomes TRH responsive near midgestation, and during the last 0.3 to 0.5 of the period of thyroid system development there is a progres-sive increase in serum T4 and free T4 levels in the face of prevailing, relatively high serum TSH concentrations. This period of increasing serum T4 concentra-tions (figure 3) reflects several simultaneous maturational events. Probably the predominant event is a progressive increase in hypothalamic TRH secretion. The pattern of maturation of extraneural tissue TRH levels and serum TRH is compatible with an effect of extrahypothalamic TRH on TSH secretion during the midperiod of thyroid system maturation (Figure 2). However, encephalectomy experiments in neonatal rat show no effect on serum TSH until the second week of life (22,23). Thus, whether extraneural TRH plays a role is not yet clear.

The ontogenesis of TSH and thyroid hormone secretion in the human fetus and newborn are shown graphically in figures 4 and 5. Here, also, the role of extraneural TRH in TSH release during midgestation is not yet clear (figure

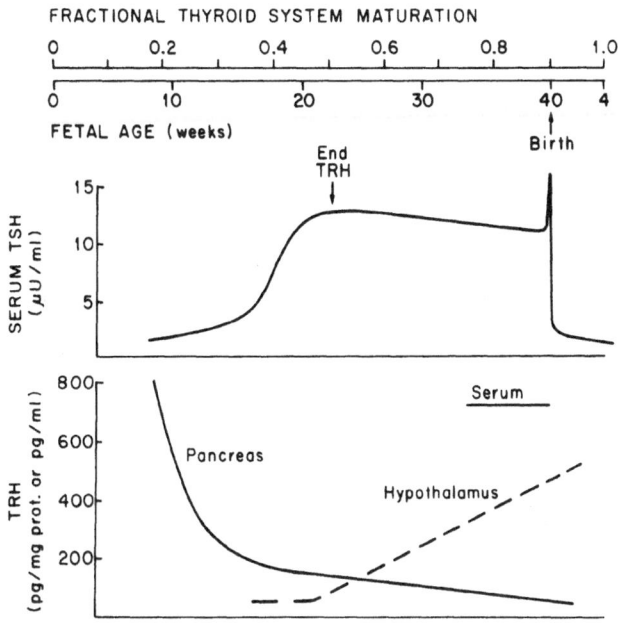

Figure 4. Maturation of TSH secretion in the human fetus and neonate. The events depicted are described in more detail in the text. Details of timing are compiled from references 1,8,34 and 37-39.

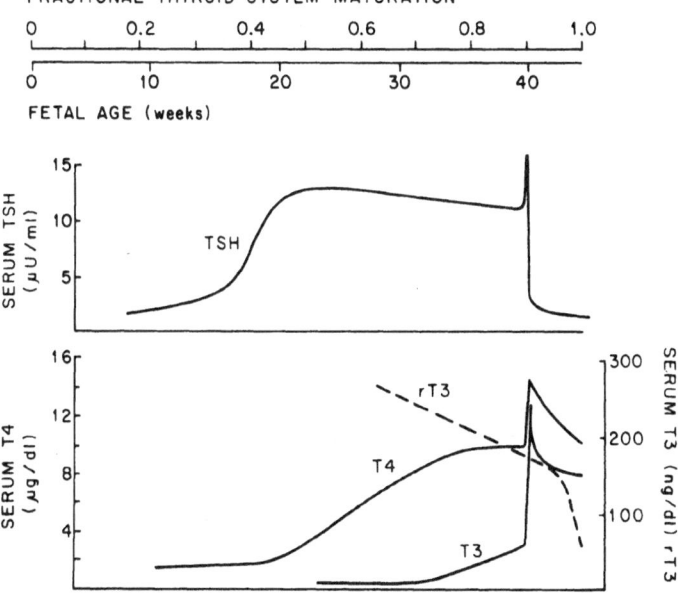

Figure 5. Maturation of thyroid hormone secretion in the human fetus and neonate. The events depicted are described in text. Details of timing are compiled from references 1,26,28.

173

4). The period of increasing fetal serum T4 concentrations between 20 and 35 weeks gestation (figure 5) represents the critical period of maturation of thyroid system control. And this period of immaturity is reflected in thyroid system function in human premature infants. Newborn premature infants have a progressively increasing prevalence of hypothyroxinemia with decreasing gestational age; the prevalence in infants <30 weeks gestation is 50% (8,35,36). This hypothyroxinemia is associated with low serum TSH values and normal TSH and T4 responses to TRH, observations characteristic of a state of hypothalamic hypothyroxinemia. The data of Tables 1 and 2 suggest that TRH binding and post receptor responsiveness are reasonably well developed at this time and are not contributing to the immature function. The progressive increase in hypothalamic TRH concentrations observed in sheep and rats would support the hypothesis of progressively maturing TRH secretion during this period of maturation (1,20,21).

There also is a progressive maturation of T3 negative feedback control of pituitary TSH release during the last half of thyroid system development. In the neonatal rat this is reflected by a progressive decrease in the dose and serum level of T3 necessary to suppress serum TSH or inhibit the serum TSH response to TRH (10). Pituitary T3 receptors and pituitary T4 to T3 conversion capacity are not rate limiting (Table 1). Hinkle and Goh (5) have shown that TRH binding and the TSH response to TRH in 10 to 12-day rat pituitary cells in culture are decreased by exposure to T3. Thyroid hormones also increase T3 receptor binding in developing pituitary gland (14). These data suggest that the maturation of pituitary sensitivity to negative feedback suppression of TSH by T3 is a complex process. Increasing T3 sensitivity with maturation seems to be due to a progressive decrease in the TRH effect mediated by a) a T3-induced decrease in TRH receptors, b) a T3-induced augmentation of T3 receptor binding, and c) T3 inhibition of hypothalamic TRH synthesis (20,40). Qualitatively, maturation of negative feedback control appears similar in the rat and human (figures 2-5).

Finally, there is a progressive maturation of the thyroid gland responsiveness to TSH. This has been most clearly evident in the fetal sheep, where a progressive increase in the T4 response to TRH-stimulated endogenous TSH release has been observed during the third trimester of gestation (9). This maturation probably involves a) an increase in TSH glycosylation, and b) an increased thyroid response to TRH. The progressive increase in serum free T4 levels in the human fetus in the face of relatively elevated and stable serum TSH concentrations would support this hypothesis (34) (Figures 4,5).

REFERENCES

1.   Fisher, D.A., Dussault, J.H., Sack, J. and Chopra, I.J. (1977) Ontogenesis of hypothalamic-pituitary-thyroid function and metabolism in man, sheep, and rat. Rec Prog Horm Res 33:59.
2.   Gershengorn, M.C. (1986) Thyrotropin releasing hormone stimulation of pituitary hormone secretion. Ann Rev Physiol 48:515.
3.   Lippman, S.S., Amr, S. and Weintraub, B. (1986) Discordant effects of thyrotropin (TSH)-releasing hormone on pre- and postranslational regulation of TSH biosynthesis in rat pituitary. Endocrinology 119:343.
4.   Morley, J.E. (1981) Neuroendocrine control of thyrotropin secretion. Endocrine Rev 2:396.
5.   Hinkle, P.M. and Goh, K.B.C. (1982) Regulation of thyrotropin releasing hormone receptors and responses by l-triiodothyronine in dispersed rat pituitary cell cultures. Endocrinology 110:1725.
6.   Kaplan, M.M. (1983) Metabolism of thyroid hormones. In: Congenital Hypothyroidism (Dussault JH and Walker P, Eds), Marcel Dekker Inc., New York, pp 11-35.

7. Banerji, A. and Prasad, C. (1982) The postnatal development of the pituitary thyrotropin releasing hormone receptor in male and female rats. Endocrinology 110:663.

8. Delange, F., Dalhem, A., Bourdoux, P., Lagasse, R., Glinoer, D., Fisher, D.A., Walfish, P.G. and Ermans, A.M. (1984) Increased risk of primary hypothyroidism in preterm infants. J Pediatr 105:462.

9. Klein, A.H. and Fisher, D.A. (1980) Thyrotropin releasing hormone stimulated pituitary and thyroid gland responsiveness and 3, 5, 3'-triiodothyronine suppression in fetal and neonatal lambs. Endocrinology 106:697.

10. Walker, P., Coulombe, P. and Dussault, J.H. (1980) Effects of triiodothyronine on thyrtropin releasing hormone induced thyrotropin release in the neonatal rat. Endocrinology 107:1731

11. Gyves, P.W., Gesundheit, N., Taylor, T., Butler, J.B. and Weintraub, B.D. (1987) Changes in thyrotropin (TSH) carbohydrate structure and response by concanavalin-A chromatography. Endocrinology 121:133.

12. Cheron, R.G., Kaplan, M.M. and Larsen, P.R. (1980) Divergent changes in thyroxine 5' monodeiodination in rat pituitary and liver during maturation. Endocrinology 106:1405.

13. Segall-Blank, M., Connolly, J.L. and Ingbar, S.H. (1982) Comparative studies of the metabolism of thyroxine on the pituitaries of pregnant sheep and their fetuses. Endocrinology 111:1996.

14. Coulombe, P., Ruel, J., Favre, R. and Dusssault, J.H. (1983) Pituitary nuclear triiodothyronine receptors during development in the rat. Am J Physiol: Endocrinol Metab 8:E81.

15. Price, D.J. and Sherwin, J.R. (1986) Autoregulation of iodide transport in the rabbit: absence of autoregulation in fetal tissue and comparison of maternal and fetal thyroid iodination products. Endocrinology 119:2547.

16. Castaign, H., Fournet, J.P., Leger, F.A., Keisgen, F., Piette, C., Dupard, M.C. and Savoie, J.C. (1979) Thyroid of the newborn and postnatal iodine overload. Arch Fr Pediatr 36:356.

17. Theodoropoulos, T., Fang, S.L., Prosky, J. and Vagenakis, A.G. (1979) Iodide induced hypothyroidism: a potential hazard during perinatal life. Science 205:502.

18. Sherwin, J.R. and Price, D.J. (1986) Autoregulation of thyroid iodide transport: evidence for the mediation of protein synthesis in iodide-induced suppression of iodide transport. Endocrinology 119:2553.

19. Oliver, C., Giraud, P. and Conte-Devolx, B. (1982) Influence of endogenous somatostatin on growth hormone and thyrotropin secretion in neonatal rats. Endocrinology 110:1018.

20. Polk, D.H., Reviczky, A.L., Lam, R.W. and Fisher, D.A. (1988) Thyrotropin releasing hormone: effect of thyroid status on tissue concentrations in fetal sheep. Clin Res 36:203A.

21. Engler, D., Scanlon, M.F. and Jackson, I.M.D. (1981) Thyrotropin releasing hormone in the systemic circulation of the neonatal rat is derived from the pancreas and other extraneural tissues. J Clin Invest 67:800.

22. Strbak, V. and Greer, M.A. (1979) Acute effects of hypothalamic ablation on plasma thyrotropin and prolactin concentrations in the suckling rat: Evidence that early postnatal pituitary-thyroid regulation is independent of hypothalamic control. Endocrinology 105:488.

23. Strbak, V. and Greer, M.A. (1981) Thyrotropin secretory response to thyrotropin releasing hormone in the hypothyroid perinatal rat: Further evidence of thyrotroph independence of the hypothalamus during early ontogenesis. Endocrinology 108:1403.

24. Theodoropoulos, T., Braverman, L.E. and Vagenakis, A.G. (1979) Thyrotropin releasing hormone is not required for thyrotropin secretion in the perinatal rat. J Clin Invest 63:588.

25. Oliver, C., Giraud, P., Lissitzky, J.C., Conte-Devolx, B. and Gillioz, P. (1981) Influence of thyrotropin releasing hormone on the secretion of thyrotropin in neonatal rats. Endocrinology 108:179.

26. Fisher, D.A. and Klein, A.H. (1981) Thyroid development and disorders of thyroid function in the newborn. New Engl J Med 304:702.
27. Fraser, M., Gunn, T.R., Butler, J.H., Johnston, B.M. and Gluckman, P.D. (1985) Circulating thyrotropin in the ovine fetus: evidence for pulsatile release and the effect of hypothermia in utero. Pediatr Res 19:208.
28. Fisher, D.A. and Odell, W.D. (1969) Acute release of thyrotropin in the newborn. J Clin Invest 48:1670.
29. Frankel, S. and Lange, G. (1980) Maturation of hypothalamic-pituitary thyroid response in the rat to acute cold. Am J Physiol (Endocrinol Metab)2:E223.
30. Theodoropoulos, T., Braverman, L.E. and Vagenakis, A.G. (1979) Circulating immunoreactive TRH in the neonatal rat. Dissociation between TRH release and TSH response following cold exposure. Endocrine Soc., Abstr. 427.
31. Sack, J., Fisher, D.A., Grawer, L.A., Lam, R.W. and Wang, C.C. (1977) The response of newborn sheep to TRH with and without somatostatin. Endocrinology 100:1533.
32. Becu, D. and Libertun, C. (1982) Comparative maturation of the regulation of prolactin and thyrotropin by serotonin and thyrotropin-releasing-hormone in male and female rats. Endocrinology 110:1879.
33. Roti, E., Robuschi, G., Emanuele, R., D'Amato, L., Gnudi, A., Fatone, M., Benassi, L., Foscolo, M.S., Gualerzi, C., and Braverman, L.E. (1983) Failure of metoclopramide to affect thyrotropin concentration in the term human fetus. J Clin Endocrinol Metab 56:1071.
34. Harris, A.R.C., Fang, S.L., Prosky, J., Braverman, L.E. and Vagenakis, A.G. (1978) Decreased outer ring monodeiodination of thyroxine and reverse triiodothyronine in the fetal and neonatal rat. Endocrinology 103:2216.
35. Klein, A.H., Oddie, T.H., Parslow, M., Foley, T.P., Jr. and Fisher, D.A. (1982) Development changes in pituitary thyroid function in the human fetus and newborn. Early Human Devel 6:321.
36. Hadeed, A.J., Asay, L.D., Klein, A.H. and Fisher, D.A. (1981) Significance of transient postnatal hypohthyroxinemia in premature infants with and without RDS. Pediatrics 68:494.
37. Koivusalo, F., (1981) Evidence of thyrotropin releasing hormone activity in autopsy pancreata from newborns. J Clin Endocrinol Metab 53:734.
38. Leduque, P., Aratan-Spire, S., Czernichow, P. and Dubois, P.M. (1986) Ontogenesis of thyrotropin releasing hormone in human fetal pancreas. J Clin Invest 78:1028.
39. Perelman, A.H., Klein, A.H. and Fisher, D.A.: Cord blood thyrotropin releasing hormone. Clin Res. 29:111A (Abstract).
40. Segerson, T.P., Kauer, J., Wolfe, H.C., Mobtaker, H., Wu, P., Jackson, I.M.D. and Lechan, R.M. (1987) Thyroid hormone regulates TRH biosynthesis in the paraventricular nucleus of the rat hypothalamus. Science 238:78.

THE FETUS AND IODINE DEFICIENCY:

MARMOSET AND SHEEP MODELS OF IODINE DEFICIENCY

Basil S. Hetzel, Mark L. Mano, and J. Chevadev*
CSIRO Division of Human Nutrition, Kintore Avenue
Adelaide, Australia
*Dept. Physiology, Monash University
Clayton, Vic. (Now Dept. Anatomy,
Mahidol University, Bangkok, Thailand)

INTRODUCTION

Iodine Deficiency Disorders (IDD) are a major international public health problem (1). The effects of iodine deficiency occur at all ages, but are particularly important during the period of fetal development. The available epidemiological evidence has been complemented by experimental studies of fetal development in animal models. These have focussed particularly on fetal brain development because of its obvious importance. Definite effects have been observed in a variety of animal models - the rat, the marmoset and the sheep. In addition studies of the mechanisms involved have been carried out which have revealed the importance of maternal thyroid function for fetal brain development.

In this paper our work on the marmoset and the sheep models is reviewed.

It is well known that the timing of brain development differs between species. In the human, maximum growth occurs at the time of parturition which is also the time of maximum growth of the pig brain (hence termed "perinatal brain developer"). In the rat and rabbit the maximum growth period is postnatal (termed "postnatal brain developer"). In the sheep and the monkey the maximum growth period is prenatal (termed "prenatal brain developer").

As Dobbing (2) has pointed out comparisons can be made between species if the stages rather than the ages of brain development are taken into account.

One major problem in these studies has been the development of a suitable low iodine diet. It was essential that the diet be satisfactory apart from the one particular nutrient, iodine, so that any non-specific effects could be eliminated. This was achieved with both species showing normal fetal growth on the control diets which consisted of the iodine deficient diet with an iodine supplement. This meant that differences between the control and iodine deficient animals could, with confidence, be attributed to the iodine deficiency.

IODINE DEFICIENCY IN THE MARMOSET

The effect of iodine deficiency on fetal development in the primate was investigated by Mano et al (3) in the common cotton-eared marmoset (Callithrix jacchus jacchus) (native of Brazil). A low-iodine diet (14.9 ± 0.8 μg iodine per kilo) was developed composed of maize (60%), peas (15%), torula yeast (10%), low-iodine dried meat meal (10%) - prepared from the slaughter of iodine-deficient sheep (4) with added maize oil, minerals, amino acids and vitamins. The marmosets were confined to cages in pairs of opposite sex, and compared to the control pairs which received the same diet to which potassium iodate was added to provide a total intake of 433 ± 34 μg iodine per kilo). The iodine-deficient pairs maintained body weight and good health but had grossly reduced plasma T_4 levels and elevated TSH levels. Thyroid iodine was reduced and histological examination of the glands revealed hyperplasia, hypertrophy and total absence of colloid material in the follicles (3). After a year on the diet, the female animals became pregnant and the newborn marmosets were studied following the first pregnancy and again following a second pregnancy.

The newborn iodine-deficient marmosets showed some sparsity of hair growth but there was no evidence of significant changes in body weight or definite evidence of skeletal retardation as indicated by delayed epiphyseal development. There was gross reduction in plasma T_4 in both mothers and offspring, increased newborn thyroid iodine weight and reduced thyroid iodine content. The differences were more obvious in the second pregnancy than in the first associated with a greater severity of iodine deficiency (Table 1).

Table 1. Body weight, Plasma thyroxine (T_4), and brain weight of newborn marmosets (mean ± SEM).*

	Number of Observations	Plasma T_4 (nmol/ml)	Body weight (g)	Brain weight (g)
Normal	10	>300	28.2 ± 0.7	3.63 ± 0.08
Control	28	>300	28.1 ± 0.4	3.53 ± 0.04
Iodine deficient - first pregnancy	14	133 ± 20[a]	26.5 ± 0.8	3.41 ± 0.07
Iodine deficient - second pregnancy	10	47 ± 7[a]	27.2 ± 1.6	3.31 ± 0.11[b]
Iodine deficient - all pregnancies	24	97 ± 15[a]	26.8 ± 0.8	3.37 ± 0.06[b]

[a] $P < 0.001$

[b] $P < 0.05$

*Modified from Mano et al (5).

178

The brain weights of the newborn marmosets were significantly reduced in the second pregnancy (Table 1). This occurred particularly in the cerebellum in which a significant reduction in cell number was observed (5).

Histological examination of the brain revealed morphological changes in the cerebellum, and cerebral hemispheres. In the cerebellum a thickened external germinal layer indicated impaired cell acquisition, a consistent feature of hypothyroidism in the developing brain. Other changes which were seen only in brains of offspring from second pregnancies were a reduction in total area and in molecular layer area and an increase in Purkinje cell linear density. The changes in the molecular layer area and the Purkinje cell linear density are indicative of reduced arborization of Purkinje cell dendrites.

In the cerebral hemispheres, there was little change in weight or cell number (as measured by the DNA content) but in the visual cortex area an increase in neuronal density was apparent in the granular band in the second pregnancy iodine-deficient newborns. Synaptic counts also were reduced in the visual cortex (Table 2).

Table 2. Counts of neuronal cells (in visual cortex) and synapses (in the band between the pia mater and supragranular band) in the cerebral hemispheres of newborn marmosets.*

	Control mean SEM n = 8	I-deficient second pregnancy mean ± SEM n = 4	% Difference from control
Neurons			
per mm^2 of supragranual band	5650 ± 234	5906 ± 131	5
per mm^2 of granular band	8977 ± 237	11264 ± 502	25[a]
per mm^2 of infragranular band	3948 ± 111	3672 ± 122	-7
Synapses			
per m^2 of section	0.296 ±0.015	0.154 ±0.013	-48[a]

n = number of observations

[a] P < 0.001

*From Mano et al (5)

These findings indicate definite effects of iodine deficiency on brain development in a primate model. The relative lack of effect in the first pregnancy indicates that severe iodine deficiency was necessary.

IODINE DEFICIENCY IN THE SHEEP

The sheep has been used as an animal model for extensive studies on iodine deficiency because of its convenience for the study of maternal-fetal function (6).

Severe iodine deficiency was produced in sheep (4) with a low-iodine diet of crushed maize and pelleted pea pollard (8-15 μg iodine/Kg) which provided 5-8 μg iodine per day. After a period of 5 months, although body weights were maintained, iodine deficiency was evident with the appearance of goitre, low plasma T_4 and T_3 values, elevated TSH levels and low daily urinary excretion of iodine. Control animals received the same diet but were supplemented with 2 mg sodium iodide administered by subcutaneous injections each week or by a single iodized oil injection (1 ml = 480 mg iodine). The ewes were mated with normal fertile rams, the dates of conception established, and fetuses delivered at 56, 70, 98 and 140 days gestation by hysterotomy. (The normal gestation period for the sheep is 150 days).

Over the whole period of some 8 years during which these studies were carried out there was a significant loss of lambs to the 69 iodine deficient ewes (21% due to abortion and stillbirths) in comparison to the 67 controls (4.3%).

Goiter was evident from 70 days in the iodine-deficient fetuses and thyroid histology revealed evidence of hyperplasia from 56 days gestation. The increase in thyroid weight was associated with a reduction in fetal thyroid iodine content and greatly reduced plasma T_4 values associated with greatly reduced plasma T_4 levels in the mother (Table 3).

The iodine-deficient fetuses at 140 days were grossly different in physical appearance to the control fetuses. They weighed less and there was a notable absence of wool growth. Goiter, varying degrees of subluxation of the foot joints, and deformation of the skull were also present. Bone maturation was slowed as shown by the delayed appearance of the epiphyses in the limbs. Examination of the viscera of the fetuses revealed reduced weight of the fetal heart and lungs (7) (8).

Lowered brain weight and brain DNA were noticed as early as 70 days (Table 3), indicating a reduction in cell number probably due to slowed neuroblast multiplication which normally occurs from 40-80 days in the sheep. Although brain protein was reduced in the deficient fetuses, the ratio of protein:DNA remained unchanged indicating no significant change in the size of the brain cells. The reduction in protein and DNA to less than normal in the 98 day and 140 day fetal brains implies that after 80 days the development of neuroglia is slowed also in iodine deficiency.

Morphological changes were observed in the cerebellum and cerebral hemispheres at 140 days gestation (Table 4). Delayed maturation of the cerebellum was shown by reduced migration of cells from the external granular layer to the internal granual layer and increased density of Purkinje cells. The greater density of Purkinje cells indicates reduced Purkinje cell arborisation within the molecular layer. In the cerebral hemispheres the cells were more densely packed in the motor and visual areas while the pyramidal neurons in the hippocampus were denser in the CA1 and CA4 regions indicating severe retardation in neuropil growth in both subfields (7).

Table 3. Effect of severe iodine deficiency on fetal brain
development in the sheep.*

Gestational Age		56 days	70 days	98 days	140 days
Maternal plasma	I-defic.	(5) 37[a]	(6) 17[a]	(5) 15[a]	(7) 19[a]
T_4 (nmol/1)	Control	(3) 126	(4) 134	(5) 141	(3) 137
Fetal plasma	I-defic.	(5) 3[b]	(6) 4[a]	(5) 4[a]	(7) 6[a]
T_4 (nmol/1)	Control	(2) 10	(4) 25	(5) 125	(3) 216
Brain weight	I-defic.	(5) 1.79	(7) 4.20[c]	(5)19.0	(7)46.4[b]
(g)	Control	(3) 1.68	(4) 5.01	(5)22.1	(6)53.8
Cell number	I-defic.	(5) 8.86	(7)14.2 [c]	(5)27.8[b]	(7)62.6[a]
(mg DNA)	Control	(3) 8.37	(4)16.2	(5)32.5	(6)74.5
Body weight	I-defic.	(5)31.7	(7) 101 [c]	(5) 662	(7)2930[b]
(g)	Control	(3)32.2	(4) 129	(5) 753	(6)3820

Number of observations shown in parentheses.
[a] $P < 0.001$ [b] $P < 0.01$ [c] $P < 0.001$ (two-tailed "t" test)
*Modified from Potter et al (7).

Table 4. Histological data at 140 days gestation of brain regions
from fetal sheep during severe dietary iodine deficiency
(mean values).*

	Control	I-deficient
Cerebral hemispheres		
Parietal cortex area density (neurons/mm^2)	1287	1726[b]
Motor cortex area density (neurons/mm^2)	973	1124
Cerebellum		
Molecular area/Total area	0.468	0.418[b]
Granular area/Total area	0.360	0.390
Medullary area/Total area	0.173	0.208[b]
Molecular area/Granular area	1.303	1.080[b]
Purkinje cell linear density (cells/mm)	15.55	17.61[b]
Granule cell area density (cells/mm^2)	7.67	7.90
External germinal cell linear density (cells/mm)	305	363
Hippocampus		
Basal dendrite depth (mm)	0.227	0.243
Apical dendrite depth (mm)	0.530	0.646
Subfield CA1 area density (neurons/mm^2)	543	707[a]
Subfield CA4 area density (neurons/mm^2)	339	404[b]
Cranial nerve		
Abducent (number of myelinated axons)	2862	2761

Number of observations is 4 for all parameters.
[a] $P < 0.01$ [b] $P < 0.05$ (Two tailed "t" test)
*From Potter et al (7).

Evidence of retarded myelination in the cerebral hemispheres and brainstem also was provided by lowered cholesterol:DNA ratios. An increased water content in the brain at 140 days was further evidence of brain retardation in iodine deficiency (7).

The effect of iodine supplementation on the retarded fetal brain development induced by iodine deficiency was investigated with a single intramuscular injection of iodized oil containing 500 mg iodine given to the pregnant sheep at 100 days gestation (9). The administered iodine restored the maternal and fetal plasma T_4 values to normal and improved the physical appearance of the fetuses. There was an increase in brain growth and to a lesser extent in body growth; and restoration of the number of cells (DNA) and myelination (cholesterol:DNA ratio) in the cerebellum and cerebral hemispheres suggested a catch-up of neuroblast development during pregnancy. Histological examination, however, revealed that counts of synapses (density) in the cerebral cortex after iodized oil were still less than those of the control fetal brains at 140 days gestation or ten days before parturition. The effects of iodine deficiency and iodized oil administration on the cerebellum and cerebral hemispheres are compared with those of the control brains in Fig. 1.

The effects of severe iodine deficiency on fetal brain and somatic development were found to be more severe but similar to those observed by McIntosh et al (10) when thyroidectomy was performed on fetal sheep at 50-60 days. Fetal thyroidectomy later in gestation at 98 days caused less severe effects (11). Further investigations by McIntosh et al (12), following earlier indications (13) that the maternal thyroid could exert an influence on fetal development in early pregnancy, revealed that a combination of maternal thyroidectomy before conception and fetal thyroidectomy at 98 days gestation produced more severe effects than those of iodine deficiency (Fig. 1).

These observations suggested that normal brain development requires the availability of both maternal and fetal thyroid hormones, a suggestion which is at variance with earlier reports (9) (14) (15) (16) that the placenta in many mammalian species is relatively impermeable to thyroid hormones and with the suggestion (17) (18) that early mammalian development takes place normally in the absence of thyroid homrones. The observations do agree however with a more recent report (19) that rat embryonic tissues are provided with T_4 and T_3 only four days after uterine implantation and well before the onset of fetal thyroid function at 17 days. They are also supported by the work of Woods et al (20) who showed that $[I^{125}]$ T_4 and $[I^{125}]$ T_3, when injected into pregnant rats, entered the rat fetus in substantial amounts early in pregnancy, but in minimal amounts in later pregnancy.

Potter et al (21) extended and confirmed their earlier observations and were able to show that surgical thyroidectomy of ewes 6 weeks prior to conception caused a reduction in fetal brain and body growth in mid-gestation compared to that of fetuses from sham-operated ewes. The differences were no longer evident in the neonatal brain at parturition after the onset of fetal thyroid function but the possibility of residual damage remains to be explored with behavioural and functional studies of brain function.

In the light of the available data and observations on the sheep, it may be concluded that the effects of severe iodine deficiency are mediated by a combination of maternal and fetal hypothyroidism, the effect of maternal hypothyroidism being earlier than the onset of fetal thyroid secretion. This would infer an effect on neuroblast multiplication which

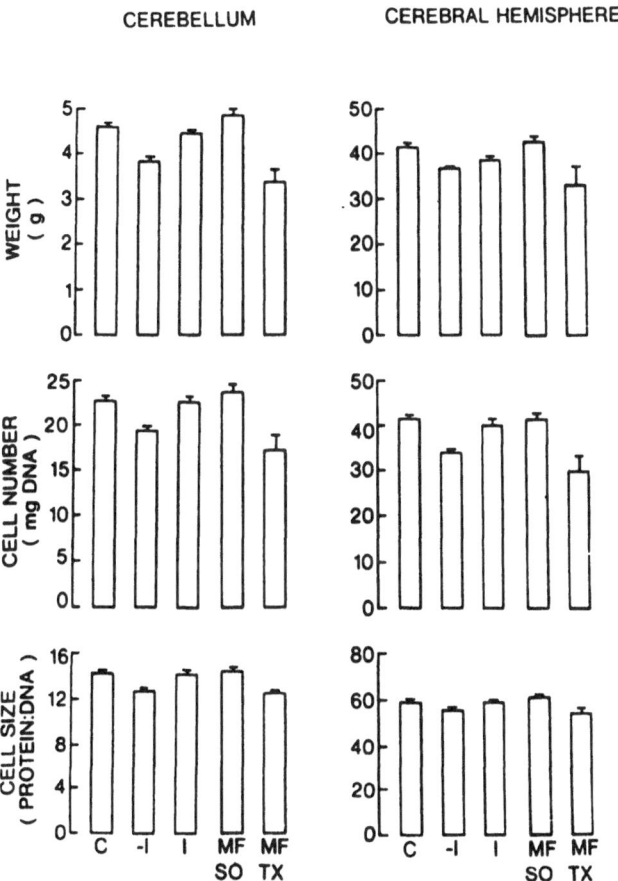

CEREBELLUM CEREBRAL HEMISPHERES

Fig. 1. Comparison of brains of sheep fetuses at 140 days' gestation.
C = Control; -I = iodine-deficient; I = iodine at 100 days;
MFSO = mother+fetus sham-operated; MFTX = mother+fetus
thyroidectomised (From Hetzel (1) with permission).

occurs from 40-80 days gestation in the sheep and 11-18 weeks in the human (2). In the rat (a postnatal brain developer) an effect of the maternal thyroid early in pregnancy is indicated by reduced weight and number of embryos, reduced brain weight and reduced transfer of maternal T_4 (19) (22).

The findings suggest that iodine deficiency has an early effect on neuroblast multiplication and if so this could be important in the pathogenesis of the neurological form of endemic cretinism (8) (23).

From current knowledge of the embryology of the brain it is suggested that the critical period when iodine deficiency affects the development of the fetus is at the mid-trimester of pregnancy at which time the neurons of the cerebral cortex and basal ganglia are formed (14-18 weeks). Formation of the cochlea which is severely affected in endemic cretinism (23) also occurs during this period (10-18 weeks) and antedates the onset of fetal thyroid function. This information and the earlier belief that iodine, but not thyroxine, could traverse the placenta led to the suggestion that iodine in elemental form exerts an influence on fetal brain development irrespective of any action mediated by the thyroid (24) (25). The work of Obregon et al (19); Woods et al (20); Morreale de Escobar et al (22); and Potter et al (21) has produced later evidence of the transfer of thyroxine from the mother to the fetus and the early effects of iodine deficiency would appear to be a reflection of the failure of maternal thyroid function in the early stages of pregnancy.

This is further indicated by evidence of the relation of maternal thyroxine levels to the risk of cretinism and the psychomotor defect from studies in Papua New Guinea (26). The subsequent lack of fetal thyroid hormones due to the inadequate supply of iodine in iodine deficiency would exacerbate the effects of maternal thyroid insufficiency and the combination of effects, which were represented experimentally in the sheep by maternal thyroidectomy before conception (21) combined with fetal thyroidectomy at 98 days (12), might be expected to produce the multiple defects of neurological cretinism.

Myxedematous cretinism would result from fetal hypothyroidism later in pregnancy - a condition that can be reversed with treatment by thyroid hormones or correction of the iodine deficiency - but the earlier damage in mid-trimester appears to be irreversible.

It has not been possible yet to reproduce in animals the full clinical syndrome of cretinism - with deaf mutism and spastic diplegia - although only limited observations have been made into the postnatal period. Further investigations are required before the pathogenesis of endemic cretinism can be fully elucidated.

SUMMARY

Animal models in the marmoset and the sheep have been developed to study the effect of severe iodine deficiency on brain development. Both these models are characterised by the production of severe maternal and fetal hypothyroidism which is associated with effects on the maturation of the cerebral cortex and cerebellum. There was a reduced brain weight with a reduced number of cells as indicated by reduced DNA, a greater density of cells in the cerebral cortex and reduced cell acquisition in the cerebellum.

Studies of the mechanisms involved have been carried out in the sheep. The findings reveal significant, though less severe, effects of fetal thyroidectomy (late gestation) and a significant effect of maternal thyroidectomy on brain development in mid gestation. A combination of maternal and fetal thyroidectomy has similar but more severe effects than iodine deficiency.

The data reviewed reveal significant effects on fetal brain development. The study of the mechanisms in the sheep reveals the role of both maternal and fetal thyroid function.

We conclude that iodine is essential for normal fetal brain development because of the critical role of the thyroid hormones. Further investigations are required into the postnatal period before the pathogenesis of endemic cretinism can be elucidated.

REFERENCES

1. Hetzel, B.S. Iodine Deficiency Disorders (IDD) and their Eradication. Lancet 2:1126-1129 (1983).
2. Dobbing, J. The later development of the brain and its vulnerability, in: "Scientific Foundations of Paediatrics", J.A. Davis and J. Dobbing, eds. Heinemann Medical, London (1974) pp. 565-577.
3. Mano, M.T., Potter, B.J., Belling, G.B. and Hetzel, B.S. Low-iodine diet for the production of severe I deficiency in marmosets (Callithrix jacchus jacchus). Brit. J. Nutrition 54: 367-372 (1985).
4. Potter, B.J., Jones, G.B., Buckley, R.A., Belling, G.B., McIntosh, G.H. and Hetzel, B.S. Production of Severe Iodine Deficiency in Sheep using a prepared Low-iodine Diet. Aust. J. Biol. Sci. 33: 53-61 (1980).
5. Mano, M.T., Potter, B.J., Belling, G.B., Chavadej, J. and Hetzel, B.S. Fetal brain development in response to iodine deficiency in a primate model (Callithrix jacchus jacchus). J. Neurological Sciences 79: 287-300 (1987).
6. Fisher, D.A., Dussault, J.H., Erenberg,A. and Lam, R.W. Thyroxine and triiodothyronine metabolism in maternal and fetal sheep. Pediatric Research 12: 894-899 (1972).
7. Potter, B.J., Mano, M.T., Belling,G.B., McIntosh, G.H., Hua, C., Cragg, B.G., Marshall, J., Wellby, M.L. and Hetzel, B.S. Retarded fetal brain development resulting from severe dietary iodine deficiency in sheep. Neuropath. appl. Neurobiology 8: 303-313 (1982).
8. Hetzel, B.S. and Potter, B.J. Iodine deficiency and the role of thyroid hormones in brain development, in: "Neurobiology of the Trace Elements" I. Dreosti and R.M. Smith. Humana Press, New Jersey (1983) pp. 83-133.
9. Potter, B.J., Mano, M.T., Belling, G.B., Martin, D.M., Cragg, B.G., Chavadej, J. and Hetzel, B.S. Restoration of Brain Growth in Fetal Sheep after Iodized Oil Administration to Pregnant Iodine-deficient Ewes. J. Neurological Sci. 66: 15-26 (1984).
10. McIntosh, G.H., Baghurst, K.I., Potter, B.J., Hetzel, B.S. Fetal thyroidectomy and brain development in the sheep. Neuropath. Appl. Neurobiol. 5: 103-114 (1979).
11. McIntosh, G.H., Potter, B.J., Hetzel, B.S., Hua, C.H., Cragg, B.G. The effects of 98 day fetal thyroidectomy on brain development in the sheep. J. Comp. Path. 92: 599-607 (1982).

12. McIntosh, G.H., Potter,B.J., Mano, M.T., Hua, C.H., Cragg, B.G. and Hetzel, B.S. The effect of maternal and fetal thyroidectomy on fetal brain development in the sheep. Neuropath. Appl. Neurobiol. 9: 215-223 (1983).

13. Potter, B.J., McIntosh, G.H. and Hetzel, B.S. The effect of iodine deficiency on fetal brain development in sheep, in: "Fetal Brain Disorders - Recent Approaches to the Problem of Mental Deficiency" B.S. Hetzel and R.M. Smith Eds. Elsevier, Amsterdam (1981) pp.119-148.

14. Erenberg, A. and Fisher, D.A. Thyroid hormone metabolism in the fetus, in: "Fetal and Neonatal Physiology" R.S. Comline, K. W. Cross, G.S. Dawes and P.W. Nathanielsz eds. Cambridge University Press, Cambridge (1973) pp. 508-526.

15. Thorburn, G.D. and Hopkins, P.S. Thyroid function in the fetal lamb, in: "Fetal and Neonatal Physiology",R.S. Comline, K.W. Cross, G.S. Dawes and P.W. Nathanielsz eds. Cambridge University Press, Cambridge (1973) pp. 488-507.

16. Fisher, D.A., Dussault, J.H., Sack, J. and Chopra, I.J. Ontogenesis of hypothalamic pituitary thyroid function and metabolism in man, sheep and rat. Rec. Prog. Horm. Res. 33: 59-116 (1977).

17. Osorio, C. and Myant, N.B. The passage of thyroid hormone from mother to fetus and its relation to fetal development. Br. Med. Bull. 16: 159 (1960).

18. Hamburgh, M. The role of thyroid and growth hormone in neurogenesis, in: "Current Topics in Developmental Biology", J. Moscona and J. Monroy eds. Academic Press, New York (1969) 4: 109.

19. Obregon, J., Mallol, R., Pastor, G., Morreale de Escobar and Escobar del Rey, F. L-thyroxine and 3,5,3'-Triiodo-L-Thyronine in Rat Embryos before onset of Fetal Thyroid Function. Endocrinology 114: 305-307 (1984).

20. Woods, R.J., Sinha, A.K. and Ekins, R.P. Uptake and metabolism of thyroid hormones by the rat fetus in early pregnancy. Clin. Sci. 67: 359-363 (1984).

21. Potter, B.J., McIntosh, G.H., Mano, M.T., Baghurst, P.A., Chavadej, J., Hua, C.H., Cragg, B.G. and Hetzel, B.S. The effect of maternal thyroidectomy prior to conception on fetal brain development in sheep. Acta Endocrinologica 112: 93-99 (1986).

22. Morreale de Escobar, G., Pastor, R., Obregon, M.J. and Escobar Del Rey, F. Effects of maternal hypothyroidism on weight and thyroid hormone content of rat embryonic tissues,before and after onset of fetal thyroid function. Endocrinology 117: 1890-1900 (1985).

23. DeLong, R. Neurological involvement in iodine deficiency disorders, in: "The Prevention and Control of Iodine Deficiency Disorders", B.S. Hetzel, J.T. Dunn and B.S. Stanbury eds. Elsevier, Amsterdam (1987) pp. 49-63.

24. Pharoah, P.O.D. Epidemiological studies of endemic cretinism in the Jimi River Valley in New Guinea, in: "Endemic cretinism", B.S. Hetzel and P.O.D. Pharoah eds, Papua New Guinea Institute of Human Biology Monograph Series (1971) 2: 109-116.

25. Hetzel, B.S. and Hay, I.D. Thyroid function, iodine nutrition and fetal brain development. Cl. Endo. 11: 445-460 (1979).

26. Pharoah, P.O.D., Connolly, K.J., Hetzel, B.S. and Ekins, R.P. Maternal thyroid function and motor competence in the child. Develop. Med. and Child Neurology 23: 76-82 (1981).

MODELS OF FETAL IODINE DEFICIENCY

Gabriella Morreale de Escobar, Carmen Ruiz de Oña,
María Jesús Obregón and Francisco Escobar del Rey

Unidad de Endocrinología Experimental, Instituto
de Investigaciones Biomédicas, C.S.I.C., Facultad de
Medicina, Arzobispo Morcillo 4, 28029 Madrid (Spain)

INTRODUCTION

An experimental model would obviously be of great importance to define the etiopathological factors leading to the neurological damage of the cretins, and to poor mental and psychomotor development of inhabitants of iodine-deficient areas. We shall only describe the rat model, as other animal models are being dealt with by Dr. Basil H. Hetzel.

The rat has been successfully used as a model of human congenital hypothyroidism since the pioneering work by Eayrs (for reviews[1,2]). Though the term "cretin" has been often applied to rats rendered hypothyroid at birth or soon thereafter, it should be realized that this may lead to considerable confusion, as such rats are not the animal counterpart of the neurological cretins from iodine-deficient areas, but of the "sporadic" congenital hypothyroid baby. Attempts to produce rats with CNS damage which could be attributed to iodine deficiency have seldom met with success. Van Middlesworth reported in 1977 [3] that 89 % of rats exposed to severe prenatal and postnatal iodine depletion suffered from severe audiogenic seizures, even if fed a normal diet after weaning as compared to 2 % in controls. The low iodine diet (LID) employed had an extremely low iodine content (0.006 µg / g) and is no longer available. Similar audiogenic seizures and partial deafness were later observed in rats born from mothers treated with a goitrogen between the end of gestation and until 16 days after birth [4]. The similarity of results suggested that the CNS lesion caused by iodine depletion could be attributed to fetal thyroid hormone deficiency prolonged during most of the suckling period, combined with maternal thyroid hormone deficiency. Unfortunately, direct data regarding fetal and neonatal thyroid hormone economy in the LID (or the goitrogen-treated rats) were not available.

More recently Chen et al. [5] fed female rats an LID with a very low iodine content (0.016 µg / g), prepared using maize and black soybean from an area of severe iodine deficiency and cretinism, or the same diet supplemented with iodine. Three months later they were mated and the first litter discarded. They were inmediately mated again and the second studied at 42 days of age. As compared to the animals on the iodine-supplemented LID, the iodine deficient progeny grew poorly, had a thin skull, disproportions in skull dimension, dry, short, scant and patchy hair, all of which are characteristic changes observed in rats rendered hypothyroid early in development. Learning ability was markedly impaired in 42 % of the offspring; there was a decrease in brain weight, DNA and protein contents, delayed myelination of the spinal cord and a decrease in the number of spines along the shaft of pyramidal neurons of the visual cortex. These rats indeed appear to be the experimental counterpart of the human endemic cretin, and the study constitutes a very important contribution.

It would be very useful for our understanding of the timing at which the damage to the CNS was produced, to obtain data regarding the degree of thyroid hormone deficiency in different fetal and neonatal tissues at different stages of development. It would also be important to define for how long the iodine-deficient situation has to be prolonged after birth for irreversible lesions to occur. We will here describe our attempts to clarify the first of these two points with respect to the fetal period of development; some data regarding postnatal development will be summarized in the contribution by Escobar del Rey et al., at this Meeting, and nave been reported in detail elsewhere [6]. We have not yet obtained cretin rats as those described by Chen at al. [5], or with the audiogenic seizures described by Van Middlesworth [3], and attribute this lack of success to the fact that our LID had a higher iodine content (0.02-0.06 μg / g).

It has been well established that neurological cretins are born in areas where the iodine intake is very low, and that correction of the iodine deficiency before, or very early in pregnancy,prevents their birth[7]. Women living in such areas are unable to raise their low serum T4 levels during pregnancy, a fact considered of causal importance for the birth of cretins[8]. The degree of maternal hypothyroxinemia has been correlated both to the frequency of congenital abnormalities, especially the birth of cretins[9,10], and to the poor psychomotor performance of the non-cretinous part of the population[11]. There were no correlations when maternal T3 levels were considered, as they are usually normal in severe iodine deficiency.

Damage caused by maternal hypothyroxinemia has usually been attributed to difficulties in maintaining adequate placental function, transfer of nutrients to the fetus, etc.[12]. This possibility is supported by the idea that the human placenta is impermeable to the iodothyronines[13,14]. However, it is also possible that maternal thyroid hormones do cross the placenta. A decreased contribution in cases of maternal hypothyroxinemia or hypothyroidism might have damaging effects on the developing embryo, especially during stages antedating onset of its own thyroid function. We shall here summarize information obtained in the rat model which might be pertinent to this problem. Special attention will be given to results which might answer to the following questions:

1) For how long does the fetus from an iodine deficient mother suffer from thyroid hormone deficiency?

2) Why is maternal hypothyroxinemia damaging to the fetus, even if maternal T3 levels are normal?

THYROID HORMONES IN EMBRYONIC TISSUES BEFORE ONSET OF FETAL THYROID FUNCTION

Work performed by Weiss and Novak [15] and by Sweney and Shapiro[16] strongly suggested the possibility that maternal thyroid hormones are available to the rat embryo before onset of fetal thyroid function, despite opinions to the contrary [17]. We decided to re-investigate this possibility by measuring T4 and T3 concentrations in embryonic and fetal tissues. We had developed specific and sensitive RIAs [18], and extensive extraction and purification

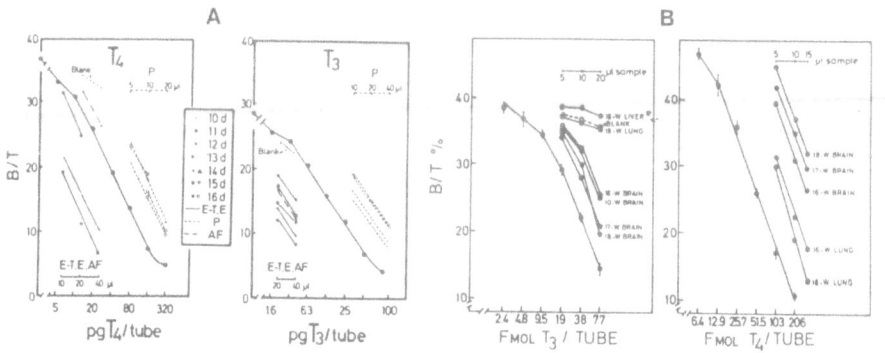

Figures 1 A and B. 1 A compares the T3 and T4 standard curves with serial
dilutions of eluates obtained from embryo-trophoblasts (E-T), embryonic (E)
placental (P) and amniotic fluid (AF) samples obtained at different gestational ages in the rat, shown in the inset. 1 B shows similar curves with
extracts from human fetal brains, between 10 and 16 weeks of gestation.
From Obregón et al.[19] and Bernal & Pekonen[27], with permission of publishers.

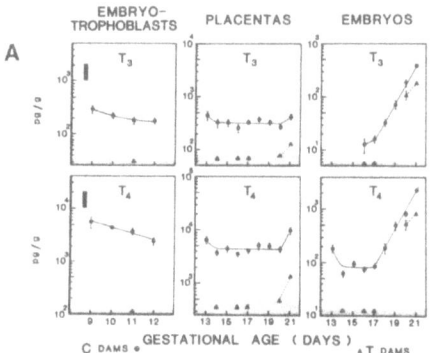

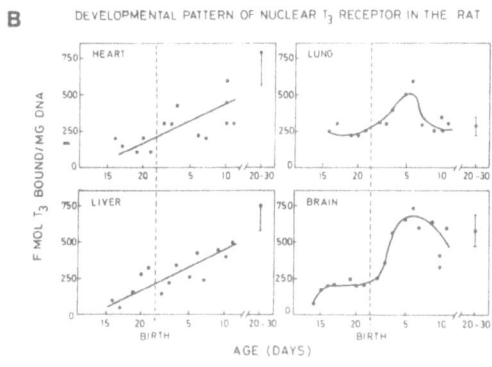

Figure 2 A and B. 1 A shows the concentrations of T4 and T3 found in 9- to
12- day old embryo-trophoblasts and in 13- to day old placentas and fetuses from C dams in black dots, and from T dams in triangles. The shaded
areas indicate the limits of detection. All differences between C and T
were significant. 1 B shows concentrations of nuclear T3 receptor in fetal tissues, also from rats. From Morreale de Escobar et al.[20] and Pérez-
Castillo et al.[26] with permission of the publishers.

procedures which permitted the determination of very small concentrations of these iodothyronines in tissues [19,20]. Different types of samples were obtained from the embryonic compartment, at different stages of development, and after extensive perfusion of the dams with phosphate buffered saline. The day of appearance of the vaginal plug was taken as gestational day (gd) = 0. With this dating, birth occurs at 22.7 ± 0.2 (SEM) gd. As summarized in Fig 1 A, all tissues obtained from the fetal compartment before onset of fetal thyroid function contained T4 and T3 [19]. They were found in molar ratios which were quite different from those of maternal plasma, thus excluding contamination with maternal blood. These results were in agreement with the simultaneous independent study by Woods et al. [21] showing transfer from mother to embryos of labeled iodothyronines early in gestation.

To confirm the maternal origin of the iodothyronines found in early embryonic samples, we determined their concentrations in fetal tissues obtained from normal (C) mothers and from dams which had been thyroidectomized (T) a few months before mating [20]. When the embryonic samples were obtained from T mothers, concentrations of both T4 and T3 were below the limits of detection in all tissues, placentas included, at least up to 17 gd (Fig. 2 A).

The extrathyroidal concentrations of T4 and T3 in whole fetuses (thyroid excluded) from normal dams do not increase until after 17 gd. This finding agrees with the report that the rat fetal thyroid starts secreting hormones at 17.5-18 gd [22]. Therefore, fetuses from T mothers develop in a thyroid hormone-deficient condition, at least up to 17.5-18 gd. Their number is markedly reduced (from 11.1 ± 0.4 to 6.8 ± 0.3), and the individual weight of the fetuses is smaller. Before onset of fetal thyroid function, fetal weight is reduced by maternal T to 59 % of C values [20]. Once fetal thyroid function starts,here appears to be a catch-up in growth, but even at 21 gd (near term) the body weight of fetuses from T dams is only 80 % [20,23] of normal. Bonet and Herrera [24] have recently described that maternal hypothyroidism during the first half of gestation results in a reduction of the pituitary growth hormone content of the fetuses to 31 % of that of fetuses from C dams, as measured at 21 gd. This decrease was only prevented by treating the T dams with T4 during the first half of gestation; treatment during the second half did not restore fetal pituitary GH. Permanent behavioral defects have been described in the progeny of T rats [25].

Studies by Pérez-Castillo et al. [26] have shown the presence of nuclear receptors for T3 in rat embryos as early as 13 gd, in brain by 14 gd, and in liver, heart and lung from 16 gd onwards (Figure 2 B). Therefore, both the hormone and its receptor are present before onset of fetal thyroid function, and it does not appear far-fetched to suggest that a biological effect might ensue, though this possibility has not yet been confirmed nor the possible effect defined.

There are similarities between the above findings in rats, and those reported for human fetuses. Bernal & Pekonen [27] have shown the presence of nuclear receptors for T3 in the brain of 10-weeks old human embryos. T3 was also found in their brain (Fig. 1 B). The receptor was later (personal communication by J. Bernal) identified in a whole 7 week-old human embryo, in which T4 was also detected; the amount of tissue was too small to attempt the detection of T3. Thus, both the receptor and the hormone are present in the human fetal brain at a stage of development when active neuroblast division is starting in the forebrain, and several weeks before onset of active thyroid function. After 10 weeks of gestation, and coinciding with the active phase of neuroblast division, there was a rapid increase in cerebral DNA; receptor concentration increased ten-fold by 16 weeks of gestation. The results show that there is maternal transfer of thyroid hormone in man before the beginning of the second trimester of pregnancy.

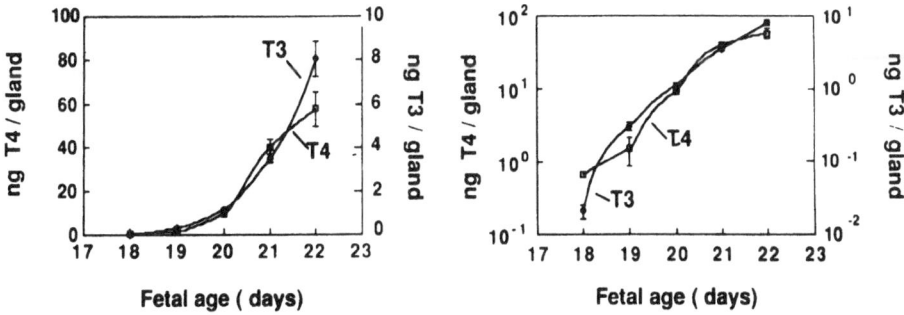

Figure 3. Concentrations of T4 and T3 in Pronase digests of fetal thyroids between 18 and 22 days of gestation. The right hand panel shows the data on a logaritmic scale. Unpublished data.

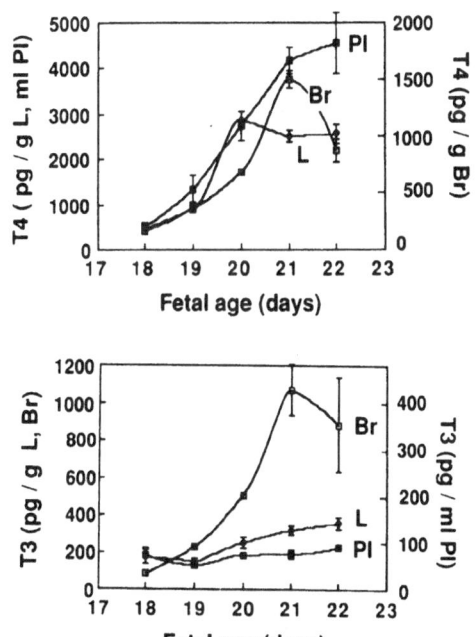

Figure 4. Concentrations of T4 and T3 in fetal plasma (Pl), liver (L) and brain (Br) between 18 and 22 days of gestation. Unpublished data.

[#] The fetus minus thyroid, blood, lung, liver, brain and interscapular brown adipose tissue pads.

Therefore, maternal hypothyroidism could result in decreased availability of thyroid hormone during the initial phases of development. It is known that maternal hypothyroxinemia results in lowered I.Q. of the progeny, even in areas without iodine deficiency[28].

THYROID HORMONES IN FETAL TISSUES AFTER ONSET OF FETAL THYROID FUNCTION

We have measured the content of T4 and T3 in the thyroid glands of fetuses from normal dams between 18 and 22 gd, using methods already described in detail[20]. Results are shown in Fig. 3. Total thyroidal content of T4 increased from 0.66 $\pm$ 0.07 (SEM) ng / gland at 18 gd to 57.7 $\pm$ 8.2 at 22 gd; data for T3 were 0.021 $\pm$ 0.005 ng / gland and 8.07 $\pm$ 0.78. The increases were about a hundred-fold for T4, and fourhundred-fold for T3. The T4 to T3 ratio was about 30 : 1 at 18 gd and 7 : 1 at 22 gd. When the mother had been T, the fetal T4 and T3 contents were the same as those of fetuses from C dams up to 20 gd, but then stopped increasing. By 22 gd the thyroidal T4 and T3 contents in fetuses from T dams were 24 % and 28 %, respectively, of those of fetuses from C dams. This finding confirms a previous observation at 21 gd [21] and suggests that the thyroid of the fetuses from T mothers is stimulated to secrete increasing amounts of hormones to compensate for the maternal contribution, which would be absent when the dam is T. Both the data and their interpretation agree with those of Gray and Galton[29].

T4 and T3 were measured in fetal plasma and tissues between 18 and 22 gd. During this period the concentrations of T4 increased almost ten-fold. In striking contrast to this, and despite the marked increase in thyroidal T3 content, the plasma concentration of T3 increased less than two-fold during the same period. Considering that low activities of outer ring iodothyronine deiodinating enzymes (5' D) had been reported in rat fetal tissues as compared to adult levels[30,31], it might have been concluded that tissue T3 levels would increase two-fold between 18 and 22 gd, whereas tissue T4 concentrations would increase ten-fold.

The approximately ten-fold increase in T4 concentration was indeed found for the brain, liver and carcass #. In contrast to this, only fetal liver T3 levels increased in parallel to plasma T3. In the brain, carcass and lung the increase was seven- to ten-fold. T4 and T3 concentrations in plasma, liver and brain are shown in Fig. 4. Linear regression coefficients were calculated for T4 and T3 concentrations over fetal age, after logarithmic transformation of the concentration data. The rates of increase were compared. There were no statistically significant differences between the rates of increase in brain, or liver T4, and plasma T4, nor between that of liver T3 and plasma T3. The rate of increase of brain T3 was significantly higher than that of plasma T3, and the same as that of plasma T4.

These results strongly suggested that the fetal brain does not depend on plasma T3 for its intracellular supply of this iodothyronine, but is deriving most of it by local generation from T4. This was confirmed when type II 5' D (5' D-II) activity was measured: there was a six-fold increase between 17 and 21 gd, with a slight decrease between 21 and 22 gd [32]. The important role of this enzyme in supplying the brain with increasing amounts of T3 during this period of development is also suggested by the fact that its activity increases in response to a decrease in fetal brain T4 concentrations as early as 17 gd; by 18 gd the increase in 5 ' D-II activity in response to hypothyroidism was three-fold[32].

The concentrations of T3 in the carcass and lung also increased at a much higher rate than plasma T3. We have tentatively attributed the findings for the carcass to its content of BAT from sources other than the interscapular

Table 1. Experimental design and treatments of the dams.

Group of dams :		MMI	T4
C	or T	-	-
C + MMI	T + MMI	+	-
C + MMI + T4	T + MMI + T4	+	+

MMI was given in drinking water, 0.02%, starting at 16 gd.
T4 was infused (osmotic micropump), 1.8 µg / 100 g BW . day,
from 17 dg.

pads of brown adipose tissue (BAT), as BAT 5' D-II is quite active in rat
fetuses[32,33]. But the rapid increase in lung T3 was totally unexpected, as
5' D-I activity is low in fetal lungs, as compared to adult levels[31].
Preliminary results, however, indicate that 5' D-I is already active in the
fetal lung, and increases about five- to ten-fold between 18 and 22 gd (Ruiz
de Oña et al., unpublished).

We have already briefly described the changes occurring in fetal
thyroidal T4 and T3 contents as a consequence of maternal T. These changes
suggest an increased secretion of T4 and T3 to compensate for the effects of
maternal T. These effects are not the same for all fetal tissues. In some of
them, T4 and T3 concentrations are lower in fetuses from T dams as compared
to those from C dams up to 21-22 gd. The brain, however, appears to be
preferentially spared from T4 and T3 deficiency throughout this period of
development, confirming previous results obtained at 21 gd[20].

CONTRIBUTION OF MATERNAL HORMONES AFTER ONSET OF FETAL THYROID FUNCTION.

We have summarized evidence that maternal thyroid hormones cross the
placenta into the fetal compartment early in gestation. Once the fetal
thyroid starts functioning, this transfer might decrease and eventually
stop, or it might continue until term. It is experimentally difficult to
investigate these possibilities. Labeled iodine-containing hormones found in
the fetal compartment after their injection into the mother do not
necessarily indicate transfer, as the fetal thyroid might have synthesized
them from labeled iodine deiodinated from the hormone administered to the
dam. Determinations of T4 and T3 in fetal tissues by RIA do not permit to
distinguish between maternal and fetal contributions. Fetal thyroid function
has to be blocked. But conclusions may then only be drawn for maternal-fetal
ralationships where the fetus is hypothyroid, and cannot be extrapolated to
normal conditions.

Table 1 outlines the experimental design which we used. Methimazole (MMI)
was used to block maternal and fetal thyroid function. Some of the
MMI-treated dams received a replacement dose of T4 given by constant
infusion, as described in detail elsewhere[23]. The main findings are
summarized in Fig. 5, which shows the T4 and T3 concentrations in fetal
carcass and brain. Comparison of the hormone levels in fetuses from C + MMI +
T4 (or T + MMI + T4) dams with those from C + MMI (or T + MMI) mothers shows
that infusion of T4 into the mothers not only ameliorates fetal deficiency
of T4, but also of T3. This occurs without an increase of fetal plasma T3,
suggesting that the fetal brain derived its T3 from local deiodination of
T4.

When the fetal thyroid is impaired (as in congenital hypothyroidism)
maternal T4 has a protective effect on the fetal brain and on other fetal
tissues[23]. This might mitigate adverse effects of fetal thyroid failure on
the developing brain. It appeared important to determine whether, or not, T3

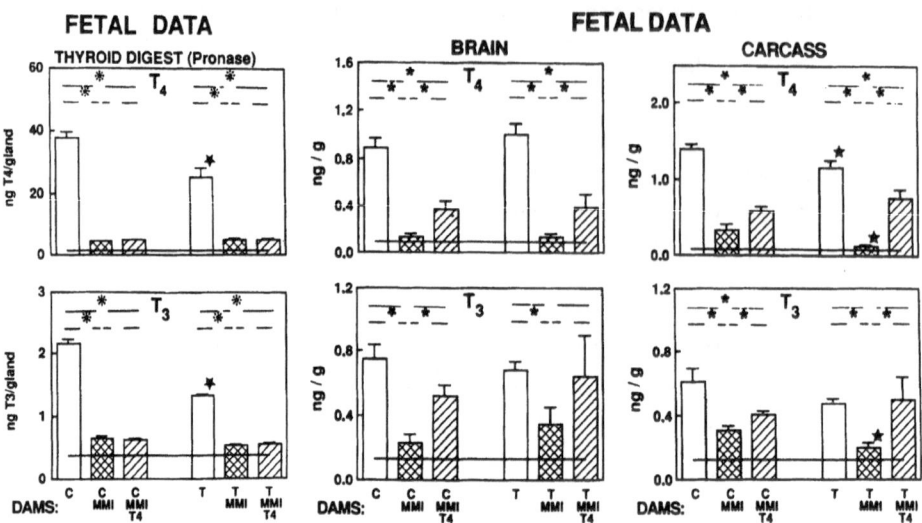

Figure 5.- Concentrations of T4 and T3 in fetal thyroid digests, and in the brain and carcass of fetuses from the six groups of dams described in Table 1. Data are means (+ SEM). Horizontal lines indicate the limits of detection. Asterisks between data bars indicate statistical significance of the difference. From Morreale de Escobar et al.[23], with permission from the publishers.

is also transferred from the mother to the fetus near term, also mitigating the effects of fetal hypothyroidism. An experiment was carried out[34] similar to the one outlined in Table 1, but with several modifications. MMI was started two days earlier, at 14 gd, and the infusion of hormone into the dams was started at 15 gd. Several MMI-treated dams were infused with T3 at a dose (0.45 μg / 100 g BW. day) equivalent to that of T4[35]. Data from the groups of dams infused with T4 confirmed those of the previous experiment[23] illustrated in Fig. 5 as regards the transfer of T4 and its effects on the T4 and T3 deficiency of fetal tissues, the brain included. The new finding was that infusion of T3 into the MMI-treated dams increased the concentration of T3 in fetal plasma (Fig. 6), and in all fetal tissues studied, except the brain (Fig. 7). MMI treatment had increased both the maternal (not shown) and fetal plasma TSH levels (Fig. 6). Infusion of both T4 and T3 decreased plasma TSH to normal levels in the mothers themselves. Infusion of T4 decreased fetal TSH, although it remained elevated as compared to normal fetuses. In contrast, the infusion of T3 had no effect

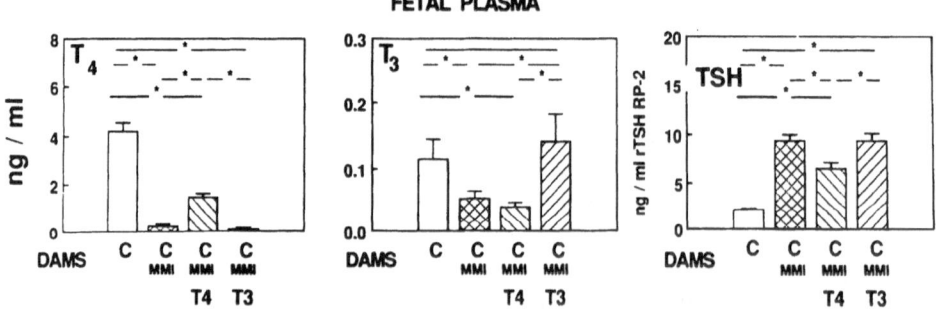

Figure 6.- Concentrations of T4, T3 and TSH in the plasma of fetuses from dams on MMI, or MMI + T4, or MMI + T3. From Morreale de Escobar et al.[34]

on fetal plasma TSH. Data shown in Figs. 6 and 7 correspond to the groups from C dams only; results from the four groups of fetuses from T dams were very similar.

Previous data summarized here have suggested an important role of local generation of T3 from T4 in providing the developing brain with increasing amounts of T3. The results of this last experiment show that the brain of rat fetuses near term is entirely dependent on T4 availability, as T3 does not enter the brain, at least at physiological plasma T3 levels. Preliminary results have been obtained in fetuses from MMI-treated dams infused with doses of T3 which were 3 and 10 times the dose used for the experiment illustrated in Figs. 6 and 7. These results show a dose-related transfer of T3, accompanied by a dose-related increase of T3 in several fetal tissues. But again, fetal plasma TSH was not affected, even at the highest T3 dose, and there was no increase in cerebral T3 above the level found in MMI-treated fetuses. Recent studies on the role of the choroid plexus in the transport of T4 and T3 into the brain show that after injection of labeled thyroid hormones, the choroid plexus and the brain avidly accumulate T4, but not T3 [36].

5' D-II activities were measured in the cerebral cortex of the fetuses corresponding to Fig. 6. Activity increased with MMI treatment, in response to thyroid hormone deficiency. Infusion of T4 into the mothers decreased enzyme activity markedly, although not to control levels, whereas infusion of T3 did not. The 5' D-II activities were inversely correlated to the T4 and T3 concentrations of the fetal brains[32], showing that the changes in thyroid hormone level were accompanied by a biological effect.

All these findings show that maternal thyroid hormones cross the placenta late in gestation also, at least when the fetal thyroid is impaired, and despite active deiodination by the placenta[37]. Maternal T4 has a more important role, as it mitigates T3 deficiency of the brain, whereas maternal T3 does not. Neither does maternal T3 decrease fetal TSH secretion. We have already discussed more extensively the possibility that these results are relevant to human pregnancies[23,38]. Transfer of maternal thyroid hormones to athyreotic fetuses would explain that in some of them cord-blood T4 levels are detectable, and then decrease rapidly. It would also explain the frequent absence of overt signs of hypothyroidism at birth. Part of the conclusion that the human placenta at term is virtually impermeable to the iodothyronines is based on studies[39,40] where normal women were given T3 near term, and some parameter related to fetal TSH secretion was measured. The data obtained in the rat suggest that this might be a misleading end-point to assess transfer of T3. Moreover, the results should not be extrapolated to T4.

THE IODINE-DEFICIENT RAT FETUS

Female rats were fed a diet with a low iodine content (LID; 0.02-0.06 μg / g diet) or the same LID supplemented with KI (LID + I; 7-10 μg I / rat.day) for 5-6 weeks, after which they were mated with normal males.

Embryonic and fetal samples were obtained before (at 11 and 17 gd) and after (21 gd) onset of fetal thyroid function[41]. Maternal plasma T4 was very low in the LID dams, and could only be measured after extraction of larger samples than used for the direct RIA. T3 levels were similar to those of LID + I dams, and decreased in both groups at the end of pregnancy. Length of gestation was unaffected, but the number of fetuses was decreased in the LID group, from (11.2 $\pm$ 0.6 to 8.8 $\pm$ 0.6). Their weight, as measured at 21 gd, was decreased, and their brains were smaller. Cerebral DNA and protein contents were lower.

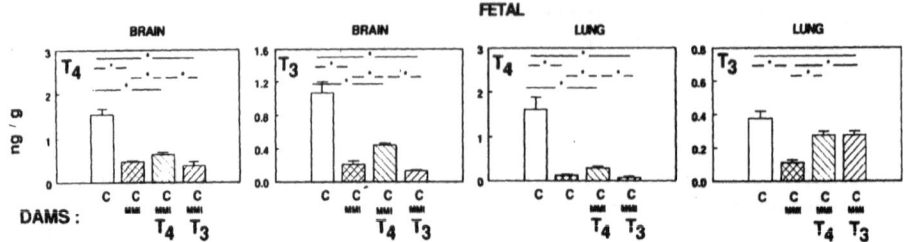

Figure 7. Concentrations of T4 and T3 in the brain and lung of the same fetuses as those of Fig. 6. Data from Morreale de Escobar et al[34].

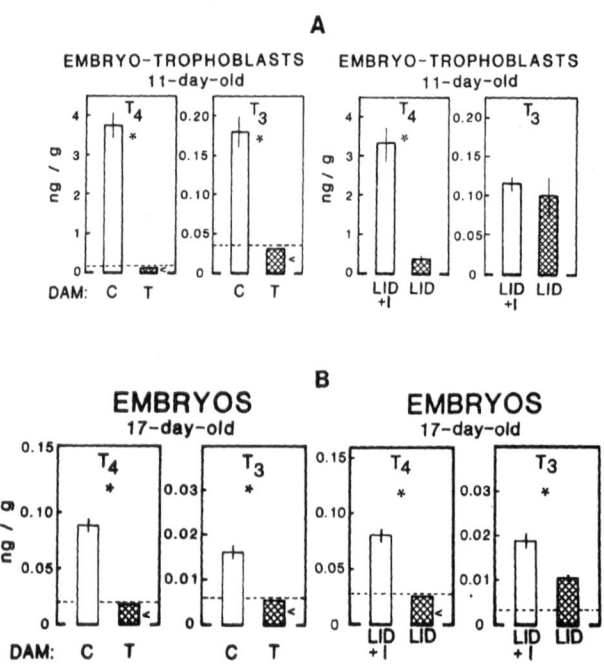

Figure 8. Concentrations of T4 and T3 in 11-day old embryo-trophoblasts and 17-day old embryos from LID, LID + I, T and C dams. Dotted horizontal lines indicate the limits of detection. Data are from Morreale et al.[20] and from Escobar del Rey et al.[6] . Figures are reproduced with permission of Plenum Press, from Frontiers in Thyroidology (1986).

The LID fetuses developed goiter, and their thyroidal total iodine was 4.7 % of that of LID + I fetuses[6]. The concentrations of T4 and T3 in different embryonic and fetal samples are shown in Figs. 8 and 9, where they are compared to data from age-paired samples obtained from C and T dams. Before onset of fetal thyroid function (11-day-old embryotrophoblasts and 17-day old fetuses) T4 concentrations are decreased by LID to a degree comparable to that of concepta taken from T mothers. T3 concentrations did not differ initially from those of the LID + I group. By 17 gd, however, T3 concentrations were lower both in the fetus, and placenta (not shown). Once fetal thyroid function starts, important differences become apparent between fetuses from T and LID dams. The activated secretion of T4 and T3 by the thyroid in fetuses from T mothers is able to compensate for previous differences related to maternal hypothyroidism, at least as far as the brain is concerned. But this is not possible for fetuses faced with a very low iodine supply, and cerebral T3 is quite low. Similar results were later obtained in another experimental series[6].

In a more recent experiment (Obregón et al., unpublished) cerebral T3 concentrations at 21 gd were reduced to 12 % of the values found for LID + I fetuses, a value which is even lower than in MMI-treated fetuses. This occurred despite a five- to ten-fold increase in cerebral cortex 5' D-II activity.

After birth, there is a relative increase of iodine availability through maternal milk which may result in an increase of plasma and cerebral T4 sufficient to maintain normal brain T3 levels throughout most of the suckling period [6] , as 5' D-II remains elevated in LID pups after birth (Obregón et al., unpublished). After the suckling period, cerebral T3 decreases again as compared to that of LID + I rats. If the iodine intake were lower, however, as in the experimental groups studied by Van Middlesworth[3] and by Chen et al.[5], this postnatal normalization of cerebral T3 levels might not occurr.

The situation found in iodine-deficient human and sheep fetuses are described at this Meeting in the contributipons by Ma Tai, Liu Jia-Liu, and B. S. Hetzel.

GENERAL COMMENTS

Implicit in many of our comments is the concept that thyroid hormones do play a role during fetal life. Direct evidence regarding a role in early brain development is still lacking. But thyroid hormones do reach the fetal compartment, and both T3 and its nuclear receptors are found in the brain early in development. We believe that statements such as "Thyroid hormone is neither available nor required during fetal life" made referring to neurogenesis[17], are no longer tenable for the rat, and probably neither for man. The early timing of the CNS damage of the neurological cretin also constitutes indirect evidence of a role of thyroid hormone in early human brain development.

It is also of interest that in fetal sheep brain, there is not only an increase of T3 receptor concentration between 50 and 100 gd, but receptor occupancy is markedly increased at 100 gd[42], when active neuroblast division is starting. Three tissues which are undergoing important maturational processes in the rat near birth, namely the brain, lung and BAT, are endowed with mechanisms to ensure an increasing supply of T3 during a period when plasma T3 is hardly increasing. When this is prevented, as in the hypothyroid rat fetus, the number of beta-adrenergic receptors of the brain decrease, as measured by Smith et al.[43] at 21 gd. The lung is maturing to face breathing at birth, and Hitchcock[44] has shown acceleration of this process by intra-amniotic injection of T4. BAT is also undergoing rapid maturation geared to the production of thermogenin. Obregón et al.[45] have found that thermogenin mRNA is expressed in fetal BAT by 21 gd, and decreases to 30 % of normal values when the fetus is hypothyroid.

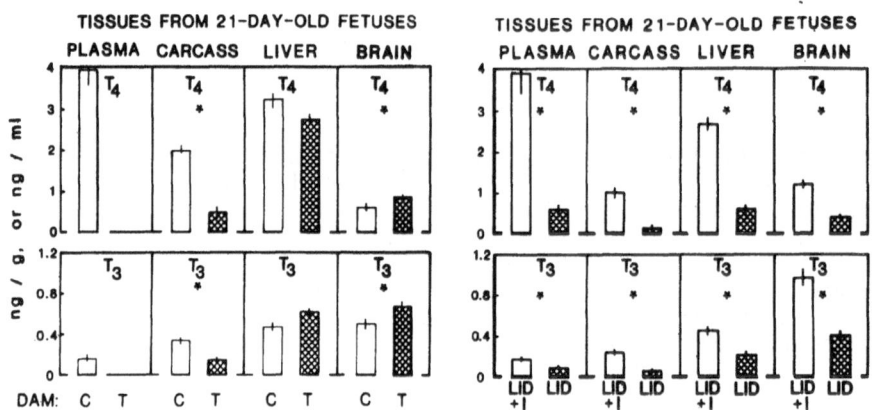

Figure 9.- Concentrations of T4 and T3 in plasma and tissues from 21-day old fetuses from the same groups of dams as in Figure 8. Data for plasma T4 and T3 in fetuses from T dams had not been obtained. Figure reproduced with permission of Plenum Press, Frontiers in Thyroidology (1986).

As regards the first of the questions outlined in the introduction, we now know that the LID rat fetus is markedly T4-deficient throughout all of gestation, to a degree comparable to that of fetuses from T mothers. T3 is initially normal, but later decreases, and the brain is T3 deficient during a period of active neurogenesis, and this might have damaging effects on fetal brain maturation

As regards the second question, results described here might afford a possible explanation for the adverse effects of maternal hypothyroxinemia on the CNS of the progeny, even if maternal T3 levels are normal. This is a frequent finding in women with severe iodine deficiency[8-10]. If production of fetal thyroid hormones is impaired because of inadequate iodine supply, and the mother also has very low plasma T4, the brain would be totally deprived of T3, despite marked increases in cerebral 5' D-II. The normal maternal plasma T3 levels cannot mitigate this cerebral T3 deficiency, because the T3 transferred from the mother does not enter the fetal brain. However, the normal maternal levels of T3 might afford some protection to other fetal tissues, which would be in a better situation than if maternal T4 and T3 were both low. This might also explain why at birth babies from such areas do not present the clinical signs usually associated with hypo-thyroidism. If the rat data may be extrapolated to the human situation, the cord-blood TSH of the iodine-deficient babies should be higher than normal, and their brain deficient in T3, even if their plasma T3 is within the normal range.

REFERENCES

1. Eayrs JT, Influence of thyroid hormone on the central nervous system, Brit Med Bull 16:122 (1960)
2. Morreale de Escobar G, Ruiz-Marcos A, Esobar del Rey F, Thyroid hormones and the developing brain, in: "Congenital Hypothyroidism", Dussault J, Walker P (eds), Marcel Dekker, New York (1983), p 85
3. Van Middlesworth, L, Audiogenic seizures in rats after severe prenatal and perinatal iodine depletion, Endocrinology 100:242 (1977)
4. Van Middlesworth L, Norris CH, Audiogenic seizures and cochlear damage in rats after perinatal antithyroid treatment, Endocrinology 106:1686 (1980)
5. Chen ZP, Chen XX, Dong L, Hu X, Zhao WD, Wang D, Ma T, The iodine deficient rat, in: "Iodine Deficiency Disorders and Congenital Hypothyroidism", Medeiros-Neto G, Maciel RMB, Halpern A (eds) ACHE Press, Sao Paolo (1986) p 46
6. Escobar del Rey F, Pastor R, Mallol J, Morreale de Escobar G, Effects of maternal iodine deficiency on the L-Thyroxine and 3,5,3'-Triiodo-L-thyronine contents of rat embryonic tissues before and after onset of fetal thyroid function, Endocrinology 118:1259 (1986)
7. Hetzel BS, Iodine deficiency disorders (IDD) and their eradication, Lancet 2:1126 (1983)
8. Choufoer M, van Rhijn M, Querido A, Endemic goiter in Western New Guinea. II. Clinical picture, incidence and pathogenesis of endemic cretinism, J Clin Endocrinol Metab 25:385 (1965)
9. Pharoah POD, Ellis SM, Ekins RP, Williams ES, Maternal thyroid function, iodine deficiency and fetal development, Clin Endocrinol 5:159 (1976)
10. Pharoah POD, Connolly K, Hetzel BS, Ekins RP, Maternal thyroid function and motor competence in the child, Develop Med Child Neurol 23:76 (1981)
11. Connolly KJ, Pharoah POD, Hetzel HB, Fetal iodine deficiency and motor performance during childhood, Lancet ii:1149 (1979)
12. McMichael AJ, Potter JD, Hetzel BS, Iodine deficiency, thyroid function and reproductive failure, in: " Endemic Goiter and Endemic Cretinism", Stanbury JB, Hetzel BS (eds), John Wiley & Sons, New York, (1980) p 445
13. Fisher DA, Dussault JK, Sack J, Chopra IJ, Ontogenesis of hypothalamic-pituitary-thyroid function in man, sheep and rat, Rec Progr Horm Res 33:59 (1977)
14. Fisher DA, Klein AK, Thyroid development and disorders of thyroid function in newborn, New England J Med 304:702 (1981)
15. Weiss RM, Noback CR, The effects of thyroxine and thiouracil on the time of appearance of ossification centers of rat fetuses, Endocrinology 45:389 (1949)
16. Sweney LR, Shapiro BL, Thyroxine and palatal development in the rat, Dev Biol 42:19 (1975)
17. Hamburgh M, The role of thyroid and growth hormone in neurogenesis, in: "Current Topics in Developmental Biology", Moscona AA, Monroy A (eds) Academic Press, New York, vol 4:109 (1969)
18. Obregón MJ, Morreale de Escobar G, Escobar del Rey F, Concentrations of triiodothyronine in the plasma and tissues of normal rats as determined by radioimmunoassay: Comparison with results obtained by an isotopic equilibration technique, Endocrinology 103:2145 (1978)
19. Obregón MJ, Mallol J, Pastor R, Morreale de Escobar G, Escobar del Rey F, Thyroxine and triiodothyronine in rat embryos before onset of fetal thyroid function, Endocrinology 114:305 (1984)

20. Morreale de Escobar G, Pastor R, Obregón MJ, Escobar del Rey, F, Effects of maternal hypothyroidism on the weight and thyroid hormone content of rat embryonic tissues, before and after onset of fetal thyroid function, Endocrinology 117:1890 (1985)
21. Woods RJ, Sinha AK, Ekins R, Uptake and metabolism of thyroid hormones by the rat foetus in early pregnancy, Clin Sci 67:359 (1984)
22. Nataf B, Sfez M, Debut du fonctionnement de la thyroide foetale du rat, Compt ren Soc Biol 156:1235 (1961)
23. Morreale de Escobar G, Obregón MJ, Ruiz de Ona C, Escobar del Rey F, Transfer of thyroxine from the mother to the rat fetus near term: Effects on brain 3,5,3'-triiodothyronine deficiency, Endocrinology 122:1521 (1988).
24. Bonet B, Herrera E 1987 Different responses to maternal hypothyroidism during the first and second half of gestation in the rat. Endocrinology 122:450 (1988)
25. Hendrich CE, Jackson WJ, Porterfield SP, Behavioral testing of progenies of Tx (hypothyroid) and growth hormone-treated Tx rats: an animal model for mental retardation, Neuroendocrinology 38:429 (1984)
26. Pérez-Castillo A, Bernal J, Ferreiro B, Pans T, The early ontogenesis of thyroid hormone receptor in the rat fetus, Endocrinology 117:2457 (1985)
27. Bernal J, Pekkonen F, Ontogenesis of the nuclear 3,5,3'-triiodothyronine receptor in the human fetal brain, Endocrinology 114:677 (1984)
28. Man EB, Serunian SA, Thyroid function in human pregnancy. IX. Development or retardation of 7-year- old progeny of hypothyroxinemic women, Am J Obst Gynecol 125:949 (1976)
29. Gray B, Galton VA, The transplacental passage of thyroxine and foetal thyroid function in the rat, Acta Endocrinol (Kbhvn) 75:725 (1974)
30. Harris ARC, Fang S, Prosky J, Braverman LE, Vagenakis AG, Decreased outer ring monodeiodination of thyroxine and reverse triiodothyronine in the fetal and neonatal rat, Endocrinology 103:2216 (1978)
31. McCann UD, Shaw EA, Kaplan MM, Iodothyronine deiodination reaction types in several rat tissues: effects of age, thyroid status, and glucocorticoid treatment, Endocrinology 114:1515 (1984)
32. Ruiz de Oña C, Obregón MJ, Escobar del Rey F, Ontogenesis of 5' deiodinase activity in fetal brain and brown adipose tissue. Effects of maternal and fetal hypothyroidism, Annales d'Endocrinologie 48: 176 A (1987)
33. Iglesias R, Fernández JA, Mampel T, Obregón MJ, Iodothyronine 5' deiodinase activity in brown adipose tissue during development, Biochim Biophys Acta 923:233 (1987)
34. Morreale de Escobar G, Obregón MJ, Ruiz de Oña C, Comparison of maternal to fetal transfer of T3 versus t4 in rats, as assessed by T3 levels in fetal tissues, Annales d'Endocrinologie 48:178 A (1987)
35. Pittman CS, Pittman JA, Relation of chemical structure to the action and metabolism of thyroactive substances, in: Handbook of Physiology, Greer ME, Solomon DH (eds)., Section 7, volume III, Thyroid,, American Physiological Society, Washington, DC, pp 233 (1974)
36. Dickson PW, Aldred AR, Menting JGT, Marley PD, Sawyer WH, Schreiber G, Thyroxine transport in choroid plexus, J Biol Chem 262:13907 (1987)
37. Roti E, Gnudi A, Braverman LE, The placental transport, synthesis and metabolism of hormones and drugs which affect thyroid function, Endocrine Rev 4:131 (1983)
38. Morreale de Escobar G, Obregón MJ, Escobar del Rey F, Fetal and maternal thyroid hormones. Horm Res 26:12 (1987)
39. Raiti S, Holzman GB, Scott RL, Blizzard RM, Evidence for the placental transfer of tri-iodothyronine in human beings. New Engl J Med 277:456 (1967)
40. Dussault JH, Row VV, Lickrish G, Volpé R, Studies of serum triiodothyronine concentration in maternal and cord blood: Transfer of triiodothyronine across human placenta. J Clin Endocrinol 29:595 (1969)

41. Escobar del Rey F, Pastor R, Mallol J, Morreale de Escobar G, Effects of maternal iodine deficiency on the L-Thyroxine and 3,5,3' -Triiodo-L-Thyronine contents of rat embryonic tissues before and after onset of fetal thyroid function, Endocrinology 118:1259 (1986)
42. Ferreiro B, Bernal J, Morreale de Escobar, Potter B, Preferential saturation of brain T3 receptor during development in fetal lambs, Endocrinology 122:438 (1988)
43. Smith RM 1981 Thyroid hormones in brain development. In: Hetzel BS, Smith RM (eds) Fetal Brain Disorders, Elsevier North Holland Publishing Co, Amsterdam p 149
44. Hitchcock KR, Hormones and the lung.I.Thyroid homones and glucocorticoids in lung development. Anat Rec 194:15 (1979)
45. Obregón MJ, Pitamber R, Jacobsson A, Nedergaard J, Cannon B, Euthyroid status is essential for the perinatal increase in thermogenin mRNA in brown adipose tissue. Biochem Biophys Res Comm 148:9 (1987)

SCREENING FOR CONGENITAL HYPOTHYROIDISM; BENEFICIAL EFFECTS ON

NEUROPSYCHOLOGICAL DEVELOPMENT

J. H. Dussault and J. Glorieux

The Quebec Network for Genetic Medicine
CHUL
Ste-Foy, Quebec

INTRODUCTION

Untreated congenital hypothyroidism is characterized by neurologic and mental retardation[1],[3]. Recent evidence has been interpreted to indicate that most of these infants can attain normal development if treatment is instituted before age 3 months. On the other hand, numerous reports have shown that some infants have some degree of impaired psychologic and neuromuscular function at later ages, even when therapy has been started early[4],[8]. The Quebec Network for Genetic Medicine has been screening every newborn infant for congenital hypothyroidism since April, 1974, using filter paper blood spot thyroxine and thyroid-stimulating hormone measurements[9], thereby allowing early detection and treatment of affected infants.

MATERIALS AND METHODS

Infants with congenital hypothyroidism were detected through the Quebec Network for Genetic Medicine Screening Program. The diagnosis was confirmed from measurements of serum TSH, T_4, and thyroxine-binding globulin (TBG) obtained at approximately 3 weeks of age and treatment was started immediately with sodium L-thyroxine (8-10 µg/kg/d). Because there was no statistical difference in their psychological global and specific scores, infants with thyroid agenesis, hypoplasia, or ectopic tissue were included in this study, the diagnosis being retrospectively confirmed by a thyroid scan at 3 years of age. Girls outnumbered boys by a ratio of two to one. Adequacy of therapy and compliance was assessed by regular physical examination and serum T_4-TSH measurements. All treated patients had subsequent normal growth, physical development, and hormonal control measurements.

Their development has been assessed by the same two psychologists from the beginning, which permitted a close relationship with the patient and their families. Only those children who strictly complied with treatment were enrolled in the study. Evaluation of mental development was assessed by the Griffiths Mental Development Scales[10], which give a global quotient and five or six specific scales that permit evaluation of and comparison between locomotion, social and personal development, audition and language, fine coordination, performance, and practical reasoning. At each level (1 1/2, 3, 5 years) a control group was constituted and selected

according to the following criteria: age, sex, rank in the family, ethnic
origin, mother's occupation, and Hollingshead Index (socioeconomic
status)[11]. At age 7 years the children were evaluated with the Wechsler
Intelligence Scales for Children (Revised)[12], using the North American
norms.

RESULTS

In absolute value, our results (Table I) shows that our two groups of
subjects (hypothyroid: H and control: C) have a mean DQ above 100 which is
considered normal. At age one there is no difference between the
hypothyroid and control children whereas starting at 18 months up to five
year differences in certain scales are appearing and are significant for
the global scores. Using the Hoteling T2 statistical test, discrimination
occurs between the two groups specifically at age five for practical
reasoning and non verbal reasoning. No differences are observed between
the two groups at age 7. Thus we can conclude that the global performance
of the hypothyroid children is very satisfying with certain restrictions in
regard to specific scales. In search of a causal factor that might explain
these differences we looked at different parameters: diagnosis (agenesis
or ectopic gland), age of initiation of therapy, the clinical index and
could not find any correlation. On the other hand looking at the T_4
concentration at birth and the bone surface at diagnosis (3-4 weeks of age)
one could predict reasonably well a sub group which is at risk of
developing manifestations of minimal brain damage as expressed by the
various scales by which the infants were tested. (Table II) In summary a
newborn with a T_4 lower than 2µg/dl with no bone surface has more chance to
have abnormal results. This predictive index is very useful since it will
determine which infant should have a psychological follow-up and its
integration and various specific stimulation programs.

Finally school integration was evaluated. Preliminary results from
45 of our patients who were admitted in primary school are not different
than those obtained by their schoolmates: 3 needed adapted schooling, 2
had to repeat their year and 9 have needed orthopedagogy in either french
or mathematic. It should be noted that out of those 14 subjects, 9 were
classified at "risk" as defined earlier.

TABLE I

Age (year)	N	Locomotion	Social	Language	Coordination	Reasoning (non verbal)	Reasoning (Practical)	Global DQ	Range
1	H: 45	119	111	111	111	117	-	115 ± 1.5	88-133
	C: 37	119	111	111	113	119	-	115 ± 1.4	102-139
1.5	H: 98	109	104*	100*	104	105*	-	105 ± 1.2*	64-122
	C: 41	112	108	108	107	114	-	110 ± 1.0	94-121
3	H: 69	102*	110	102*	99	105*	90*	101 ± 2.3	55-140
	C: 40	111	118	128	107	116	103	114 ± 2.2	82-139
5	H: 51	100	114	106	101	100*	85*	101 ± 1.8*	60-123
	C: 45	105	116	112	105	110	93	107 ± 1.4	83-128
7	H: 44			92		102		97 ± 2.2	49-128
	C: -			100		100		100	

*: p < .01

TABLE II

Global WISC-R IQ at the age of 7 years

	Number of children with IQ	
	< 90	> 90
B.S. < 0.05 cm² and T4 < 2 µg/dL	10	6
B.S. < 0.05 cm² and/or T4 > 2 µg/dL	3	24
Total	13	30

From: "Useful parameters to predict the eventual mental outcome of hypothyroid children", Pediatric Research in press with permission.

DISCUSSIONS

Smith et al.[13] first noted the importance of initiating treatment within the first few months of age in infants with hypothyroidism to maximize the probability of achieving a normal IQ. They reported improved mental prognosis in congenital hypothyroidism when treatment was started before 7 months of age; almost 50% of the infants had an IQ >90. Raiti and Newns[1] and Klein et al.[2] found an even better result when infants received treatment before 3 months of age, with 85% achieving an IQ >85. With the advent of screening programs, which permitted initiation of therapy before 1 month of age, these results were corroborated in prospective studies; almost 90% of affected infants achieved an IQ >90 at 3 to 4 years of age[14],[15].

Our data confirm these results in children who have reached school age. Only 10% have DQ <90, with the performance and practical reasoning scales being the most discriminant.

At age 7 years the mean quotients are not different from the accepted norm. In our patients, two factors may have been influential in the good results: professional guidance was started as soon as problems were noticed, and the group of patients with the lowest scores have not yet attained the age of 5 or 7 years.

Throughout the last decade numerous groups have also been involved in similar research programs. Publications from the New England Collaboration Study have shown that in their cohort of patients there was no difference between the IQ of both groups with no correlation between the biological parameters and the various psychological indexes[16],[17]. It is the only group that has such an optimistic outlook on the eventual outcome of these patients. On the other hand Rovet et al.[19],[20] come to similar conclusions as our own using siblings as controls. In Europe, through the European Society of Pediatric Endocrinology numerous questionnaires were sent to various countries and an effort made to collate all the results. Although the results confirm the efficiency of screening, there are so many diverse methods of evaluation and ages of evaluation that no firm conclusions can be drawn from that collaborative study.

In summary, we can state that screening for congenital hypothyroidism has been very beneficial but there is still much work to be done to come to a final conclusion. We think that the recommendations stated at the last International Symposium on screening are very appropriate[22]. "The

observation that there still exists a relationship between factors
associated with hypothyroidism and developmental measures, however,
suggests that children are still affected, but to a much lesser degree.
This implies there is still a need for further research in this area.
Issues that should be studied include identifying the important risk
factors, studying the children at school-age, determining optimal treatment
levels. By assessing different screening and management practices in terms
of outcome, guidelines for the optimal development of these children could
be provided."

REFERENCES

1. S. Raiti and G.H. Newns, Early diagnosis and its relation to mental
   prognosis, Arch. Dis. Child. 46:692 (1971).
2. A.H. Klein, S. Meltzer, and F.M. Kenney, Improved prognosis in
   congenital hypothyroidism treated before age 3 months, J. Pediatr.
   89:912 (1972).
3. New England Congenital Hypothyroidism Collaborative, Effects of
   neonatal screening for hypothyroidism: Prevention of mental
   retardation by treatment before clinical manifestations, Lancet
   2:1095 (1981).
4. J. Maenpaa, Congenital hypothyroidism: Etiological and clinical
   aspects, Arch. Dis. Child. 47:914 (1972).
5. P.J. Collipp, S.A. Kaplan, M.D. Kogut, W. Tasem, F. Plachte, V.
   Schlamm, D.C. Boyle, S.M. Ling, and R. Koch, Mental retardation in
   congenital hypothyroidism: Improvement with thyroid replacement
   therapy, Am J. Ment. Defic. 80:432 (1965).
6. H.J. Andersen, in: "Endocrine and genetic diseases of childhood,"
   ed. 2, W.B. Saunders Co., Philadelphia, p. 238 (1975).
7. G.A. Von Harnack and H. Wallis, Zur psychopatholgie der
   hypothyreose ' im Kindesalter, Monatsschr Kinderheilkd 108:373
   (1960).
8. S. Zabransky, R. Richter, F. Hanefeld, B. Weber, and H. Helge, Zur
   prognose der angeborenen hypothyreose. Psychopatholgische befunde
   bein 30 langzeitbehandelten kindern, Monatsschr Kinderheilkd
   123:475 (1975).
9. J.H. Dussault, J. Morissette, J. Letarte, H. Guyda, and C. Laberge,
   Modification of a screening program for neonatal hypothyroidism, J.
   Pediatr. 92:274 (1978).
10. R. Griffiths, "The abilities of young children," Somerset: Young & Son
    (1970).
11. A.B. Hollingshead, "Two factor index of social position." Yale
    University Press, New Haven, Conn. (1965).
12. D. Wechsler, "Wechsler Intelligence Scales for Children (Revised,"
    Psychological Corp., New York (1974).
13. D.W. Smith, R.M. Blizzard, and L. Wilkins, The mental prognosis in
    hypothyroidism of infancy in childhood, Pediatrics 19:1011 (1957).
14. J.H. Dussault, J. Glorieux, J. Letarte, and H. Guyda, Preliminary
    report on psychological development at age one of treated
    hypothyroid infants detected in the Quebec Screening Network for
    Metabolic Disorder, Pediatr. Res. 12:412 (1978).
15. New England Congenital Hypothyroidism Collaborative. Characteristics
    of infantile hypothyroidism discovered on neonatal screening, J.
    Pediatric. 104:539 (1984).
16. R.Z. Klein, Infantile hypothyroidism then and now: the results of
    neonatal screening, Curr, Probl. Pediatr. 15:1 (1985).
17. New England congenital hypothyroidism collaboration, Screening for
    congenital hypothyroidism, Lancet 403 (1986).
18. J.F. Rovet, D.L. Westbrook, and R.M. McErlich, Neonatal thyroid
    deficiency: early temperamental and cognitive characteristics,
    J. Am. Acad. Child. Psychiatry 1:10 (1984).

19. J. Rovet, R. Erlich, and D. Sorbara, Intellectual outcome in children with fetal hypothyroidism, J. Pediatr. 5:700 (1987).

20. J. Rovet, D.L. Sorbara, and R. Erlich, The intellectual and behavioral characteristics of children with congenital hypothyroidism identified by neonatal screening in Ontario, in: "The Toronto prospective study, Genetic Disease Screening and Management," Alan R. Liss Inc., New York, P. 281 (1986).

21. R. Illig, R.H. Largo, Q. Qin, T. Torresani, P. Rochiccioli, and A. Larsson, Mental development in congenital hypothyroidism after neonatal screening, Arch. Dis. Child. 62:1050-1055 (1987).

22. J. Rovet, J. Glorieux, and S. Heyerdahl, Summary of presentation and discussion of the psychological follow-up of congenital hypothyroid children identified by newborn screening, in: "Advances in Neonatal Screening," B.L. Therrel Jr. Eds: Elsevier Science Publishers B.V., Amsterdam - New York, P. 71 (1987).

PLACENTAL DEIODINATION OF THE THYROID HORMONES

Lewis E. Braverman

University of Massachusetts Medical School
Worcester, Mass.

Deiodination of the thyroid hormones in peripheral tissues is a well established phenomenon which affects both the tissue and plasma concentrations of these hormones and their action at the cellular level. For example, the rate of 5' or outer ring PTU sensitive type I deiodination of T4 is a major factor in the regulation of the plasma triiodothyronine (T_3) concentration in the euthyroid state and PTU insensitive Type II 5'-deiodination of T_4 plays a major role in the generation of nuclear T_3 in brain, pituitary and brown adipose tissue (BAT).[1] Inner ring or 5-deiodination of T_4 and T_3 is present in highest concentrations in the placenta and brain and serves primarily to inactivate T_4 and T_3 by generating iodothyronines with little or no known metabolic activity.[1] Since the placenta is intimately related to the transport of iodothyronines from mother to fetus and receives a relatively large proportion of the fetal cardiac output, it seemed likely that the placenta would play important roles in regulating the passage of thyroid hormones across the placenta, in fetal thyroid hormone metabolism and action, and in determining the concentration of thyroid hormones in amniotic fluid. In 1981 and 1982, we first reported that the rat and human placentas are rich in 5-deiodinase activity[2,3] and have continued to be concerned with the possible importance of this activity in affecting maternal and fetal thyroid hormone metabolism and those factors which might regulate the enzyme's activity.

In the human, the concentration of reverse T_3 (rT_3), an iodothyronine generated from 5-deiodination of T_4, in amniotic fluid (AF) far exceeds its concentration in maternal serum.[4] Rat models of fetal hypothyroidism, maternal hypothyroidism and combined maternal-fetal hypothyroidism were employed to explore the source(s) of AF rT_3 since it had earlier been suggested that AF rT_3 concentration might be used as a means of diagnosing fetal hypothyroidism in utero.[5] These studies indicated that AF rT_3 content was influenced by maternal and not fetal thyroid function (Table 1) and that rT_3 itself was poorly transported from maternal serum into the amniotic cavity. Fetal serum rT_3 concentration is dependent upon both maternal and fetal thyroid function. Suzuki et al., have also reported that changes in rat AF rT_3 concentration are mainly dependent upon maternal thyroid function.[6] These findings suggested to us that tissues in contact with both maternal serum and fetal serum, or maternal tissues and the amniotic cavity could deiodinate maternal T_4 to generate rT_3. Since the membranes of hemochoroidal placentas are in contact with maternal and fetal blood and the chorioamniotic membranes are in contact with maternal decidua and AF, these tissues were tested for deiodinase activity.

Table 1. Amniotic fluid (AF) reverse T_3 (rT_3) in rat models of maternal-fetal hypothyroidism (MFH), maternal hypothyroidism (MH) and fetal hypothyroidism (FH)

Group	AF rT_3, % of Control Group		
	Exp. 1	Exp. 2	Exp. 3
MFH	34.7*	39.8*	20.1*
n	14	11	11
MH	34.7*	58.4*	26.0*
n	19	8	10
FH	103.6	95.2	79.6
n	13	9	12

*$p < .01$ vs control group

MATERIALS AND METHODS

Outer ring labeled ^{125}I-T_4, ^{125}I-T_3 and ^{125}I-rT_3 were obtained from New England Nuclear (Boston, MA). The specific activities of these labeled iodothyronines were from 942 to 1250 uCi/ug. When labeled iodothyronines were incubated with tissue homogenates, the incubation products were analyzed by paper chromatography using a hexane-tertiary amyl alcohol-NH_3 system modified from that described by Bellabarba et al.[7] The paper chromatograms were analyzed by autoradiography and counting of 1 cm sections of the lanes. In other experiments, stable T_4 was incubated with homogenates and the rT_3 generated from T_4 measured by radioimmunoassay. Placental 5'-deiodinase activity was measured by the formation of ^{125}I when ^{125}I-rT_3 was incubated with placental homogenates.[8] Sprague Dawley rats were obtained from Charles River (Boston, MA) and mated in our laboratory. Timed pregnant guinea pigs were obtained from Elm Hill Breeding Labs (Chelmsford, MA). Guinea pig placentas were perfused in situ at term, after removal of a single fetus, with ^{125}I-T_3 and perfusates analyzed by HPLC for ^{125}I-labeled iodothyronine metabolites.

Iodothyronine Deiodination by the Placenta

Homogenates of rat and human placentas actively deiodinate T_4 and T_3 in the tyrosyl (5) ring, generating rT_3 from T_4 and $3,3'$-T_2 and $3'$-T_1 from T_3 (Figure 1). rT_3 generated from T_4 or rT_3 added directly to the homogenates was not further deiodinated. This 5-deiodinase activity was in the microsomal fraction and was protein, pH, time, and DTT dependent. The effects of PTU, iodothyronines and other agents on placental microsomal 5-deiodinase activity were assessed.[9] The apparent Michaelis-Menton (K_m) for the deiodinase in human placental microsomes was 1.2×10^{-7}m. T_3, $3,3'$-T_2, iopanoic acid (IA), iodoacetic acid, diamide and propranolol, but not PTU, exhibited dose-dependent inhibition of human and rat placental 5-deiodinase activity in the presence of 10 mM DTT (Figure 2). When the DTT concentration was lowered to 0.25 mM, PTU inhibited deiodinase activity by approximately 71%. These studies may be relevant since PTU, propranolol and iopanoic acid may be administered to pregnant women.

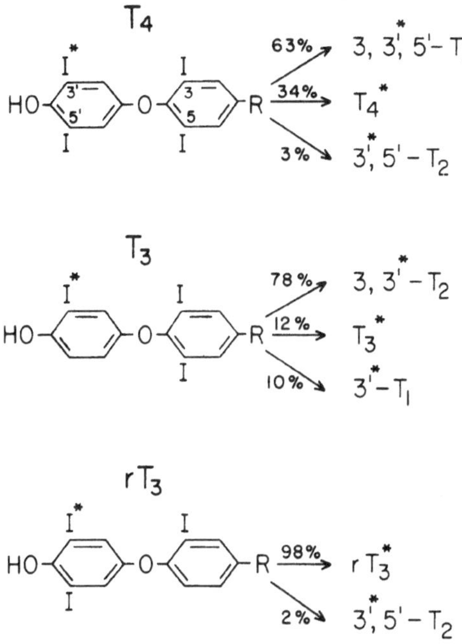

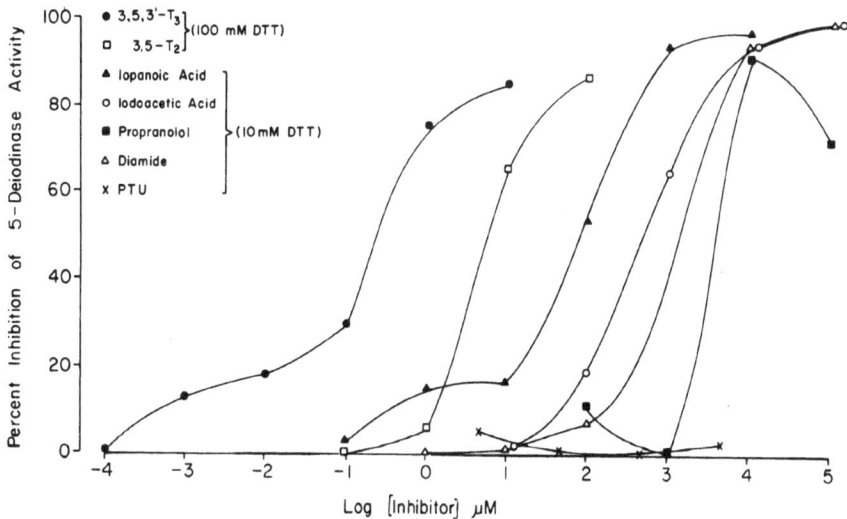

Figure 1. The metabolism of ^{125}I-labeled T_4, T_3 and rT_3 by human placental homogenates.

Figure 2. The inhibitory effect of iodothyronines (3,5,3'-T_3 and 3,5,-T_2), iopanoic acid, propranolol, and the sulfhydryl oxidizing agents (iodoacetic acid, diamide, and PTU) on inner ring iodothyronine deiodinase activity in the microsomal fraction of human placenta. The T_4 concentration was 6.4 x 10^{-7} M. The incubation time was 60 min.

In addition to iodothyronine 5-deiodinase activity, placental homogenates[10] and cultures of human chorionic decidual cells[11] contain T_4 and rT_3 5'-deiodinase activity but this activity is far less than that observed for 5-deiodinase.

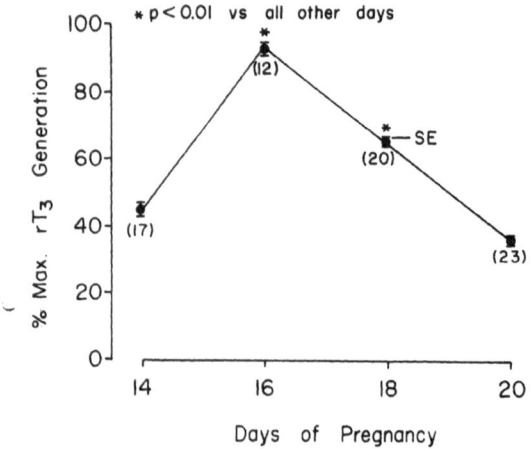

Figure 3. The ontogenesis of rat placental homogenate 5-deiodinase. Results are expressed as a percentage of the maximal value of rT_3 generated from T_4 from days 14-20 of pregnancy.

In order to further determine the relationship between AF rT_3 concentrations and placental 5-deiodinase activity, the ontogenesis of these 2 parameters were studied during pregnancy in the rat. Sufficient placental tissue was obtained to measure 5-deiodinase activity on all but the 12th day of gestation. The highest level was observed on day 16. 5-deiodinase activity on days 14, 18, and 20 was 52%, 77%, and 41%, respectively, of that observed on day 16 (P<0.01, day 16 vs. all other days) (Figure 3). Amniotic fluid rT_3 concentrations were highest on day 18 and were 61% and 64%, respectively, of that observed on day 18 (P<0.01, days 16 and 20 vs. day 18). These studies indicate that there are age-dependent changes in placental deiodinase activity in the rat and emphasize the importance of gestational age in studies of placental inner ring iodothyronine deiodinase. AF rT_3 concentrations appear to reflect these changes in deiodinase activity. Yoshida et al., have reported a significant negative correlation between the net placental rT_3 generation from T_4 and gestational age in humans.[12]

Since thyroid status and fasting have profound effects on iodothyronine deiodinases in other tissues, studies were performed to determine if these perturbations affected 5-deiodinase activity. Control and treated rats were mated and killed near term on the 20th day of gestation. 5-deiodinase activity was determined in placenta homogenates enriched with dithiothreitol by measuring the conversion of T_4 to rT_3. In four of 5 studies, 5-deiodinase activity was similar in dams that underwent thyroidectomy (Tx) on day 7 of gestation and sham Tx dams. 5-deiodinase activity was not altered in dams that were treated with methimazole (MMI) to induce maternal and fetal hypothyroidism. Treatment of dams with supraphysiological doses of T_4, beginning on the seventh day of gestation, did not significantly affect 5-deiodinase activity. In three of 4 studies, 5-deiodinase activity was similar in fed dams compared to values in dams fasted for the last 5 days of pregnancy. Placenta iodothyronine 5'-deiodinase activity was also measured in some studies. 5'-deiodinase activity was not decreased in Tx rats and was modestly decreased in MMI-treated rats. However, the effect of MMI was not reversed by the administration of supraphysiological doses of T_4, suggesting that MMI itself may decrease 5'-deiodinase activity in the placenta. Tx, MMI treatment, and fasting all decreased hepatic T_4 5'-deiodinase activity in the pregnant rats. These results strongly suggest that thyroid status and fasting do not alter placenta 5-deiodinase activity. Suzuki et al. also reported that placental 5-deiodinase activity is not affected by maternal thyroid status.[6]

Since all studies of placenta deiodinase activity had been carried out in broken cell preparations, it was important to determine whether the intact placenta could also deiodinate the iodothyronines. To address this question, studies have been performed employing the perfused guinea pig placenta. At about 60 days gestation, pregnant guinea pigs were anesthetized, a fetus removed and one of the placenta perfused through an umbilical artery using a technique modified from Kihlstrom and Kihlstrom.[13] The perfusion effluent was returned to the same reservoir from which the perfusion fluid was drawn. This recirculating reservoir system was, therefore, analogous to the fetal vascular pool. In a temperature-controlled chamber (37'C), the fetal side of the placenta was perfused through the umbilical artery at a rate of 1 ml/min with 3% bovine serum albumin Krebs-Henseleit buffer containing 0.14 nM outer ring labeled ^{125}I-T_3. Placenta effluent fractions were collected at timed intervals from the umbilical vein canula throughout a 120-min perfusion period. The contents of the perfusion buffer and the various effluent fractions were analyzed for their iodothyronine content by HPLC. In 5 experiments, the percent composition of ^{125}I-labeled iodothyronines in the perfusion buffer and placenta effluent was 95.3 ± 1.0 (mean $\pm$ SE) and 70.2 ± 2.1 for T_3 ($P<0.01$), 2.5 ± 0.7 and 20.1 ± 1.8 for 3,3'-T_3 ($P<0.01$), and 0 and 8.2 ± 0.9 for 3'-T_1 (Figure 4). There was no difference between the percent ^{125}I-iodide in the perfusion buffer and in the placenta effluents. When placentas were perfused with IA and ^{125}I-T_3, after perfusion with ^{125}I-T_3 alone, there was a significant increase ($P<0.01$) in the percent ^{125}I-T_3 in the placenta effluents, and a significant decrease in ^{125}I-3,3'-T_2 ($P<0.01$) and ^{125}I-3'-T_1 ($P<0.01$) (Figure 5). In contrast, PTU did not affect the composition of labeled iodothyronines in the placenta effluents, despite the fact that the addition of PTU significantly ($P<0.001$) inhibits the inner-ring deiodination of ^{125}I-T_3 in human or guinea pig placenta microsomes in the presence of low (0.25 mM) concentrations of dithiothreitol. These studies demonstrate that T_3 is actively deiodinated in the inner ring to 3,3'-T_2 by the intact guinea pig placenta. A portion of 3,3'-T_2 is further deiodinated in the inner ring to generate 3'-T_1. No outer ring deiodination of T_3 was seen under the conditions employed. IA, but not PTU, inhibits T_3 deiodination in the placenta perfused in situ. We conclude that the placenta is probably a site for fetal T_3 metabolism.[14] Cooper et al.[15] have reported T_4 inner ring deiodination in the perfused guinea pig placenta.

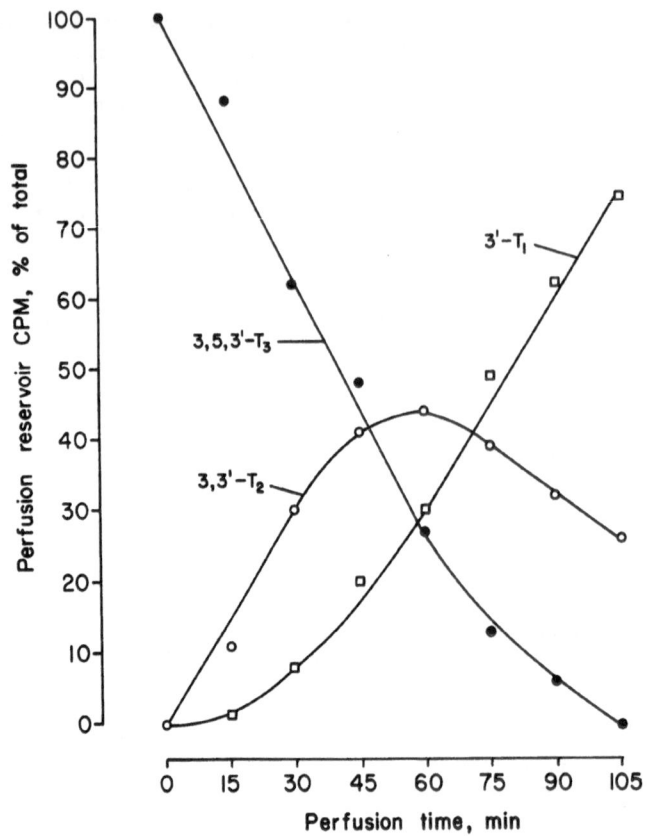

__Figure 4.__ Perfusion of the guinea pig placenta using a recirculating reservoir system.

Inner Ring Deiodination by Fetal Membranes

Indirect evidence, based on injection of thyroxine (T_4) into the amniotic cavity of humans, and maternal thyroidectomy in the rat, suggests that fetal membranes might be capable of converting T_4 to rT_3 by virtue of inner ring iodothyronine deiodinase activity. The present study was undertaken to provide direct evidence that human fetal membranes contain inner ring iodothyronine deiodinase activity directed toward T_4 and T_3. Homogenates of human fetal membranes were incubated with ^{125}I-labeled T_4, rT_3, and T_3, and with stable T_4. Conversion of ^{125}I-T_4 to ^{125}I-rT_3 was noted in chorion and amnion. ^{125}I-T_3 was converted to ^{125}I-3,3'-diiodothyronine (T_2) in chorion and amnion. ^{125}[I]-rT_3 was stable in fetal membranes under the incubation conditions employed. Time, temperature, pH, and protein content-dependent conversion of stable T_4 to rT_3 was found in fetal membranes. Iodothyronine metabolism did not occur in the absence of dithiothreitol. These studies indicate that human fetal membranes contain an inner ring deiodinase enzyme. Because of their intimate contact with the amniotic cavity, this fetal membrane enzyme may generate a portion of the rT_3 found in amniotic fluid.[16]

It can be postulated that placental deiodinase activity has two effects on maternal-fetal thyroid economy. The first is that the placenta is a site for "peripheral deiodination" in the maternal and/or fetal circulation. The meaning of "peripheral deiodination" in this discussion is restricted to the situation in which products of deiodination are returned to the compartment from which the precursor

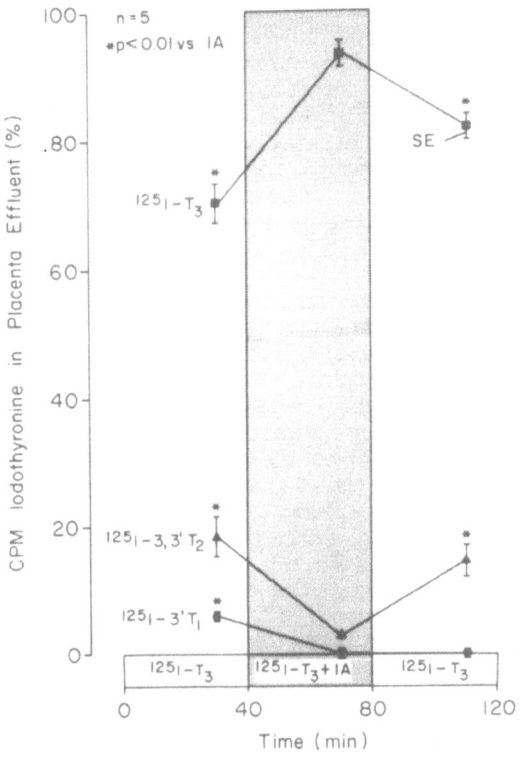

Figure 5. Effect of IA on the metabolism of ^{131}I-T_3 in the perfused guinea pig placenta.

originated. For example, inner ring deiodination of fetal serum T_4 by the placenta would generate fetal serum rT_3. The evidence that the placenta is an important site for peripheral deiodination, while indirect, is compelling. The placenta is perfused with approximately 25-50 percent of the fetal cardiac output, more than any other organ.[17] Many organs besides the placenta are undoubtedly responsible for peripheral deiodination in the fetus.[18,19] Comparative studies of the deiodinase activities of different fetal organs, including the placenta, are required in order to quantitate the relative importance of the placenta as a site for peripheral deiodination in the fetus.

Differences in placentation among species may determine whether the placenta is also a site for peripheral deiodination of maternal iodothyronines. In the hemochorial placenta, maternal and fetal blood is in direct contact with placental tissue.[20] Therefore, in these placentas, iodothyronines from both the maternal and fetal circulations are likely to provide substrates for inner ring deiodinase activity. Species with hemochorial placentation include the rat, guinea pig, and human.[20] In the sheep, which has endotheliochorial placentation, only fetal blood is in contact with placental tissue.[20] For this reason the placental deiodination of maternal iodothyronines in the sheep may be limited.

The second possible effect of placental deiodinase activity on maternal-fetal thyroid economy is that substrates from one compartment are deiodinated and the products transferred into a second compartment which also borders the deiodinating tissues. For example, maternal plasma T_4 may be deiodinated by the placenta and a portion of the rT_3 generated transferred to the fetal vascular pool. According to this schema, fetal T_4 could also be the source of some rT_3 in maternal serum. In addition, as discussed earlier, maternal rT_4 could contribute to AF rT_3 by virtue of deiodinase activities in the chorion and amnion.

Free T_4 and T_3 concentrations are much lower in the fetal circula-tion than in the maternal circulation during most of gestation.[21] These relatively low concentrations of T_3 and T_4 in fetal blood are probably optimal for fetal development. Placental dysfunction could conceivably result in fetal hyperthyroxinemia and hypertriiodothyroninemia by permitting T_4 and T_3 to be freely transferred from the mother to the fetus. An important question is whether transfer of T_4 and T_3 between the maternal and fetal circulations is limited by placental deiodinase activity. To prove this hypothesis it must be shown that inhibition of the deiodinase enzyme increases the transfer of intact T_4 and T_3 from mother to fetus. At the present time, the evidence that placenta inner ring deiodinase limits the transfer of intact hormone is indirect. It consists largely of the studies described above which suggest that maternal T_4 is a source of fetal serum rT_3. However, it would be an over simplification to equate enhanced deiodinase activity with decreased transfer from the placenta. rT_3 is both resistant to deiodination and is poorly transferred across the placenta. One speculation is that access of iodothyronines to the active site on the placental 5-deiodinase and movement of iodothyronines into and across the placental membrane cells are related.

Bonet and Herrera[22] and Escobar et al[23] have recently reported that thyroid hormones cross the placenta from dam to fetus and that maternal hypothyroidism does affect fetal growth and development and brain T_3 content. Further studies in the rat and other species, including man, will further clarify this question.

References

1. J. Kohrle, G. Brabant, and R.D. Hesch, Metabolism of the Thyroid Hormones, Hormone Res 26:58 (1987).
2. E. Roti, S.L. Fang, K. Green, C.H. Emerson, and L.E. Braverman, Human placenta is an active site of thyroxine and 3,3',5-triiodothyronine tyrosyl ring deiodination. J. Clin. Endocrinol. Metab. 53:498 (1981).

3. E. Roti, S.L. Fang, L.E. Braverman, and C.H. Emerson: Rat placenta is an active site of inner ring deiodination of thyroxine and 3,3'5-triiodothyronine. Endocrinology 110:34 (1982).

4. I.J. Chopra and B.F. Crandell, Thyroid hormones and thyrotropin in amniotic fluid. N. Engl. J. Med. 293:740 (1975).

5. I.J. Chopra, D.R. Hollingsworth, S.L. Davis, R.P. Belin, and M.C. Reid, Amniotic fluid 3,3'- and 3,5'-diiodothyronine in fetal hypothyroidism in sheep. Endocrinology 104:596 (1979).

6. M. Suzuki, K. Yoshida, T. Sakurada, N. Kaise, K. Kaise, H. Fukazawa, T. Nomura, Y. Itagaki, K. Yonemitsu, M. Yamamoto, S. Saito, and K. Yoshinaga, Effect of changes in thyroid state on metabolism of thyroid by rat placenta, Endocrinol. Japon 33:37 (1986).

7. D. Bellabarba, R.E. Peterson, and K. Sterling, An improved method for chromatography of iodothyronines. J. Clin. Endocrinol. Metab. 28:305 (1968).

8. T.J. Visser, J.L. Leonard, M.M. Kaplan, and P.R. Larsen, Kinetic evidence suggesting two mechanisms for iodothyronine 5'-deiodination in rat cerebral cortex. Proc. Natl. Acad. Sci. USA 79:5080 (1982).

9. M. Fay, E. Roti, G. Wright, L.E. Braverman, and C.H. Emerson, Effect of propylthiouracil, iodothyronines, and other agents on thyroid hormone metabolism in human placenta. J. Clin. Endocrinol. Metab. 58:280 (1984).

10. K. Banovac, L. Bzik, T. Tislaric, and M. Selso, Conversion of thyroxine to triiodothyronine and reverse triiodothyronine in human placenta and fetal membranes. Horm. Res. 12:253 (1980).

11. J.T. Hidal, and M.M. Kaplan, Characteristics of thyroxine 5'-deiodination by cultured human placenta cells. Regulation by iodothyronines. J. Clin. Invest. 76:947 (1985).

12. K. Yoshida, M. Suzuki, T. Sakurada, O. Shinkawa, T. Takahashi, N. Furuhashi, N. Kaise, K. Kaise, H. Kitaoka, H. Fukazawa, T. Nomura, Y. Itagaki, M. Yamamoto, S. Saito, and K. Yoshinga, Human placental thyroxine inner ring deiodinase in complicated pregnancy, Metabolism 34:535 (1985).

13. I. Kihlstrom, and J.E. Kihlstrom, An improved technique for perfusion of the guinea pig placenta in situ giving viable conditions demonstrated by placental transport of amino acids (L- and D-alanine). Biol. Neonate 39:150 (1981).

14. M.I. Castro, L.E. Braverman, S. Alex, C.F. Wu, and C.H. Emerson, Inner rng deiodination of 3,5,3'-triiodothyronine in the in situ perfused guinea pig placenta. J. Clin. Invest. 76:1921 (1985).

15. E. Cooper, M. Gibbens, C.R. Thomas, C. Lowy, and C.W. Burke, Conversion of thyroxine to 3,5'-5'-triiodothyronine in the guinea pig placenta in vivo. Endocrinology 112:1808 (1983).

16. E. Roti, S.L. Fang, K. Green, L.E. Braverman, and C.H. Emerson, Inner ring deiodination of thyroxine and 3,5,3'-triiodothyronine by human fetal membranes. Am. J. Obst. Gyn. 147:788 (1983).

17. B.S. Nuwayhid, Fetal hemostasis. In: Principles and Practice of Obstetrics and Perinatology. Ed. L. Iffy, H. Kaminetzky. Wiley, New York p. 272 (1981).

18. A.R.C. Harris, S.L. Fang, J. Prosky, L.E. Braverman, and A.G. Vagenakis, Decreased outer ring monodeiodination of thyroxine and reverse triiodothyronine in the fetal and neonatal rat. Endocrinology 103:2216 (1978).

19. S.Y. Wu, A.H. Klein, I.J. Chopra, and D.A. Fisher, Alterations in tissue thyroxine 5'-monodeiodinating activity in the perinatal period. Endocrinology 103:235 (1978).

20. E.M. Ramsey, and M.W. Donner, Placental Vasculature and Circulation. Saunders, Philadelphia p. 85 (1980).
21. D.A. Fisher, J.H. Dussault, J. Sack, and I.J. Chopra, Ontogenesis of hypothalamic-pituitary-thyroid function and metabolism in man, sheep, and rat. Recent Prog. Horm. Res. 33:59 (1977).
22. B. Bonet and E. Herrera, Different response to maternal hypothyroidism during the first and second half of gestation in the rat, Endocrinology 122:450 (1988).
23. G. Morreale de Escobar, M.J. Obregon, C. Ruiz de Ona, and F. Escobar del Rey, Transfer of thyroxine from the mother to the rat fetus near term, Effects on brain 3,5,3'-triiodothyronine deficiency, Endocrinology 122:1521 (1988).

ENDEMIC CRETINISM. AN OVERVIEW

François M. Delange

Depts. of Pediatrics and Radioisotopes [×]
University Hospital Saint-Pierre
University of Brussels, Belgium.

INTRODUCTION

The term endemic cretinism applies to individuals born and living in areas of severe iodine deficiency and endemic goiter exhibiting irreversible anomalies of intellectual and physical development not explained by other causes than the environmental factors responsible for goiter. The prevalence of the disorder can reach 10 % of the whole population in severely affected areas and cretinism constitutes the most serious complication of endemic goiter.

The diagnosis of endemic cretinism is descriptive and mostly made on epidemiological grounds because its etiopathogenesis is only partly understood and because information on its pathology is scanty. In 1986, a study group of the Pan American Health Organization (1) formulated the following definition of endemic cretinism :

"The condition of endemic cretinism is defined by three major features :
1. Epidemiology. It is associated with endemic goiter and severe iodine deficiency.
2. Clinical manifestations. These comprise mental deficiency, together with either :
   a) a predominant neurological syndrome including defects of hearing and speech, squint, and characteristic disorders of stance and gait of varying degree; or
   b) predominant hypothyroidism and stunted growth.
   Although in some regions one of the two types may predominate, in other areas a combination of the two syndromes will occur.
3. Prevention. In areas where adequate correction of iodine deficiency has been achieved, endemic cretinism has been prevented."

Several extensive review papers on endemic cretinism with comprehensive bibliography are available (2-6).

In this chapter, the author will summarize present knowledge in the field and express his personal views on what is clear and unclear in the various aspects of endemic cretinism.

[×] Correspondance : F. Delange M.D.
Dept. of Radioisotopes
Hôpital Saint-Pierre
322, rue Haute, 1000 - Brussels, Belgium.

1. CLINICAL ASPECTS AND EPIDEMIOLOGY

We know that there are two extreme types of endemic cretinism, neurological and myxedematous, that there are regional variations in the frequency of the two types, that there are mixed types in some individuals and areas and, finally, that there is neurointellectual impairment in non cretin individuals in severe endemic areas (1-12).

We do not know the reasons for the regional variations in the frequency of the two types or the reasons for the predominance of neurological impairment of thyroid failure in individuals with the mixed type of cretinism. There is no doubt that the focus of interest and the training of the investigator play a role. Endocrinologists will take more notice of signs of thyroid failure and neurologists of signs of impairment of the nervous system (13). The question also arises as to what extent neurological signs in myxedematous cretins are a consequence of longstanding untreated hypothyroidism (14-16) or whether they are the consequence of encephalopathy of the type found in neurological cretins.

2. PATHOLOGY

a. Brain

We know from early reports by De Quervain and Wegelin in Switzerland (17) and, from more recent pathological and radiological studies made in Ecuador (18) and China (19) that the brain of neurological cretins is atrophic, especially in the pontomesencephalic and cerebellar areas, and that there is dilatation of the ventricles and subarachnoid spaces. There is also microscopic evidence of disorganisation of the interneuronal relationship and lack of development of the pyramidal and Purkinje cells. We also know from Chinese studies (20. See also Chapter 4) that human fetuses in regions of severe endemic goiter have cerebral atrophy with increased cell density and an increased percentage of non-differentiated neuroblasts.

What is needed now is a comparison of brain pathology in neurological cretinism and cerebral palsy and in myxedematous cretinism and sporadic congenital hypothyroidism; there is also a need for study of the brain in endemic mental retardation.

b. Ear

We know very little about the pathology of the ear. And yet, hearing loss is a major component of cretinism. Old data summarized by Koenig (21) indicated that neurological endemic cretins present middle ear abnormalities characterized by hypertrophic bone changes of the promontorium, deformation of the ossicles, distorsion of the oval and round windows and thickening of the mucous membrane of the tympanic cavity. No data are available for the inner ear.

What we need is more precise information, obtained with modern methods of investigation, on the structure and ultrastructure of the middle and inner ear in neurological and myxedematous cretins. Such findings should then be compared with data concerning sporadic congenital hypothyroidism and congenital infections. There is undoubtedly scope for further work in audiometry and auditory evoked potentials in endemic and sporadic cretinism.

c. Thyroid

Swiss authors (2, 17) reported that neurological cretins have hyperplastic, colloid, multinodular goiters similar to those seen in non cretin individuals in the same area. They showed that myxedematous endemic cretins have thyroid atrophy and sclerosis. As a consequence of their thyroid failure, myxedematous cretins also have hyperplastic pituitary glands.

Human fetuses in regions of severe endemic goiter have hyperplastic colloid goiters with no evidence of thyroid atrophy or destruction (22).

We need more information on the structure and ultrastructure of the thyroid in the two types of cretinism. We need to know whether there are specific aspects of thyroid morphology in hypothyroid newborns and still-births in endemic areas. We also need to know whether there are age-related changes in the thyroid of hypothyroid infants and children. In other words, how valid is the hypothesis that only progressive atrophy of the thyroid will result in permanent hypothyroidism and myxedematous cretinism in some individuals while the development of compensatory thyroid hyperplasia explains the transient character of hypothyroidism in others (5). Finally, is there any morphological evidence of an autoimmune process within the thyroid ?

3. ETIOLOGY

a. Iodine deficiency

The role of iodine deficiency in the etiology of endemic cretinism is well established 1) by the correlation between the degree of iodine deficiency and the frequency of cretinism 2) by the prophylactic action of iodine on the incidence of cretinism and 3) by the emergence of cretinism in previously unaffected populations as a consequence of iodine deficiency of recent onset, as observed in the Jimi valley in New Guinea after replacement of natural rock salt rich in iodine with low iodine industrial salt (23).

Finally, iodine deficiency during gestation in animals results in thyroid deficiency in the offspring. All the models used mimic the myxedematous type of cretinism; none was able to produce the neurological type (16).

We do not know why it is only in certain individuals that the thyroid, and possibly the brain, are unable to adapt themselves to iodine deficiency.

b. Naturally occurring goitrogens

We know that thiocyanate, a naturally occurring goitrogen resulting from chronic consumption of poorly detoxified cassava, aggravates the effects of iodine deficiency on the incidence of congenital hypothyroidism (24) and that thiocyanate overload is more severe in myxedematous cretins than in euthyroid controls of the same area (25). We do not know whether this last finding is the cause or the consequence of cretinism. We have no information on the possible etiological role of other goitrogens in the presence of iodine deficiency, such as the waterborne humic substances described by Gaitan et al. (26).

c. Congenital infections and perinatal anoxia

The fact that there are some similarities between neurological cretinism and cerebral palsy (18) raises the question of whether congenital infections or perinatal anoxia play a role in the etiology of cretinism. For example, in non endemic areas, as much as 12 % of congenital sensori-neural hearing loss is due to congenital infection caused by cytomegalovirus (27-29).

On the other hand, the seasonal variations in the incidence of sporadic congenital hypothyroidism (30) are not due to infectious processes (31).

More information is needed on the infectious profiles and obstetrical history in populations with endemic cretinism.

d. Thyroid autoimmunity

On the strength of findings reported initially in sporadic goiter (32), it has been suggested that goiter in endemic areas could be due to thyroid-growth-promoting antibodies, TGI (33-38).

In addition, there is now a keen controversy over the possible role of thyroid-growth-blocking antibodies of maternal origin in the etiology of sporadic congenital hypothyroidism (39-41).

Finally, the role of antithyroglobulin antibodies in the etiology of endemic goiter has been suspected in Argentina (42) but has never been conclusively confirmed, either in endemic goiter and cretinism (43) or in sporadic congenital hypothyroidism (44).

We definitely need more information about the possible relationship between iodine deficiency, goitrogens and thyroid autoimmunity.

e. Oligoelements

The serum concentration of Selenium is lower in inhabitants of the former goitrous area than in those of the non-goitrous area of Idjwi Island in Zaire (45). It has been speculated that Selenium deficiency could cause oxygen derivatives such as H_2O_2 to be toxic for the thyroid, and could be one of the factors responsible for thyroid damage resulting in myxedematous cretinism.

Manganese deficiency in rats aggravates the alterations of the vestibular function induced by PTU (46). On the other hand, supplements of manganese to iodine deficient rats partly protect them against the antithyroid action of iodine deficiency (47). The question arises as to whether manganese deficiency in man could aggravate the goitrogenic action of iodine deficiency.

Finally, Zinc deficiency enhances the hepatic deiodination of T_4 (48). Among the several nutritional disorders associated with endemic goiter, it would be of interest to investigate the specific role of Zinc nutrition.

4. PATHOGENESIS

a. Fetal hypothyroidism

Chinese data have indicated that hypothyroidism is present in human fetuses in regions of severe iodine deficiency from the fourth month of gestation (22). Thyroid failure at birth has been evidenced in several areas of the world, such as Zaire (49, 50), India (51, 52), Algeria (53), Tanzania (54), and even some parts in Europe such as Sicily (55), Southern Germany (56) and Greece (57).

We know that thyroid deficiency occurring during the initial period of brain development results in irreversible brain damage (Review in 58) and especially in case of thyroid deficiency of prenatal onset (59) probably even if, based on neontal screening (60), adequate substitutive therapy is initiated soon after birth (61, 62).

There is some suggestion that infants with sporadic congenital hypothyroidism have hearing defects (63).

Finally, it has been shown experimentally in animals that fetal hypothyroidism induces damage of the inner ear (64-68), which is possibly responsible for the audiogenic seizures reported in the offspring of extremely severely iodine deficient rats (69).

We do not know the reason for the varying degrees of fetal and neonatal hypothyroidism in the presence of uniform iodine deficiency in a given population, the frequency and the long-term consequences of transient fetal or neonatal hypothyroidism in endemic goiter and the possible role, if any, of fetal hypothyroidism in the etiology of neurological cretinism.

b. Maternal hypothyroxinemia

Maternal hypothyroxinemia is rare in non endemic areas and can result in impaired neurointellectual development in offspring (2, 70). In contrast, maternal hypothyroxinemia is extremely frequent in endemic areas (50, 71-73). It is associated with increased mortality and morbidity in offspring (72, 74) and increased incidence of hypothyroidism in neonates (49, 50). In addition, data collected in rats showed that maternal hypothyroxinemia results in fetal hypothyroxinemia and brain damage during early pregnancy, even before the onset of fetal thyroid function (75-81).

What we would like to know is the rate of transfer, if any, of thyroid hormones across the placenta in humans before and after the onset of fetal thyroid function; the presently accepted concept is that there is no transfer or only a very minimal transfer of thyroid hormones and TSH across the placenta in humans (Review in 82). What we would also like to know are the specific consequences of hypothyroxinemia and iodine deficiency for fetal development in humans. On the basis of presently available clinical and experimental data, DeLong (13) has developed the concept that all the neurological deficits in endemic cretinism result from iodine deficiency mediated through hypothyroidism and that the difference between the purely neurological and myxedematous types of cretinism are explained by timing and, perhaps, the degree of maternal and/or fetal hypothyroidism. The neurological type would result from second-trimester hypothyroxinemia of mother and fetus and the myxedematous type from third-trimester hypothyroxinemia of the fetus, possibly because of a more adequete maternal thyroid hormone contribution during early gestation. The hypothesis is extremely attractive; the problem is that it has been clearly established that maternal hypothyroxinemia is much more severe in areas with the myxedematous type (50) than in areas with the neurological type (71), not the reverse

c. Lack of elemental iodine

The data suggesting the direct role of elemental iodine on brain development is the observation that correction of iodine deficiency in mothers prevents the emergence of neurological cretinism only if correction takes place before or during early gestation, thus before the onset of fetal thyroid function (83). What we would like to know in greater detail is which parameter is corrected in the human fetus when maternal iodine deficiency is corrected before the onset of fetal thyroid function ? Is it the fetal deficiency in iodine, in thyroid hormones or in both ?

5. PREVENTION

We know that endemic cretinism can be prevented by the normalization of iodine intake in the population, for example by the introduction of iodized salt. It can also be prevented by giving injections of iodized oil to mothers during early or even late pregnancy in the case of myxedematous cretinism (Reviews in 3-6). We also know that newborns are more sensitive than children and adults to the antithyroid effects of both iodine deficiency and iodine excess (84).

What we would like to know more precisely is the duration of the preventive effect of a single injection of iodized oil in women on the incidence of cretinism in the offspring. More information is also needed on the possible adverse effects of iodized oil. The possibility that this therapy could induce a Wolff-Chaikoff effect in the fetus has been raised (85) but never conclusively demonstrated. The hypothesis has also been proposed (86), but never confirmed, that an acute increase in iodine supply could trigger thyroid autoimmunity.

Finally, we would like to know a little more about the minimum and maximum iodine intakes to be recommended during pregnancy in humans.

In conclusion, a lot is known about the clinical characteristics, pathology, etiopathogenesis and prevention of endemic cretinis, but probably a lot more is unknown. It may be expected that a lot will be learned over the next few years thanks to recent developments in clinical and experimental thyriodology and easier access to the field.

REFERENCES

1. F. Delange, S. Bastani, M. Benmiloud, E. DeMaeyer, M.G. Isayama, D. Koutras, S. Muzzo, H. Niepomniszcze, C.S. Pandav and G. Riccabona, Definitions of endemic goiter and cretinism, classification of goiter size and severity of endemias, and survey techniques, in : "Towards the eradication of endemic goiter, cretinism and iodine deficiency", J.T. Dunn, E. Pretell, C.H. Daza and F.E. Viteri eds., Pan American Health Organization Publ., Washington. PAHO Sc. Publ. n° 502 (1986) pp. 373-376.
2. M.P. König, Die Kongenitale Hypothyreose und der Endemische Kretinismus. Springer – Verlag Publ., Berlin (1968) pp. 1-175.
3. B.S. Hetzel and A. Querido, Iodine deficiency, thyroid function, and brain development, in : "Endemic goiter and endemic cretinism", J.B. Stanbury and B.S. Hetzel eds., John Wiley Publ., New York (1980), pp. 461-472.
4. P. Pharoah, F. Delange, R. Fierro-Benitez and J.B. Stanbury, Endemic cretinism, in : "Endemic goiter and endemic cretinism. Iodine nutrition in health and disease", J.B. Stanbury and B.S. Hetzel eds., John Wiley Publ., New York (1980), pp. 395-421.
5. F. Delange, Disorders of iodine deficiency. Endemic cretinism, in : "The thyroid", S.H. Ingbar and L.E. Braverman eds., J.B. Lippincott Publ., Philadelphia (1986), pp. 722-734.
6. J.B. Stanbury, Iodine deficiency-related endemic retardation, in : "Iodine nutrition, thyroxine and brain development", N. Kochupillai, M.G. Karmakar and V. Ramalingaswami eds., Tata McGraw-Hill Publ., New Dehli (1986), pp. 18-27.
7. Z.F. Shi, G.H. Zeng, J.X. Zhang, X.T. Li, M.T. Hou, T.Z. Lu, J.B. Wu, G.X. Wang, Z.Z. Tian, J.L. Liu, Z.J. Liu, S.H. Yang, S.Y. Nie, S.G. Li, D.M. Kong and X.Y. Zhu, Endemic goiter and cretinism in Gui Zhou. Clinical analysis of 247 cretins, Chinese Med. J. 97 : 689-697 (1984).
8. H.M. Wang, T. Ma, X.T. Li, X.M. Jiang, Y.Y. Wang, B.Z. Chen, F.R. Wang, S.M. Gao, L.Y. Ma and M.Y. Su, A comparative study of endemic myxedematous and neurological cretinism in Hetian and Luopu, China, in : "Current problems in thyroid research", N. Ui, K. Torizuka, S. Nagataki and K. Hiyai eds., Excerpta Medica Publ., Amsterdam (1983) pp. 349-355.
9. X.Y. Zhu, Endemic goiter and cretinism in China with special reference to changes of iodine metabolism and pituitary-thyroid function two years after iodine prophylaxis in Gui-Zhou, in : "Current problems in thyroid research", N. Ui, K. Torizuka, S. Nagataki and K. Hiyai eds., Excerpta Medica Publ., Amsterdam (1983), pp. 13-18.
10. N. Bleichrodt, I. Garcia, C. Rubio, G. Morreale de Escobar and F. Escobar del Rey, Developmental disorders associated with severe iodine deficiency, in : "The prevention and control of iodine deficiency disorders", B.S. Hetzel, J.T. Dunn, J.B. Stanbury eds., Elsevier Publ., Amsterdam (1987), pp. 65-84.
11. M. Mehta, C.S. Pandav and N. Kochupillai, Intellectual assessment of schoolchildren from severely iodine deficient villages, Indian Pediat. 24 : 467-473 (1987).

12. S. Muzzo, L. Leiva and D. Carrasco, Possible etiological factors and consequences of a moderate iodine deficiency on intellectual coefficient of schoolage children, in : "Frontiers of thyroidology", G.A. Medeiros-Neto and E. Gaitan eds., Plenum Press Publ., New York (1985), pp. 1001-1005.

13. B. Delong, Neurological involvment in iodine deficiency disorders, in : "The prevention and control of Iodine Deficiency Disorders", B.S. Hetzel, J.T. Dunn and J.B. Stanbury eds., Elsevier Publ., Amsterdam (1987), pp. 49-63.

14. S.N. Nickel and B. Frame, Neurologic manifestations of myxedema, Neurology 8 : 511-517 (1958).

15. J.W. Swanson, J.J. Kelly and W.M. McConahey, Neurologic aspects of thyroid dysfunction, Mayo Clin. Proc. 56 : 504-512 (1981).

16. B.S. Hetzel and B.J. Potter, Iodine deficiency and the role of thyroid hormones in brain development, in :"Neurobiology of the trace elements," I.E. Dreosti and R.M.Smith eds., Humana Press Publ., New Jersey (1983), pp. 83-133.

17. F. De Quervain and C. Wegelin, Der Endemische Kretinismus, Springer-Verlag Publ., Berlin (1936), pp. 1-206.

18. I. Ramirez, M. Cruz and J. Varea, Endemic cretinism in the Andean region : New methodological approaches, in : "Cassava toxicity and thyroid : research and public health issues", F. Delange, R. Ahluwalia eds., International Development Research Center Publ., Ottawa (1983), pp. 73-76.

19. T. Ma, T.Z. Lu, Y.B. Tan and B.Z. Chen, Neurological cretinism in China, in : "Iodine nutrition, thyroxine and brain development", N. Kochupillai, M.G. Karmakar and V. Ramalingaswami eds., Tata McGraw-Hill Publ., New Dehli (1986), pp. 28-33.

20. J.L. Liu, Y.B. Tan, Z.L. Zuang, X.Y. Zhu and B.Z. Chen, Morphologic study of development of cerebral cortex of therapeutically aborted fetuses in endemic goiter region, Gui-Zhou Province, China, in : "Current problems in thyroid research", N.Ui, K. Torizuka, S. Nagataki and K. Miyai eds., Excerpta Medica Publ., Amsterdam (1983), pp. 390-393.

21. M.P.Koenig and M. Neiger, The pathology of the ear in endemic cretinism, in : "Human development and the thyroid gland. Relation to endemic cretinism", J.B. Stanbury and R.L. Kroc eds., Plenum Press Publ., New York (1972), pp. 325-333.

22. J.L. Liu, Y.B. Tan, Z.J. Zhuang, Z.F. Shi, Y.L. Wang, X.T. Li, L.X. Xu, X.B. Yang, J.B. Wu, B.Z. Chen, L.Y. Ma, J.X. Zhang, D.M. Kong and X.Y. Zhu, Morphologic studies of therapeutically aborted fetus thyroid in an endemic goiter region of Guizhou Province, Chinese Med. J. 95 : 347-354 (1982).

23. P.O.D. Pharoah and R.W. Hornabrook, Endemic cretinism of recent onset in New Guinea, Lancet ii : 1038-1041 (1973).

24. F. Delange, P. Bourdoux, C. Colinet, P. Courtois, P. Hennart, R. Lagasse M. Mafuta, P. Seghers, C. Thilly, J. Vanderpas, Y. Yunga and A.M. Ermans, Nutritional factors involved in the goitrogenic action of cassava, in : "Cassava toxicity and thyroid : research and public health issues", F. Delange and R. Ahluwalia eds., International Development Research Centre Publ., Ottawa (1983), pp. 17-25.

25. R. Lagasse, P. Bourdoux, P. Courtois, P. Hennart, G. Putzeys, C. Thilly, M. Mafuta, Y. Yunga, A.M. Ermans and F. Delange, Influence of the dietary balance of iodine/thiocyanate and protein on thyroid function in adults and young infants, in : "Nutritional factors involved in the goitrogenic action of cassava", F. Delange, F.B. Iteke and A.M. Ermans eds., International Development Research Centre Publ., Ottawa (1982), pp. 34-39.

26. G. Gaitan, R.C. Cooksey and R.H. Lindsay, Factors other than iodine deficiency in endemic goiter : goitrogens and protein-calorie malnutrition (PCM), in : "Towards the eradication of endemic goiter, cretinism and iodine deficiency", J.T. Dunn, E.A. Pretell, C.H. Daza and F.E. Viteri eds., Pan American Health Organization, Scientific Publication n° 502, Washington (1986), pp. 28-45.

27. J.B. Hanshaw, On deafness, cytomegalovirus, and neonatal screening, Am. J. Dis. Child. 136 : 886-887 (1982).

28. W.D. Williamson, M.M. Desmond, N. Lafevers, L.H. Taber, F.I. Catlin and G. Weaver, Symptomatic congenital cytomegalovirus. Disorders of language, learning and hearing, Am. J. Dis Child. 136 : 902-905 (1982).

29. C.S. Peckham, O. Stark, J.A. Dudgeon, J.A. Martin and G. Hawkins, Congenital cytomegalovirus infection : a cause of sensorineural hearing loss, Arch. Dis. Child. 62 : 1233-1237 (1987).

30. K. Miyai, J.F. Connelly, T.P. Foley Jr., M. Irie, R. Illig, S.O. Lie, J. Morissette, H. Nakajima, P. Rochiccioli and P.G. Walfish, An analysis of the variation of incidence of congenital dysgenetic hypothyroidism in various countries, Endocrinol. Jpn. 31 : 77-81 (1984).

31. K. Miyai, N. Hata, T. Kurimura, O. Nose, T. Harada, T. Tsuruhara, S. Kusuda, R. Satake, H. Mizuta, N. Amino and K. Ichihara, Antiviral antibodies in congenital hypothyroidism, in : "Frontiers of thyroidology", G.A. Medeiros-Neto and E. Gaitan eds., Plenum Press Publ., New York (1986), pp. 1225-1229.

32. H.A. Drexhage, G.F. Botazzo, D. Doniach, L. Bitensky and J. Chayen, Evidence for thyroid growth stimulating immunoglobulins in some goitrous thyroid diseases, Lancet ii : 287-292 (1980).

33. R.D. Van Der Gaag, H.A. Drexhage, W.M. Wiersinga, R.S. Brown, R. Docter, G.F. Botazzo and D. Doniach, Further studies on thyroid growth-stimulating immunoglobulins in euthyroid non endemic goitre, J. Clin. Endocrinol. Metab. 60 : 972-979 (1985).

34. H. Schatz, R. Bär, F. Müller, J.A. Nickel and H. Stracke, Growth promoting effects of TSH and EGF in isolated porcine thyroid follicles and occurrence of thyroid growth stimulating immunoglobulin-in euthyroid goitre patients from an iodine-deficient endemic goitre area, in : "The thyroid and autoimmunity", H.A. Drexhage and W.M. Wiersinga eds., Elsevier Publ., Amsterdam (1986), pp. 207-208.

35. A. Halpern, G. Medeiros-Neto and L.D. Kohn, Thyroid growth promoting activity in endemic goiter, in : "The thyroid and autoimmunity", H.A. Drexhage and W.M. Wiersinga eds., Elsevier Publ., Amsterdam (1986), pp. 209-210.

36. B. Grubeck-Loebenstein, H. Kassal, P.P.A. Smyth, K. Krisch and W. Waldhäusl, The prevalence of immunological abnormalities in endemic simple goitre, Acta Endocrinol. 113 : 508-513 (1986).

37. G. Medeiros-Neto, A. Halpern, Z.S. Cozzi, N. Lima and L.D. Kohn, Thyroid growth immunoglobulins in large multinodular endemic goiters : effect of iodized oil, J. Clin. Endocrinol. Metab. 63 : 644-650 (1986).

38. P.A. Wadeleux and R.J. Winand, Thyroid growth modulating factors in the sera of patients with simple non-toxic goitre, Acta Endocrinol. 112 : 502-508 (1986).

39. R.D. Van Der Gaag, H.A. Drexhage and J.H. Dussault, Role of maternal immunoglobulins blocking TSH-induced thyroid growth in sporadic forms of congenital hypothyroidism, Lancet i : 246-250 (1985).

40. J.H. Dussault and D. Bernier, ^{125}I uptake by FRTL5 cells : a screening test to detect pregnant women at risk of giving birth to hypothyroid infants, Lancet ii : 1029-1031 (1985).

41. C. Marcocci, L. Giusti, L. Chiovato, M. Ciampi, G. Fenzi and A. Pinchera, Screening for risk of delivery of a hypothyroid baby, Lancet ii : 403-404 (1986).
42. R.J. Soto, B. Imas, A.M. Brunengo and D. Goldberg, Endemic goiter in Misiones, Argentina : pathophysiology related to immunological phenomena, J. Clin. Endocrinol. Metab. 27 : 1581-1587 (1967).
43. A.M. Ermans, Disorders of iodine deficiency. Endemic goiter, in : "The thyroid", S. Ingbar and L.E. Braverman eds., Lippincott Publ., Philadelphia (1986), pp. 905-721.
44. J.H. Dussault, J. Letarte, H. Guyda and C. Laberge, Lack of influence of thyroid antibodies on thyroid function in the newborn infant and on a mass screening program for congenital hypothyroidism, J. Pediatr. 96 : 385-389 (1980).
45. P. Goyens, J. Golstein, B. Nsombola, H. Vis and J.E. Dumont, Selenium deficiency as possible factor in the pathogenesis of myxedematous endemic cretinism, Acta Endocrinol. 114 : 497-502 (1987).
46. J.W. Oliver, Interrelationships between athyreotic and manganese deficient states in rats, Am. J. Vet. Res. 37 : 597-600 (1976).
47. A.M. Ermans, J. Kinthaert, M. Van Der Velden and P. Bourdoux, Studies of the antithyroid effects of cassava and of thiocyanate in rats, in : "Role of cassava in the etiology of endemic goitre and cretinism", A.M. Ermans, N.M. Mbulamoko, F. Delange and R. Ahluwalia eds., International Development Research Centre Publ., Ottawa (1980), pp. 93-110.
48. J.W. Oliver, D.S. Sachan, P. Su and F.M. Applehans, Effects of zinc deficiency on thyroid function, Drug-Nutrient Interactions 5 : 113-124 (1987).
49. C.H. Thilly, F. Delange, R. Lagasse, P. Bourdoux, L. Ramioul, H. Berquist and A.M. Ermans, Fetal hypothyroidism and maternal thyroid status in severe endemic goiter, J. Clin. Endocrinol. Metab. 47 : 354-360 (1978).
50. F. Delange, C. Thilly, P. Bourdoux, P. Hennart, P. Courtois and A.M. Ermans, Influence of dietary goitrogens during pregnancy in humans on thyroid function of the newborn, in :"Nutritional factors involved in the goitrogenic action of cassava", F. Delange, F.B. Iteke and A.M. Ermans eds., International Development Research Centre Publ., Ottawa (1982), pp. 40-50.
51. K. Kochupillai, M.M. Godbole, M.G. Karmakar, C.S. Pandav, K. Vasuki and M.M.S. Ahuja, TSH, T_4, T_3 and rT_3 in cord blood of newborns from areas of differing endemicity for goitre, in : "Neonatal screening", H. Naruse and M. Irie eds., Excerpta Medica Publ., Amsterdam (1983), pp. 32-33.
52. N. Kochupillai and C.S. Pandav, Neonatal chemical hypothyroidism in iodine-deficient environments, in : "The prevention and control of Iodine Deficiency Disorders", B.S. Hetzel, J.T. Dunn and J.B. Stanbury eds., Elsevier Publ., Amsterdam (1987), pp. 85)93.
53. M.L. Chaouki, F. Delange, R. Maoui and A.M. Ermans, Endemic cretinism and congenital hypothyroidism in endemic goiter in Algeria, in : "Frontiers of thyroidology", G.A. Medeiros-Neto and E. Gaitan eds., Plenum Press Publ., New York (1986), pp. 1055-1060.
54. W. Wachter, M.G. Mvungi, E. Triebel, D. Van Thiel, I. Marschner, W.G. Wood, J. Haberman, C.R. Pickardt and P.C. Scriba, Iodine deficiency, hypothyroidism and endemic goitre in Southern Tanzania, J. Epidemiol. Com. Health 39 : 263-270 (1985).
55. L. Sava, F. Delange, A. Belfiore, F. Purrello and R. Vigneri, Transient impairment of thyroid function in newborn from an area of endemic goiter, J. Clin. Endocrinol. Metab. 59 : 90-95 (1984).
56. F. Delange, P. Heidemann, P. Bourdoux, A. Larsson, R. Vigneri, M. Klett, C. Beckers and P. Stubbe, Regional variations of iodine nutrition and thyroid function during the neonatal period in Europe, Biol. Neonate 49 : 322-330 (1986).

57. C. Beckers, C. Cornette, A. Georgoulis, J. Stonfouris, G.D.
Piperingos and D.A. Koutras, Neonatal hypothyroidism in areas with
moderate iodine deficiency, in : "Thyroid Research VIII", J.R.
Stockigt and S. Nagataki eds., Australian Academy of Science Publ.,
Canberra (1980), pp. 24-26.

58. J. Legrand, Thyroid hormone effects on growth and development,
in : "Thyroid hormone metabolism", G. Henneman ed., M. Dekker Publ.,
New York (1986), pp. 503-534.

59. R. Wolter, P. Noel, M. Craen, C.H. Ernould, F. Verstraeten, J. Simons,
S. Mertens, N. Vanbroeck and M. Vanderschueren-Lodewyckx, Neuro-
psychological study in treated thyroid dysgenesis, Acta Paediatr. Scand.
Suppl. 227 : 41-46 (1979).

60. G.N. Burrow and J.H. Dussault, Neonatal Thyroid Screening, Raven
Press, New York (1980), pp. 1-322.

61. J. Rovet, R. Ehrlich and D. Sorbara, Intellectual outcome in children
with fetal hypothyroidism, J. Pediatr. 5 : 700-704 (1987).

62. J. Glorieux, J.H. Dussault, J. Morissette, M. Desjardins, J. Letartre
and M. Guyda, Follow-up at ages 5 and 7 years on mental development in
children with hypothyroidism detected by Quebec screening program,
J. Pediatr. 107 : 913-915 (1985).

63. M. Vanderschueren-Lodeweyckx, F. Debruyne, L. Dooms, E. Eggermont and
R. Eeckels, Sensorineural hearing loss in sporadic congenital hypothyroidism, Arch. Dis. Child. 58 : 419-422 (1983).

64. A. Uziel, A. Rabie and M. Marot, The effects of hypothyroidism on
the onset of cochlear potentials in developing rats, Brain Res. 182 :
172-175 (1980).

65. A. Uziel, C. Legrand and A. Rabie, Corrective effects of thyroxine on
cochlear abnormalities induced by congenital hypothyroidism in the
rat. I. Morphological study, Dev. Brain Res. 19 : 111-122 (1985).

66. A. Uziel, M. Marot and A. Rabie, Corrective effects of thyroxine on
cochlear abnormalities induced by congenital hypothyroidism in the rat.
II. Electrophysiological study, Dev. Brain Res. 19 : 123-127 (1985).

67. R. Hebert, J.M. Langlois and J.H. Dussault, Permanent defects in rat
peripheral auditory function following perinatal hypothyroidism :
determination of a critical period, Dev. Brain Res. 23 : 161-170
(1985).

68. D. Dememes, C. Dechesne, C. Legrand and A. Sans, Effects of hypothyroidism on postnatal development in the peripheral vestibular system,
Dev. Brain Res. 25 : 147-152 (1986).

69. M. Van Middlesworth and C.H. Norris, Audiogenic seizures and cochlear
damage in rats after perinatal antithyroid treatment, Endocrinology
106 : 1686-1690 (1980).

70. E.B. Man, W.S. Jones, R.H. Holden and E.D. Hellits, Thyroid function in
human pregnancy. VIII. Retardation of progeny aged 7 years; relationship to maternal age and maternal thyroid function, Am. J. Obstet.
Gynecol.111 : 905-910 (1971).

71. J.B.Stanbury, Cretinism and fetal-maternal relationship, in : "Human
development and the thyroid gland. Relation to endemic cretinism",
J.B. Stanbury and R.L. Kroc eds., Plenum Press Publ., New York (1972),
pp. 487-505.

72. P.O.D. Pharoah, S.M. Ellis, R.P. Ekins and E.S. Williams, Maternal
thyroid function, iodine deficiency and fetal development, Clin.
Endocrinol. 5 : 159-166 (1976).

73. P.O.D. Pharoah, K.J. Connolly, R.P. Ekins and A.G. Harding, Maternal
thyroid hormone levels in pregnancy and the subsequent cognitive and
motor performance of the children, Clin. Endocrinol. 21 : 265-270
(1984).

74. C. Thilly, R. Lagasse, G. Roger, P. Bourdoux and A.M. Ermans, Impaired fetal and postnatal development and high perinatal death-rate in a severe iodine deficient area, in : "Thyroid Research VIII", J.R. Stockigt, S. Nagataki, E. Meldrum, J.W. Barlow and P.E. Harding eds., Australian Academy of Science Publ., Canberra (1980), pp. 20-23.

75. M.J. Obregon, J. Mallol, R. Pastor, G. Morreale de Escobar and F. Escobar del Rey, L-thyroxine and 3,5,3' triiodo-L-thyroxine in rats embryos before onset of fetal thyroid function, Endocrinology 114 : 305-307 (1984).

76. R.J. Woods, A.K. Sinha and R. Ekins, Uptake and metabolism of thyroid hormones by the rat fetus in early pregnancy, Clin. Sci. 67 : 359-363 (1984).

77. G. Morreale de Escobar, R. Pastor, M.J. Obregon and F. Escobar del Rey, Effect of maternal hypothyroidism on the weight and thyroid hormone content of rat embryonic tissues, before and after onset of fetal thyroid function, Endocrinology 117 : 1890-1900 (1985).

78. G. Morreale de Escobar, F. Escobar del Rey, R. Pastor, J. Mallol and M.J. Obregon, Effects of maternal thyroidectomy or iodine deficiency on T4 and T3 concentrations in rat concepta, before and after onset of foetal thyroid function, in : "Iodine nutrition, thyroxine and brain development", N. Kochupillai, M.G. Karmakar and V. Ramalingaswami eds., Tata McGraw-Hill Publ., New Dehli (1986), pp. 109-117.

79. F. Escobar del Rey, R. Pastor, J. Mallol and G. Morreale de Escobar, Effects of maternal iodine deficiency on T4 and T3 contents of rats concepta, both before and after onset of fetal thyroid function, in "Frontiers of thyroidology", G.A. Medeiros-Neto and E. Gaitan eds., Plenum Press Publ., New York (1986), pp. 1033-1038.

80. R.P. Ekins, A.K. Sinha and R.J. Woods, Maternal thyroid hormones and development of the foetal brain, in : "Iodine nutrition, thyroxine and brain development", N. Kochupillai, M.G. Karmakar and V. Ramalingswami eds., Tata McGraw-Hill Publ., New Dehli (1986), pp. 222-245.

81. B.J. Potter, G.H. McIntosh, M.T. Mano, P.A. Baghurst, J. Chavadej, C.H. Hua, B.G. Cragg and B.D. Hetzel, The effect of maternal thyroid-ectomy prior to conception on foetal brain development in sheep, Acta Endocrinol. 112 : 93-99 (1986).

82. D.A. Fisher and A.M. Klein, Thyroid development and disorders of thyroid function in the newborn, N. Engl. J. Med 304 : 702-712 (1981).

83. P.O.D. Pharoah, I.H. Buttfield and B.S. Hetzel, Neurological damage to the fetus resulting from severe iodine deficiency during pregnancy, Lancet i : 308-310 (1971).

84. F. Delange, P. Bourdoux and A.M. Ermans, Transient disorders of thyroid function and regulation in preterm infants, in : "Pediatric thyroidology", F. Delange, D. Fisher and P. Malvaux eds.,Karger, Basel (1985), pp. 369-393.

85. N. Kochupillai, M.M. Godbole, C.S. Pandav, A. Mithal and M.M.S. Ahuja, Environmental iodine deficiency, neonatal chemical hypothyroidism (NCH) and iodized oil prophylaxis, in "Iodine nutrition, thyroxine and brain development", N. Kochupillai, M.G. Karmakar and V. Ramalingaswami eds., Tata McGraw-Hill Publ., New Dehli (1986), pp. 87-93.

86. D.A. Koutras, K.S. Karaiskos, K. Evangelopoulu, M.A. Bouxis, G.D. Piperingos, J. Kitsopanides, D. Makriyannis, J. Mantsos, J. Stonfouris and A. Souvatzoglou, Thyroid autoantibodies after iodine supplementation, in : "The thyroid and autoimmunity", H.A. Drexhage and W.H. Wiersinga eds., Elsevier Publ., Amsterdam (1986), pp. 211-212.

OBSERVATIONS ON THE NEUROLOGY OF ENDEMIC CRETINISM

G. Robert DeLong

Chief, Division of Pediatric Neurology
Department of Pediatrics
Duke University Medical Center
Durham, North Carolina

Several points emerge from a neurological analysis of endemic cretinism:

1) Endemic cretinism presents a coherent clinical picture, though varying in severity and emphasis.

2) The clinical neurological deficits primarily implicate impairment of cerebral cortex, cochlea and basal ganglia.

3) The major motor disorder in neurological cretinism is extrapyramidal rather than pyramidal and is characterized by rigidity, with an added measure of spasticity. It is consistent with a putamino-pallidal lesion, in addition to other elements.

4) The pattern of damage, and other information, indicates that the critical effect on brain occurs during the second trimester of intrauterine life. It seems likely that the irreversible effect on brain results from impairment of neuron production.

I. CLINICAL PICTURE

I have previously described the neurological features of endemic cretinism (1,2) and will summarize them briefly here. The three major features are deafmutism, mental deficiency and motor disorders. Deafness, if complete, results from cochlear dysfunction. Partial hearing loss may be associated with limited dysarthric speech, probably reflecting an element of cerebral cortical dysfunction.

Intellectual deficiency, shown by concreteness, lack of abstraction, primitive language and arithmetic skills, poor visuomotor integration and difficulties in praxis is ascribable to cerebral neocortical dysfunction. In very severe cases, frontal release of primitive sucking and grasping reflexes, autistic vacuity and poor visual attention can be attributed to dysfunction of frontal, temporo-limbic, and occipito-parietal cortex respectively. Association cortex functions are impaired more than those of primary analyzers, probably reflecting plasticity of function with limited overall cortical capacity.

Motor dysfunction is dominated by a proximal and axial plastic rigidity and flexion dystonia most consistent with dysfunction of the basal ganglia, particularly putamen and globus pallidus. On this is superimposed a varying degree of spasticity, again predominantly proximal, attributable to cortico-spinal dysfunction, including premotor cortex. In some severely involved individuals, leg muscles are thin and undergrown, and the feet are weak. This may reflect loss of anterior horn cells.

It is important to note those neurological systems that are apparently spared: They include the visual system; autonomic and vegetative functions including those of hypothalamus; personality and memory functions associated with the temporo-limbic system (except in a few severely deficient individuals); and cerebellar function. Seizures are not a feature.

Within the broad syndrome of neurological endemic cretinism, the severity of particular features may vary widely. For example, deafness may be minimal or may be complete; spasticity may be prominent or may be subtle; speech may be impossible, or may be functional and only mildly dysarthric, etc. Within sibships, the overall pattern tends to be more similar than is seen if unrelated individuals are compared. These observations argue that the genetic substrate influences the expression of endemic cretinism. This has recently been clearly demonstrated by Held, et al. (3)

II. PATHOPHYSIOLOGY

What can the neurological data contribute to understanding of the pathophysiology of the brain in endemic cretinism? Can neurological observations tell which systems are involved and when that involvement occurred?

The systems mainly involved are auditory, motor and intellectual. The language deficit results from a combination of deafness and intellectual deficit, as well as defective motor control of speech. The motor deficit is characterized by predominantly proximal rigidity and spasticity that implies extra-pyramidal more than pyramidal dysfunction.

The intellectual deficit deserves comment. Detailed psychological data will be presented by others; here I will make some comments from the point of view of the neurologist. The intellectual deficit in neurological cretinism appears to be generalized: it affects most, if not all, areas of mental function, including language, reasoning, visual perception and visuomotor integration, motor planning and praxis. Mental function, when it can be assessed accurately in individuals with some hearing and language, is simple, concrete and quite limited. The deficits shown by cretins are best illustrated by their crude efforts in drawing, copying, performing imitative motor actions, naming and counting. An exception to the generalized nature of the intellectual deficits may be memory, which seems to be relatively preserved in neurologic cretins of moderate severity. The generalized defect in intellectual functions is best explained by a diffuse disorder of the neocortex, particularly of high level association areas. It is a general rule of the developing brain that defects are compensated by a plastic reorganization of remaining tisue, so there is a strong tendency to integrate function utilizing the remaining brain tissue - even though, after injury, the re-integrated function may be at a low level.

After emphasizing the pervasiveness of intellectual deficits in cretins, which probably are accounted for by neocortical damage, we should comment on the generally better preservation of those functions important to personality, including emotionality and also memory. We have elsewhere noted the tendency of cretins to be equable, sociable, cheerful, tractible, able and motivated to work at simple tasks and able to sustain attention and remember at a level adequate for work-a-day activities. This suggests a relatively good preservation of limbic systems of brain which are important to personality organization. It is true by the same token that very severely defective cretins do show an absence of personality, of social responsiveness, and of motivated activity that can best be described as autism, and which implies severe deficits not only of neocortical structures but also of limbic systems.

The movement disorder in endemic cretinism is distinctive. It is not primarily a cortico-spinal or pyramidal dysfunction although in a minority there are clasp-knife spasticity and extensor plantar responses. More characteristic, however, is the rigidity seen in axial structures (trunk and neck) and in the proximal limb girdles, both at the hips and thighs, and in the shoulders and upper arms. Other cretin individuals have marked flexion dystonia with plastic rigidity and eventual contractures at the hips, knees, shoulders and elbows, with preserved free movements of the hands and wrists. This pattern of axial and proximal rigidity is classically extrapyramidal, similar to the pattern of rigidity and bradykinesia seen in Parkinsonism and in other lesions of the basal ganglia (particularly the globus pallidus). Reminiscent of Parkinsonism are cretins with marked truncal rigidity who can't flex at the hips to sit on a chair - they remain stiffly extended and slide down. Their arms are held adducted over their abdomen, with the fingers extended and immobile. Tremor is not seen, however. The lesion in Parkinson's disease primarily affects the substantia nigra, but through the nigro-striatal pathways the striatum is strongly affected.

Other disorders showing analogous motor findings include progressive supranuclear palsy, the rigid childhood form of Huntington's disease, and cases of striatal necrosis seen in young children. All these diseases have in common a lesion of the basal ganglia, particularly the putamen or globus pallidus or both. In the case of Huntington's disease, Byers et al (4) found in all 13 rigid children, the globus pallidus or the ansa lenticularis was severely involved, a finding not commonly reported in adults. Extensive bilateral striatal lesions, for example in children with Leigh's disease or acute striatal necrosis, produce a similar picture of proximal limb and trunk rigidity. Another clinical analogy is that of progressive supranuclear palsy in which there is marked rigidity especially in a cape-like distribution in the neck, shoulders and proximal upper extremities very similar to that in endemic cretinism, with sparing of wrists and hands (5). The major pertinent lesions in progressive supranuclear palsy are in the globus pallidus and substantia nigra, though other extrapyramidal areas are also involved.

The extreme form of the motor syndrome is instructive, just as the extreme form of mental deficiency in cretinism is instructive because it produces an autistic state (implicating the limbic system). The severe form of motoric dysfunction (which does not necessarily run parallel to the mental dysfunction) is characterized by marked flexion dystonia and by release of "thalamic postures." These are obligatory and stereotyped postures originally described in monkeys whose entire telencephalon above the thalamus had been destroyed. They consist of flexion of the limbs on one side, and extension of those on the other, when the animal is laid on its side on a firm surface; they reverse if the animal is laid on the other side, and represent fragments of righting reflexes. The existence of reflex responses closely reminiscent of these in cretins suggests that certain critical motor systems above thalamic level may be destroyed or non-functioning and is consistent with putaminal and globus pallidal pathology. These findings were seen only in severely affected, non-ambulatory individuals. Interestingly, despite these severe gross motor deficits, three of these same individuals had preserved use of their hands to the extent that they could throw or manipulate a ball or point with a finger.

In summary, one can make a reasonable hypothesis that implicates lesions of the basal ganglia, particularly putamen and globus pallidus, as playing an important role in the motoric defect in neurological endemic cretinism. This hypothesis must be tested by further research in pathology and neurophysiology. Obviously other systems, and other nuclei concerned with motility, including cerebral cortex, must be affected in endemic cretinism. However, the predominating finding is usually rigidity.

III. DEVELOPMENTAL TIMING

The neurological and neuropathological findings in endemic cretinism put some limits on the timing of the intrauterine insult to the developing brain. We know the consequences of other insults at specific times in brain development and can use this information to estimate the time at which damage is critical in cretinism. Of course, the exact biological mechanism of injury must vary widely with different noxious agents, but the fact remains that different insults occuring at a particular time in development tend to produce similar effects, presumably because they affect systems which are at the most active and critical stage of development. The minimal facts are as follows:

The basic organogenetic plan of brain is intact in endemic cretinism. The processes of neural tube closure, formation of eyes, midline cleavage and pairing of forebrain structures, formation of the corpus callosum, formation of the outlet foramina of the fourth ventricle, and formation of the subarachnoid pathways for cerebrospinal fluid flow – all these are normal in any cretin brain that has come to my attention. These features are the result of the main actions of the first twelve weeks of human brain development. This constitutes a strong argument that the first trimester of human brain development, if not independent of iodine and thyroid hormone action, is at least not critically impacted by the conditions obtaining in endemic cretinism. By contrast, there are close parallels between the abnormalities in endemic cretinism and those found after other insults during the second trimester of human gestation. The formation of the cochlea of the inner ear occurs during the second trimester, specifically between the 10th and 18th week, and it is particularly vulnerable in iodine deficiency (6). The time of the major burst of proliferation of neurons destined for the cerebral cortex occurs between fourteen and eighteen weeks of gestational age (7). The time of maximal generation and migration of neurons destined for the basal ganglia corresponds closely to that of the cerebral cortex, that is during the 12th to 18th weeks of gestation (14).

Thus, we can suggest with some confidence that these organs - cochlea, cortex and basal ganglia - undergo their major formative events during the second trimester and are selectively affected in cretinism. To this extent the data from clinical neurology and that from developmental neuroanatomy are congruent. They are also entirely congruent with the well-known data on iodine repletion, which show that iodine must be given by the end of the first trimester of pregnancy to prevent cretinism (8), demonstrating that the critical iodine-dependent events determining neurological cretinism begin at that time.

IV. CRITICAL PROCESSES AFFECTED

What is the critical process? Is it failure of cell proliferation or migration? Is it failure of dendritic development, or of myelin? The timing of endemic cretinism may indirectly suggest answers: There is a narrow window of vulnerability - in the second and perhaps third trimesters. This is the time of neuron generation and migration. By contrast, formation of dendrites and synapses and myelination occur later and continue over a long period. This suggests that cell proliferation and perhaps migration is the critical event - that if too few cells are generated for cerebral cortex and basal ganglia, further developmental events even if they proceed normally are subject to the constraint of a limited number of neurons. The Chinese have human fetal brain material from endemic areas that is thyroid deficient at the fourth through the eighth month of pregnancy (9). Careful quantitative study of this material may permit an answer to the question of adequacy of cell production.

What is the cause of this second-trimester defect in brain development in endemic cretinism? Is it hypothyroidism? In 1986 I wrote that "no firm data were available" on the question whether severe iodine deficiency caused fetal hypothyroidism in man (2). I must apologise for that statement, because such data were available and had been published in China, documenting hypothyroidism in fetuses in iodine-deficient areas. I believe it is fair to say that the fact of human fetal hypothyroidism in iodine deficiency was not known, or not widely appreciated, at that time outside of China. One of the basic purposes of the present meeting is to bring that Chinese experience to the attention of the rest of the world. In fact, however, the knowledge that iodine deficiency produces fetal thyroid dysfunction, shown by low levels of thyroid hormones and thyroid hypertrophy as early as four months of gestation (10) does much to clarify the pathogenesis of neurological cretinism in man.

VI. NEUROLOGICAL CRETINISM WITHOUT IODINE DEFICIENCY: A CASE

I wish to describe an exceptional case that may throw some light on the pathogenesis of endemic cretinism. Only in one instance have I seen a clinical picture similar to endemic cretinism in the United States, in a young man I examined when he was 21 years old. He had moderately severe mental deficiency, moderate hearing loss and quite limited dysarthric speech, and the motor deficits typical of endemic cretinism (proximal hypertonia and mild flexion contractures, rigidity, exaggerated adductor and knee jerks, absence of distal spasticity, disinhibited facial emotional expressions and strabismus.) In every respect, the clinical picture was that of endemic cretinism and scarcely compatible with any other clinical syndrome I know. His history is instructive: At birth he was recognized as cretinous and was treated with thyroxine subsequently with normal growth. Studies showed no functional thyroid tissue. During her pregnancy with our patient, the mother became tired, constipated, and had dry skin from the third month, but she was only diagnosed as hypothyroid with Hashimoto's thyroiditis and treated after giving birth. In retrospect, she had a history of hyperthyroidism ten years previously, for which she was briefly treated with propylthiouracil. It is difficult to escape the conclusion that her infant suffered in utero from a combination of maternal hypothyroidism and fetal athyreosis reproducing the conditions commonly present in iodine deficiency, and resulting in an irreversible insult to the fetal brain and cochlea indistinguishable from that seen in endemic cretinism. In this case, the cause of the fetal athyreosis is unknown, but it is tempting to speculate that the fetal thyroid was damaged by maternal anti-thyroid antibodies.

Blizzard et al (11) observed the simultaneous occurrence of thyroid antibodies in both mother and newborn in many human cases of cretinism. They postulated the transplacental transfer of a thyrocytotoxic factor. Goldsmith et al (12) described a family with autoimmune thyroid disease (Hashimoto's type), in which congenital thyroid suppression (partially reversible) was present in all six offspring of one member with autoimmune thyroiditis and hypothyroidism. Of these six children, all were hypothyroid at birth, four were obviously cretinous, and 3 of 4 tested had high titers of antithyroid antibodies. The four surviving children were all mentally retarded and undergrown despite replacement thyroid treatment. More recently, van der Gaag et al(13) found 15 mothers of infants with sporadic congenital hypothyroidism had circulating immunoglobulins that blocked thyroid growth (thyroid growth inhibiting immunoglobulin). All the mothers were euthyroid. These results support the view that at least some infants with sporadic congenital hypothyroidism have a maternal thyroid autoantibody mediated cause for their thyroid dysgenesis.

The reason for describing our case is that his neurological deficits were indistinguishable from neurological endemic cretinism, and were irreversible despite early treatment, presumably because there was the added factor that the mother was hypothyroid during pregnancy.

VII. SUMMARY

Clinical data indicate that the major deficits in neurological cretinism are intellectual deficiency, deafness, and motor rigidity, and indicate that the parts of the nervous system most affected are the cerebral neocortex, the cochlea, and the basal ganglia. The findings implicating basal ganglia disease have generally received little attention.

The pathogenesis of this insult to brain development in neurological cretinism has been unclear. Only recently has the existence of fetal hypothyroidism in man at the critical stages of gestation been demonstrated in studies in iodine-deficient endemic areas in China (10). This finding greatly strengthens the presumption that this is the cause of insult to the developing brain.

Knowledge of the framework of developmental events in brain makes it possible to correlate the clinical deficit in neurological cretinism with the timing of the insult with some precision. The events of the first trimester, encompassing the basic embryonic organization of the brain, are completed without error. This is consistent with the fact that iodine treatment is not necessary before the end of the first trimester. The systems associated with the major triad of symptoms - the cerebral cortex the cochlea and the basal ganglia - all undergo major formative events during the second trimester, and all may plausibly be vulnerable at that time. Also, we know that neurological cretins may not be severely hypothyroid at birth and their deficits cannot be reversed by treatment after birth. This is consistent with the finding that systems having major formative events after birth are not notably affected, for example the cerebellum and the dentate gyrus of hippocampus.

The clinical observation of a case with all the features of neurological cretinism occuring in a non-endemic area and despite treatment beginning shortly after birth helps clarify our understanding of pathogenesis. In this case both fetal athyreosis and maternal hypothyroidism were demonstrated, suggesting that in man the combination of fetal and maternal hypothyroidism is necessary to produce neurological cretinism.

The lesions of neurologic cretinism make sense in anatomic space and developmental time. They promise to yield much more information in the future about human brain development and function.

REFERENCES

1. DeLong GR, Stanbury JB, Fierro-Benitez R. Neurological signs in congenital iodine-deficiency disorder (endemic cretinism). Dev. Med. Child Neurol. 27:317-324, 1985.

2. DeLong GR. Neurological involvement in iodine deficiency disorders. In: "The Prevention and Control of Iodine Deficiency Disorders", Hetzel, BS,Dunn JT and Stansbury JS eds. Elsevier, Amsterdam (1987), pp 49-63.

3. Held KR, Cruz ME, Moncayo F. The genetics of endemic cretinism. Prog. Clin. Biol. Res. 200:207-218, 1985.

4. Byers RK, Gilles FH, Fung C: Huntington's disease in children. Neurology (Minn) 23:561, 1973.

5. Lees AJ. The Steele-Richardson-Olszewski syndrome (progressive supranuclear palsy). In: Movement Disorders 2. Marsden CD and Fahn S (eds), Butterworth, London, 1987, pp 272-287.

6. Konig MP, Neige M. The pathology of the ear in endemic cretinism. In: Human development and the thyroid gland: relation to endemic cretinism. Stanbury JB and Kroc RL (eds). New York, Plenum Press, 1972, pp 325-333.

7. Davison AN, Dobbing J. The developing brain. In: Applied Neurochemisty. Oxford: Blackwell 1968, pp 253-286.

8. Pharoah POD, Buttfield IH, Hetzel BS. Neurological damage to the fetus resulting from severe iodine deficiency during pregnancy. Lancet 1971; 1:308-310.

9. Liu JL, Zhuang ZJ, Tan YB, Shi ZF, et al. Morphologic study on cerebral cortex development in therapeutically aborted fetuses in an endemic goiter region in Guizhou. Chinese Medical Journal. 97:67-72, 1984.

10. Liu JL, Tan YB, Zhuang ZJ, Shi ZF, et al. Morphologic studies of therapeutically aborted fetus thyroid in an endemic goiter region of Guizhou province. Chinese Medical Journal 95:347-354, 1982.

11. Blizzard RM, Chandler RW, Landing BW, Pettit MD and West CD. Maternal autoimmunization to thyroid as probable cause of athyrotic cretinism. New Eng J. Med. 1960,263:327-336.

12. Goldsmith RE, McAdams AJ, Larson PR, Mackenzie M and Hess EV. Familial autoimmune thyroiditis: Maternal-Fetal relationship and the role of generalized autoimmunity. J. Clin Endocrin Metab. 1973, 37:265-275.

13. Gaag RD van der, Drexhage HA, Dussault JH. Role of maternal immunoglobulins blocking TSH induced thyroid growth in sporadic forms of congenital hypothyroidism. Lancet 1985, 1:246-250.

14. Cooper ERA. The development of the human red nucleus and corpus striatum. Brain 1946, 69:34-44.

NEUROLOGICAL ASPECTS OF CRETINISM IN QINGHAI PROVINCE

Halpern J-P[1], Morris JGL[1], Boyages S[2], Maberly
G[2], Eastman C[2], Jin C[3], Wang Z-L[3], Lim J-L[3] Yu
D[4], and You C[4]
Neurology[1] and Endocrine[2] Units, Department of
Medicine, Westmead Hospital, Sydney, Australia
Institute of Endemic Diseases[3] and Qinghai
Provincial Peoples' Hospital[4]

SUMMARY

We have examined 69 cretins from the Province of
Qinghai in the Peoples' Republic of China, an endemia where
both myxoedematous and neurological cretinism occur.   A
characteristic neurological syndrome was observed in both
euthyroid and hypothyroid individuals.   The neurological
findings were similar to those which have previously been
reported: intellectual impairment, deaf-mutism, pyramidal
signs, gait disorder, primitive reflexes and squint.   The
gait disorder was not typical of the 'spastic diplegia' of
previous descriptions; much of it could be attributed to
laxity and deformity of the joints of the lower limbs.

In this endemia there was no clear distinction between
'neurological' and 'myxoedematous' cretins on the basis of
neurological signs. It appears that a similar neurological
insult has occurred in both types of cretin.

INTRODUCTION

There have been few detailed accounts of the
neurological aspects of endemic cretinism. McCarrison's
description (1) of the 'nervous' type of cretinism in the
Himalayan valleys of Chitral and Gilgit was the first and
remains an impressive feat of clinical observation. He noted
a 'knock-kneed spasticity' of the lower limbs and summarised
his findings as follows: 'In short, the condition is one of
cretinous idiocy with associated cerebral diplegia'. Since
then, many studies have mentioned deaf-mutism, mental
retardation, gait disturbance and squint but usually without
detailed descriptions. Like McCarrison, Choufoer and his
colleagues (2) felt that the gait of cretins (in New Guinea)
was that of a spastic diplegia but noted that the tone in
the lower limbs was often normal or even reduced when the
subjects relaxed.   Hornabrook (3) emphasised the evidence of
corticospinal damage in the cretins he studied in New

Guinea. The most detailed account of the neurology of
endemic cretinism in recent times is provided by DeLong (4)
on endemias in Zaire, Ecuador and China.

In this study we describe the neurological features of
69 cretins from the Province of Qinghai in the Peoples'
Republic of China, an endemia where both myxoedematous and
neurological cretinism occur.

BACKGROUND AND METHODS

Qinghai Province lies on the Tibetan plateau and has a
population of 4 million comprising Han Chinese, Tibetans and
the Wee people. The majority of the population relies on
subsistence agriculture for their livelihood and the area is
severely iodine deficient. The prevalence of cretinism has
been estimated as exceeding 10% by the Ministry of Public
Health. Iodized salt prophylaxis was introduced in 1978 and
the cretin rate has fallen greatly since then, particularly
in the eastern counties of the Province.

69 individuals, selected by the Chinese health
authorities to include both myxoedematous and neurological
cretins were studied at the Institute of Endemic Diseases at
Xining, the capital of Qinghai Province, in November 1986
and May 1987. A wide range of assessments and tests were
performed. The results of the neurological and radiological
examinations are the subject of this communication.
Patients are divided into two groups on the basis of current
thyroid status: euthyroid (TSH < 5 mIU/L) or hypothyroid
(TSH > 5 mIU/L). Total serum thyroxine (T4) was also
measured. An Intelligence Quotient was derived for each
patient using the Hiskey Nebraska Test of Learning Ability
and the Griffiths Mental Developmental Scales. The details
of the endocrine, psychological and neurophysiological
studies are the subject of separate communications.

RESULTS

The ages of the 69 patients ranged from 4-53 years.
There were 37 males (mean age 27.1 years) and 32 females
(mean age 26 years). 32 patients were euthyroid and 37
hypothyroid. The endocrine features of the 2 groups of
patients are summarised in Tables 1a and 1b. It was clear
that features of hypothyroidism were not uncommon in
patients who were currently euthyroid as well as in those
who were currently hypothyroid on TSH testing.

Gait

A characteristic gait was observed in both euthyroid
and hypothyroid patients. Its most obvious feature was
broadening of the base and knock-knees. The feet were flat
and everted, the knees flexed and the hips adducted. The
arms did not swing and were held in a curious posture: the
shoulders abducted, the elbows and wrists flexed. The
cretins walked, shoulders swaying, in a stiff shuffling
manner and turned with difficulty, effecting this manoeuvre
in a series of small steps. The trunk was tilted in flexion
in severely affected patients. Some required assistance to

AGE, HEIGHT, TSH AND T4 LEVELS OF THE TWO GROUPS OF PATIENTS				
	Currently euthyroid (n=32)		Currently hypothyroid (n=37)	
	mean	SD	mean	SD
Age (years)	21.9	9.63	30.6	11.96
Standing height (cm)	137	18.2	132.8	15.8
Sitting height (cm)	76	9.8	74.7	8.7
TSH (mIU/L)	2.7	1.34	104.9	112.32
Total T4	118	24.2	66	40.4

Table 1b

CLINICAL SIGNS OF HYPOTHYROIDISM IN THE TWO GROUPS OF PATIENTS		
	No of patients (%)	
Sign	Euthyroid (n=31)	Hypothyroid (n=35)
Goitre	4 (13%)	4 (11%)
Myxoedema	3 (10%)	16 (44%)
Delayed ankle jerk relaxation	5 (17%)	15 (41%)
Loss of outer portion of eyebrow	15 (47%)	20 (68%)
Abnormally soft cartilage	19 (61%)	22 (63%)
Saddle nose deformity	24 (77%)	21 (60%)

walk. A spectrum of abnormality was present, flat feet
being the most consistent finding (see Table 2). Two
patients had a hemiparetic gait and in a further eight the
gait was high-stepping.

Table 2

CLINICAL FEATURES OF THE CRETIN GAIT	
Feature	% of patients (n=69)
Flat feet	97
Reduced arm swing	66
Everted feet	59
Knocked-kneed	47
Wide-based	31
Knees flexed	31
Hips flexed	20

Pyramidal signs

Pyramidal signs were present in most patients (see
Table 3a) regardless of current thyroid status (see Table
3b). Brisk reflexes and increased tone in the proximal
muscles of the lower limbs were the most striking findings.
The pyramidal signs were asymmetrical in nearly one third of
patients.

Other neurological signs

Primitive reflexes and a non-paralytic squint, with
failure of abduction of one or both eyes on lateral gaze,
were present in many patients (see Table 4). Half the
patients were deaf to a greater or lesser extent. We found
no signs of cerebellar dysfunction or of a peripheral
neuropathy. The only evidence of an extrapyramidal
disturbance was mildly abnormal posturing of the hands and a

Table 3a

SIGNS OF A PYRAMIDAL LESION	
Brisk reflexes	% of patients (n=69)
Pectorals	18
Triceps	38
Biceps	50
Supinator	71
Knee	91
Hip adductors	38
Ankle	33
Extensor plantars	25
Asymmetrical signs	29
Increased tone	
Elbow	22
Wrist	17
Knee	43
Ankle	15

Table 3b

ABNORMALITIES OF THE PYRAMIDAL SYSTEM AND OF THE HIP JOINTS IN THE TWO GROUPS OF PATIENTS		
Feature	% of patients	
	Euthyroid (n=32)	Hypothyroid (n=37)
Increased tone at the knees	22	22
Brisk knee jerks	43	49
Extensor plantars	19	6
Increased hip laxity	22	26

positive glabellar tap sign. The pupils were normal. The cretins were friendly and co-operative in a childlike way.

Table 4

OTHER NEUROLOGICAL FEATURES	
Feature	% of patients (n=69)
Deafness	51
Strabismus	47
Glabellar tap	78
Grasp reflex	62
Mild hand dystonia	21

Joint abnormalities (Table 5)

Nearly half the patients had abnormal laxity of the hips and ankles. It was surprising to be able to dorsiflex the foot, so that it almost touched the shin, in patients with obvious spasticity of the proximal muscles of the leg. Fixed deformities of the spine were present in a few patients. X-rays of the hips showed coxa vara and valga deformities, shortening of the femoral neck, deformities of the head of the femur and flattening of the acetabula.

Table 5

JOINT ABNORMALITIES	
Clinical feature	% of patients (n=69)
Lax upper limb joints	12
Lax hips	49
Genu recurvatum	18
Lax ankles	51
Kyphosis	7
Scoliosis	15
Lordosis	6
Radiological feature	(n=53)
Coxa varus	25
Coxa valgus	23

CT Scans

50 patients had CT scans of their heads. Calcification of the basal ganglia was present in 15, mild atrophy of the cerebral hemisphere in 4 and cerebellar atrophy in 3. Four patients had evidence of focal atrophy.

DISCUSSION

The main neurological findings of mental retardation, deafness, gait disorder, spasticity, primitive reflexes and squint are similar to those of DeLong and other investigators (1,2,3,4). These features result in a characteristic neurological syndrome with a wide range of expression from those with minimal disability to those who are totally dependent on others.

The unexpected aspect of these neurological findings was that they were as common in currently hypothyroid cretins as they were in euthyroid (neurological) cretins. We did not see myxoedematous cretins without neurological signs. McCarrison (1), upon whose observations the current classification of endemic cretinism is based, did not comment on the neurological abnormalities of the myxoedematous cretins. In the current definition of myxoedematous cretinism (5) it is stated that deaf mutism and spastic diplegia are infrequent though it is also emphasised that mixed forms are common. It is significant however that in a careful study of myxoedematous cretins in the Ubangi region of Zaire, DeLong (4) found that one third of the patients had neurological signs. Ibbertson and his colleagues (6) also emphasised in their study in Nepal that the current classification 'may serve to obscure what is in reality a continuing spectrum of physical, mental and functional abnormality'.

While it has proved useful, in descriptive terms, to divide cretinism into myxoedematous and neurological types it is clear from this and other studies that this classification is not ideal for all endemias. Such a classification implies different aetiologies for the two types. Our experience in Qinghai leads us to believe that endemic cretinism is the result of two variables: lack of thyroid hormone in utero, which predominantly affects the developing nervous system, and continuing hypothyroidism after birth which causes stunting of growth, failure of sexual maturation, bone deformities and other somatic effects. Both variables play a part, to a greater or lesser degree, in most, if not all cretins. On this basis, one would expect signs of continuing hypothyroidism, such as stunting of growth, in many neurological cretins. This was certainly the case in Qinghai. The paucity of neurological signs in myxoedematous cretins in some studies may reflect the fact that a cretin in whom both variables are present to a high degree would be unlikely to survive for long after birth.

The term 'spastic diplegia' does not seem an appropriate description of the gait observed in this study. Spastic diplegia implies a bilateral pyramidal lesion and typically the gait is stiff-legged with circumduction and adduction of the hips. The feet are inverted causing 'scissoring' (7). This was not the gait that we observed. There was evidence of spasticity in the proximal muscles of the legs and the hips assumed an adducted posture. The most striking feature however was a broadening of the base and knock-knees. Increased joint laxity was noticeable at the hips, knees and ankles. Marked deformities of the hips were evident on x-ray; these had the features of an epiphyseal dysplasia (8). It seems likely that these joint abnormalities contributed to the gait. Genu recurvatum and hip dislocation are also described in spasticity of the lower limbs associated with cerebral palsy (9), the bones failing to grow normally in the presence of abnormal muscle traction associated with the neurological deficit.

In this endemia there was little evidence of an extrapyramidal disturbance of the type described by DeLong in Ecuador (4) apart from a positive glabellar tap sign and mild abnormal posturing of the hands. The squint, which was present so commonly, was probably due to a failure to develop stereoscopic vision and conjugate gaze, for it was non-paralytic. This feature is common in mental handicap of any type. Considering the degree of neurological disability in so many of these patients, the findings on CT scanning of the brain were unremarkable. This contrasts with the experience in Ecuador where cerebral atrophy was severe (10).

The subject of endemic cretinism has been bedevilled over the years by the fact that iodine deficiency disorders appear to express themselves very differently from country to country. We hope that those observations on the cretins of Qinghai Province add one more piece to the immense jigsaw puzzle of endemic cretinism.

REFERENCES

1.   McCarrison R. Observations on endemic cretinism in the Chitral and Gilgit valleys. Lancet 1908;ii:1275-1280.
2.   Choufoer JC, Van Rihn M, Querido A. Endemic goiter in Western New Guinea. II. Clinical picture, incidence and pathogenesis of endemic cretinism. J Clin Endocrinol Metab. 1965; 25:385-402.
3.   Hornabrook RW. Neurological aspects of endemic cretinism in Eastern New Guinea. In: Hetzel BS, Pharoah POD. Eds. Endemic Cretinism. Papua, New Guinea. Institute of Human Biology monograph Series No.2. 1971;105-107.
4.   DeLong R. Neurological involvement in iodine deficiency disorders. In: Hetzel BS, Dunn JT, Stanbury JB. Eds.: The prevention and control of iodine deficiency disorders. Elsevier Science Publishers BV. 1987;49-63.

5. Querido A, Delange F, Dunn T, Fierro-Benitez R, Ibbertson HK, Koutras DA, Perinetti H. Definitions of endemic goiter and cretinism, classification of goiter size and severity of endemias and survey techniques. In: Dunn JT, Medeiros-Neto GA. Eds: Endemic goiter and cretinism: continuing threats to world health. Pan American Health Organisation, Scientific Publication 292, Washington DC. 1974;267-272.

6. Ibbertson HK, Tait JM, Pearl M, Lim T, McKinnon J, Gill MB. In: Hetzel BS, Pharoah POD. Eds: Endemic cretinism. Papua, New Guinea. Institute of Human Biology Monograph Series No. 2. 1971;71-88.

7. Adams RD, Victor M. Principles of Neurology. Third Edition. McGraw-Hill Book Co. New York. 1985;93.

8. Wynne-Davies R, Hall CM, Apley AG. Atlas of skeletal dysplasias. Churchill Livingstone, Edinburgh. 1985;19.

9. Simon SR, Deutsch SD, Nuzzo RM, Mansour MJ, Jackson JL, Koskinen M, Rosenthal RK. Genu recurvatum in spastic cerebral palsy. J of Bone and Joint Surgery 1978;7:882-894

10. Ramirez I. Cruz M, Varea J. Endemic cretinism in the Andean region: new methodological approaches. In: Delange F, Ahluwalia R. Eds.: Cassava toxicity and thyroid: research and public health issues. Ottowa:IDRC. 1983;73-76.

INFLUENCE OF IODINE DEFICIENCY ON HUMAN FETAL THYROID GLAND AND BRAIN

Liu Jia-Liu*, Tan Yu-Bin**, Zhuang Zhong-Jie*,
Shi Zhon-Fu*, Chen Bin-Zhon** , and Zhang Jia-Xiu*

* Guiyang Medical College, Guiyang
** Tianjin Medical College, Tianjin, China

INTRODUCTION

In order to evaluate the influence of prenatal iodine deficiency on the thyroid gland and brain in man, we studied therapeutically aborted fetuses collected from a severely iodine deficient endemic goiter and cretinism area before and after 5 year iodized salt utilization. Fetuses from a non-endemic area served as the normal controls[1,2,3].

MATERIAL AND METHODS

Material

Fetuses were obtained by therapeutic abortion as a birth control measure by the hydrostatic-bag method. Fetal age was estimated by the available clinical data of last maternal menstrual history and fetal crown-rump length.

1. The normal control group consisted of 20 fetuses from a non-endemic area: 4th month 5, 5th month 6, 6th month 4, 7th month 3, 8th month 4.

2. The iodine deficient group consisted of 30 fetuses from an endemic area before iodine supplement: 4th month 7, 5th month 6, 6th month 6, 7th month 6, 8th month 5.

3. The iodine treated group consisted of 32 fetuses from the endemic area after 5 year iodine supplement (with iodized salt 1/50,000 KI). The mothers also received one single injection of iodized oil (475 mg/ml) at the end of the first year: 4th month 6, 5th month 5, 6th month 7, 7th month 6, 8th month 8.

Laboratory Techniques

Urine iodine was determined by Barker's modified alkaline incineration technique and catalytic reduction of ceric ion by arsenite salt. A random afternoon urine sample was collected and assayed. Urine iodine was expressed in µg/gm creatinine.

Serum throxine (T4) was determined by a modified protein binding technique with the T4 kit supplied by the Atomic Energy Research Institute, Academia Sinica, Beijing.

Serum triiodothyronine (T3) was determined by RIA (separated by resin technique), with antibodies labelled by the Isotope Labelling Laboratory of the Atomic Energy Research Institute, Academia Sinica, Beijing.

Serum human thyrotropin (hTSH) was radioimmunoassayed with hTSH and antisera from Calbiochem, California.

Morphologic Study of the Thyroid Gland

For light microscopy, the fetal thyroid gland was fixed in 10% buffered formalin solution after weighing. Tissue blocks were embedded in paraffin. Sections were stained with HE and PAS. Frozen sections were stained for peroxidase activity. In addition to the routine microscopic examination, the maximal diameter of the follicles and the heights of the epithelial cells and their nuclei were measured by micrometer. In each case, 100 cells were randomly measured, the average values were obtained and the nucleus-cytoplasm index (N/C) was calculated.

For electronmicroscopy, small fragments of fresh thyroid tissue were fixed in 25% glutaldehyde. Postfixation with osmic acid was followed by dehydration in a graded series of alcohols and the specimens embedded in epon 812. Ultrathin sections were stained with lead hydroxide and studied under a JEM-100CX electron microscope.

Morphometric Study of the Brain

The brain was fixed in situ by perfusion with 10% buffered formalin and removed 1 week later. After weighing, thin pieces of cerebral tissue (0.5mm or less in thickness) were taken from certain regions of the frontal lobes and fixed again in 10% buffered formalin for 24 hours. Specimens were subsequently embedded in celloidin. All sections were stained with fast cresyl violet and Weil's methods for myelin. Blocks from similar areas were treated with Golgi's silver impregnation.

Cytoarchitectonic studies of the cerebral tissue were made on the basis of the following criteria: the number of nerve cells (including undifferentiated neuroblasts and differentiated neurons) and astrocytes in each $100 \times 100 \times 100 \ \mu^3$ volume; the percentage of differentiated neurons in the total nerve cell population; the number of heterotopic nerve cells in the layer considered; the development of dendrites and their bulbs and spines; and the degree of myelinization.

RESULTS

Iodine Nutrition Status

The average values of the urine iodine, and 24 hour thyroid 131^I uptake as the parameters for iodine nutrition status of the inhabitants are listed in Table 1.

Pituitary and Thyroid Axis Functional Status
of the Mother and Fetus

1. The serum T4 levels of the maternal, cord and fetal ventricular blood of the three groups are listed in Table 2. Both the maternal and cord serum T4 values in the iodine deficient group were significantly lower in comparison with the normal control group, while the serum T4

Table 1. Iodine Nutrition Status of Inhabitants in
Various Areas

Parameters	Normal	I-defic.	I-treated
Morbid. rate of goiter (%)	0.9 (4320)	31.5* (3790)	4.5 (4180)
Morbid. rate of cretinism (%)	0 (63)	0.2* (53)	0 (64)
Urine iodine (μg/g creat.)	63.1±1.4 (53)	14.5±2.2 (75)	91.2±4.1 (62)
24 hr thyroid 131_I uptake (%)	43.0 (93)	74.3* (83)	30.5 (104)

Numerals in parenthesis represent inhabitants examined
* $p < 0.01$ compared to the normal controls

level of the fetal ventricular blood in the iodine deficient group was
nearly the same as that of the normal control. In the iodine treated group
the maternal and fetal ventricular serum T4 values reached the normal
range, but the cord serum T4 was still lower than that of the normal.

The maternal serum T4 values in all three groups were significantly
higher than those of the cord and fetal ventricular blood. In both the
iodine deficient and the iodine treated groups there was no significant
difference between the serum values of the cord and fetal ventricular
blood, but in the normal control group, the serum T4 values showed
significant difference.

2. The serum T3 levels are shown in Table 3. Both the maternal and
cord T3 levels of the iodine deficient group were lower compared to the
normal control group. The fetal heart T3 of the iodine deficient group was
higher than of either the normal control or the iodine treated group. The

Table 2. Serum T4 Levels of the Maternal, Cord and Fetal
Heart Blood (μg/dl) ($\bar{x}$ ± S.D.)

Group	Maternal Blood	Cord Blood	Fetal Heart Blood
Normal control	14.5±2.1 (24)	10.3±5.3 (13)	5.0±1.8## (12)
I-deficient	8.7±3.2* (29)	5.1±1.9*# (25)	4.2±1.6## (29)
I-treated	14.1±2.3 (21)	6.5±2.7*## (22)	6.3±2.5## (22)

Numerals in parentheses: number of fetuses examined
* $p < 0.01$ compared to the normal control
# $p < 0.01$, ## $p < 0.05$ compared to the maternal blood

Table 3. Serum T3 Levels of the Maternal, Cord and Fetal Heart Blood (ng/dl) ($\bar{x} \pm$ S.D.)

Group	Maternal Blood	Cord Blood	Fetal Heart Blood
Normal control	141.0±3.9 (21)	49.3±23△ (12)	39.1±16.2△ (13)
I-deficient	124.5±21.6* (30)	42.4±15.6*△ (26)	46.5±18.1△* (30)
I-treated	136.8±24.4 (22)	28.8±10.0*△ (22)	25.8±6.1△* (21)

Numerals in parentheses: No. of fetuses examined
* p<0.01 compared to the normal control
△ p<0.01 compared to the maternal blood

maternal serum T3 of the iodine treated group reached the normal range. The cord and fetal heart serum T3 in the iodine treated group were lowered markedly in comparison with the normal control and iodine deficient groups. In all three groups, the maternal serum T3 values were significantly higher in comparison with the cord and fetal heart blood, the serum T3 values of the latter two showing no significant difference between each other.

3. As shown in Table 4, the serum TSH values of the maternal, cord and fetal heart blood of the iodine deficient group were significantly higher compared to the normal control and the iodine treated groups. The serum TSH value of the cord blood of the iodine treated group was lower than that of either the normal control or the iodine deficient group. In the normal control group, the difference between the serum TSH values of the fetal heart and the cord blood was significant, but in the iodine

Table 4. Serum TSH Levels of the Maternal, Cord and Fetal Heart Blood (uU/ml)

Group	Maternal Blood $\bar{x}\pm$S.D.(log)	G	Cord Blood $\bar{x}\pm$S.D.(log)	G	Fetal Heart Bl. $\bar{x}\pm$S.D.(log)	G
Normal control	0.56±0.28 (22)	3.6	0.78±0.57 (14)	6.7△	0.30±0.31 (13)	2.0
I-deficient	0.65±0.35 (29)	4.5**	1.09±0.38 (26)	12.3*△	1.08±0.43 (30)	12.0*△
I-treated	0.49±0.27 (22)	1.9*	0.33±0.24 (21)	1.7*	0.37±0.23 (18)	2.3

Numerals in parentheses: Fetuses examined
* p<0.01, ** p<0.05 compared to normal and I-treated
△ p<0.01, △△ p<0.05 compared to maternal

deficient and the iodine treated groups, these values did not show any
significant difference between each other.

Morphology of the Thyroid Gland

Thirteen of 30 iodine deficient fetuses showed marked enlargement of
the thyroid gland, the weight of which was considerably above the upper
limit ($\bar{x} \pm 2$ S.D.) of the weight of the normal control group. The thyroid
gland weights of these goitrous fetuses as well as the fetuses of the
iodine treated and the normal control groups are listed in Table 5.

Histologic study shows that the histologic features of the thyroid
gland of the iodine treated fetuses were comparable to normal control
fetuses. In the 4th - 5th month fetuses, the thyroid gland showed distinct
formation of follicles lined with small cuboidal epithelial cells and
containing scanty colloid. In the 6th - 8th month fetuses, the thyroid
tissue showed more mature differentiation. The follicle size and the
amount and staining density of the colloid increased with the fetal age.
The epithelial cells were also cuboidal in shape. In the 13 goitrous
iodine deficient fetuses, the thyroid gland showed marked hyperplasia
irrespective of the fetal age. There were papillary tufts formed. The
follicular lumen contained no colloid. Hypertrophy of the epithelial cells
which became high columnar with large and irregular nuclei was noted.
Prominent hyperemia of the perifollicular capillaries was seen. Histologic
examination of the 17 nongoitrous iodine deficient thyroid specimens showed
that, in 10 of them, the thyroid histologic picture assumed no significant
difference from the histologic picture of the normal control thyroids of
the same age group, but in the remaining 7, the thyroid gland assumed
hyperplasia similar to that seen in the goitrous fetuses but of milder
severity.

Electronmicroscopy of the thyroid tissue in the goitrous fetuses
confirmed the observation made by light microscopy. The high columnar
epithelial cells showed well developed mitochondria and rough-surfaced
endoplasmic reticulum with dilated cisternae suggesting an increase in the
function of protein synthesis. The microvilli were underdeveloped and
colloid droplets were scarce, indicating poor functioning of colloid
absorption and hormonal secretion. These changes were in accordance with
the hypofunctional state of the thyroid gland.

Table 5. Weight of the Thyroid Gland (mg)

Age Group (month)	Normal Control		I-deficient		I-treated	
	N	$\bar{x} \pm$ S.D.	N	$\bar{x} \pm$ S.D.	N	$\bar{x} \pm$ S.D.
4	5	187±38	3	460±30*	6	164±36
5	4	484±66	3	1090±332*	5	465±57
6	4	597±80	2	1955±75*	7	583±67
7	3	838±137	3	1967±982*	6	865±124
8	4	1350±237	2	5970±240	8	1234±310

13 of 30 iodine deficient fetuses showed enlargement of the
thyroid gland
* p<0.01 compared to the normal controls of same age group

Table 6. Weight of the Brain (g)

Age Group	Normal Control		I-treated		I-deficient	
	N	$\bar{x}$ ± S.D.	N	$\bar{x}$ ± S.D.	Case 1	Case 2
6th-month	3	155±5	7	152±12	126	133
8th-month	3	448±46	8	432±36	243	228

Morphometry of the Brain

The weights of the brains of the fetuses are listed in Table 6. As compared with the fetuses of the normal group, two of the 6th-month fetuses and two of the 8th-month fetuses in the iodine deficient group showed decreased brain weight.

Cytoarchitectonic study. 6th-month fetuses: The cerebral cortices of various areas were not well differentiated or developed, only three layers being distinguishable. Cortex nerve cells consisted almost entirely of undifferentiated neuroblasts. No mitoses were found. With Golgi's silver impregnation, apical dendrite budding was seen, but not the basal dendrites. No myelination was revealed in the cortex with Weil's stain. Two of the iodine deficient fetuses showed a significant increase in cell population compared to the other two groups (above the upper limit of the normal range). The cell count is listed in Table 7.

8th-month fetuses: The cerebral cortex of all areas examined was clearly differentiated into 6 layers. The cell elements were readily distinguished into neuroblasts, pyramidal cells, granular cells and astrocytes, etc. Heterotopic nerve cells were found in the cortical layers and underlying white matter. On Golgi preparations, the dendrites of pyramidal cells were developed. Weil's stain revealed no myelination in the cortex yet. In respect of the cell count, two of the iodine deficient

Table 7. Cerebral Cortex Cell Count in the 6th-month Fetuses (Number per each 100X100X100 u^3)

Area	Layer	Normal (3 cases) $\bar{x}$ ± S. D.	I-treated (7 cases) $\bar{x}$ ± S. D.	I-deficient Case 1	Case 2
Precentral	I	1279±69	1330±72	1380	1395
gyrus	II	3170±115	3200±125	3430	3433
	III	2184±87	2200±68	1865	2565
Frontal	I	1779±45	1850±45	1865	1915
pole	II	3465±96	3428±78	3775	3675
	III	2349±88	2731±92	2606	2615
Temporal	I	1880±42	1784±57	2050	1935
lobe	II	3288±137	3375±126	3484	3565
lob	III	2858±72	2759±47	3085	2305
Occipital	I	1537±65	1724±72	1655	1665
pole	II	3397±104	3425±98	3670	3660
	III	2663±115	2554±104	2440	2485

Table 8. Precentral Gyrus Cell Count in 8th-month Fetuses (Number per each 100X100X100 u^3)

Group	Layer	Nerve Cells $\bar{x} \pm$ S. D.	Undif. Neuroblast %	Heterotopic Nerve Cells $\bar{x} \pm$ S. D.
Normal	I	363±59	31	78±18
Control	II	2792±257		147±21
(3 cases)	III	503±107	30	162±8
	IV	953±36		212±84
	V	417±18	32	
	White matter			173±3
Iodine	I	370±45	30	69±19
treated	II	2562±231		162±9
(8 cases)	III	489±121	31	168±58
	IV	968±42		168±58
	V	403±25	32	
	VI	545±27	38	
	White matter			182±3
Iodine deficient				
Case 1	I	465	25	95
	II	2785		141
	III	715	48	240
	IV	1135		375
	V	615	41	
	VI			
	White matter			195
Case 2	I	455	245	95
	II	2814		90
	III	705	63	140
	IV	1055		480
	V	450	44	
	VI	580	47	
	White matter			225

fetuses showed increased cell density in all layers of various areas compared to the normal control and the iodine treated groups, as shown in Table 8.

DISCUSSION

Urine iodine excretion is the best single index of organism iodine nutrition. Although we were unable to detect the urine iodine content of the fetuses, we could estimate their iodine nutrition status indirectly. The fetuses collected from the endemic area before iodine supplementation were born in an area where the inhabitants were severely deficient in iodine and the mothers were suffering with goiter and hormonal hypothyroidism (indicated by lower T4 and higher TSH). The fetuses collected from the same endemic region after 5 year iodized salt prophylaxis were considered in normal iodine nutrition status, because the iodine deficiency of the inhabitants including the mothers had been corrected by iodine treatment as shown by the values of the urine iodine and thyroid ^{131}I uptake, as well as the serum T4 and TSH. These values were comparable to the non-endemic normal control fetuses.

Our study on the maternal and fetal serum hormones showed that in the iodine deficient group, the maternal, cord and fetal heart serum T4 levels were lowered and TSH raised. These values however, returned to normal after correction of iodine deficiency with iodized salt prophylaxis. In all the three groups, the maternal serum T4 and T3 levels were significantly different from the comparison with either the cord or the fetal serum levels, indicating that placental transfer of iodothyronines is severely limited in both the maternal to fetal and fetal to maternal direction. This is also in accordance with previous studies[3]. Our data strongly support that the fetal pituitary-thyroid axis is independent of the mother's. These facts are against the concept that maternal hypothyroidism plays a role in the pathogenesis of endemic cretinism. The cord and fetal heart serum T3 levels were significantly lower than those of the mothers in all the three groups. A high maternal-fetal gradient was apparent and this may be due to the inability of T3 to cross the placental barrier. The possibility exists that a compensatory increase in fetal production of T3 in the iodine deficient fetuses might occur as Pretell has suggested[4].

Our observations on the thyroid gland showed goiter was found in 43.3% of the fetuses collected from the endemic region before iodine supplement. The pathologic changes were substantially the early changes in the hyperplastic stage seen in acquired endemic goiter. It is presumed that both conditions are the results of chronic overstimulation of the thyroid by TSH.

It has been reported that the incidence of congenital goiter in neonates is high in severe endemic goiter regions[4,5,6]. Some authors considered the incidence an important index in deciding the severity of the prevalence of this disease. Our finding of high incidence of congenital goiter in the fetuses collected from the endemic region before iodine prophylaxis might serve to demonstrate the severity of this disease in that region. However, after 5 year iodized salt utilization, with correction of iodine deficiency in the inhabitants, the aborted fetuses showed no more goiter changes.

The present investigation showed that a certain number of the iodine deficient fetuses had decreased brain weight as compared to the normal control fetuses of the same age. This was considered as one of the criteria for retardation of brain development. The morphometric study of the cerebral cortex in the iodine deficient fetuses also revealed certain features indicating retardation of brain development. In case of the 6th-month fetuses, the retardation was manifested by an increase in the cell population per unit of brain volume which, according to Rabinowich[7], is one of the morphologic criteria for immaturity due to scarcity of cytoplasm and intercellular substances. In case of the 8th-month fetuses, the retardation of development was manifested by some additional features besides the increase of cell density. The percentage of undifferentiated neuroblasts was increased suggesting depression of their differentiation; the number of heterotopic neurons was increased indicating a slower migration rate of some neurons.

Although the materials studied in the present paper and the above mentioned differences were relatively small, the morphometric changes in the brains of certain iodine deficient fetuses are definite. The results suggest that prenatal iodine deficiency exhibits certainly a depressive influence on the brain in some of the fetuses in the endemic goiter and cretinism areas retarding its development. This might account for the neurological deficit of cretinism or cretinoid syndrome.

REFERENCES

1. Liu Jia-Liu, et al., Morphologic studies of therapeutically aborted
    fetus thyroid in an endemic goiter region of Guizhou Prov. Chinese
    Med J. 95:347 (1982).

2. Liu Jia-Liu, et al., Morphologic study of development of cerebral
    cortex of therapeutically aborted fetuses in an endemic goiter
    region, Guizhou Province, China. in: "Current Problems in Thyroid
    Research (Proceedings of the Second Asia and Oceania Thyroid
    Association Meeting, Tokyo)," Nobuo Ui, et al., ed., ISBN Elsevier
    Science Publishing Co., New York (1982).

3. Chen Bin-Zhon, et al., Survey of endemic goiter and cretinism in Heba
    Commune, Guizhou, China. Changes of pituitary-thyroid axis
    functional status after 3 year iodized salt prophylaxis, Chinese J
    Endemic Dis. 3:97 (1984).

4. E.A. Pretell, et al., Iodine deficiency and the maternal-fetal
    relationship, in: "Endemic Goiter and Cretinism", J.T. Dunn, et
    al., ed., PAHO, Washington (1974).

5. U. Tezuka et al., Congenital goiter found in a district of Omuro,
    Kochi, Sikoku, Japan, Endocrinol. Jpn. 18:281 (1971).

6. D.B. Jones, Congenital goiter in North America, Am J Pathol. 27:85
    (1971).

7. T.H. Rabinowich, The cerebral cortex of the premature infant of the
    8th-month, in: "Growth and Maturation of the Brain", D.P. Purpura
    ed., Elsevier Publishing Co., New York (1964).

NEUROPSYCHOLOGICAL STUDIES IN IODINE DEFICIENCY AREAS IN CHINA

T. Ma, Y.Y. Wang, D. Wang, Z.P. Chen, and S.P. Chi

Institute of Endocrinology
Tianjin Medical College
Tianjin, China

There are two types of cretin in China. The major type is neurological and the minor type is myxedematous. Neuropsychological findings such as mental retardation, deaf-mutism and neuromuscular abnormalities are more prominent among the neurological cretins and mild among myxedematous cretins. The difference is both quantitative and qualitative. The neuropsychological difference between the two types has been compared not only in different endemias, but also in the two types of patients in the same endemia.

Besides the typical clinical cretin, we have found a subclinical type of cretin in various endemias in China. The characteristics of the "subcretin" are summarized and the diagnostic criteria as well as the method of assessment discussed in detail.

The presence of "subcretins" may be more significant than the presence of cretins in China in terms of the impact on society and productivity.

Iodine deficiency disorders are still a significant problem in China. More than 1/3 of our one billion population lives in iodine deficient areas. However, 90% of them are under iodization programs at present, but there still remain 11 million goiter patients. We haven't found any apparent psychological defects in goiter patients. The neuropsychological defects are found only among the offspring of iodine deficient parents.

There are two types of cretinism in China[1]. The major type is neurological cretinism which is found in almost all the cretin endemias of China; the myxedematous cretin is the minor type which is found in the north-west part of China. The latter are always mixed with the neurological cretin; there is no clear cut demarcation between these two types of cretins. One could consider the different kinds of cretin arranged on a spectrum: the typical neurological cretin is on one extremity of the spectrum; the typical myxedematous is on the other extremity, and most cretins fall between.

I. Characteristics of cretins in China

Chengde is an endemia of neurological cretinism in China. Ma and Chen[2] examined the cretins there. Only 2 cases had frank myxedema in addition to their neurological signs. Mental defects (idiot and imbecile)

Table 1. Comparison of Vestibular Function Between
Neurological & Myxedematous Cretins

| | Vestibular Function | |
	Normal	Abnormal
Neurological Cretin (102)	5(4.9%)	97(95.1%)
Myxedematous Cretin (13)	12(92.3%)	1(7.7%)
Control Children (193)	193(100%)	0(0%)

were found in all of the cretins in Chengde. Hearing and speech defects
were also quite significant. 95% of them had hearing defects in different
degree and 97.5% had different degrees of speech defects.

The defect in vestibular function is usually parallel to the defect
in hearing in neurological cretins. Wang[3] found 95.1% of neurological
cretins had abnormal vestibular response.

Neuromuscular defects in neurological cretins in Chengde showed
chiefly pyramidal signs such as increased tendon reflexes as well as the
presence of pathological reflexes. Cranial motor nerve defects were not
so common in Chengde but quite common in Guizhou[4]. As a rule, most
patients showed truncal rigidity with their forearms adducted over the
trunk and fingers extended and immobile, similar to the posture of
parkinsonism. Some patients showed rigidity more prominent on proximal
than the distal part of extremities. Giving L-DOPA to 8 neurological
cretins, four of them showed remarkable release of rigidity, and the
neurological involvement of these four patients was moderate; no
remarkable improvement was seen in either severely or mildly affected
patients. These findings indicate the presence of extrapyramidal
involvement. No definite evidence of muscular atrophy caused by a lower
motor neuron lesion was seen. The neurological involvement of the mixed
type cretins was quite similar to those of the neurological type except
for the presence of some signs of hypothyroidism. Psychomotor tests, such
as choice reaction time and tapping are all performed much more poorly by
cretins as compared with the control children; the myxedematous type
cretins do somewhat better than the neurological cretins.

Table 2. Comparison of IQ of Different Types of Cretin

	N	$\bar{X}$	S	P
Neurological	12	21.3	9.1	<0.02
Myxedematous	12	37.9	18.9	

Table 3. Comparison of Tapping Ability of Different
         Types of Cretins

	N	$\bar{X}$	S	P
Neurological	9	45.5	10.9	<0.01
Myxedematous	11	59.7	9.8	
School Children in IDD Endemia	52	67.5	8.5	<0.01
School Children in Non-endemia Area	64	78.0	12.4	

Table 4. Comparison of Steadiness of Different Types
         of Cretins

	N	$\bar{X}$	S	P
Neurological	3	0.187	0.052	<0.05
Myxedematous	11	0.246	0.075	

Table 5. Comparison of Choice Reaction Time (m sec) of
         Different Types of Cretins and School Children

	N	$\bar{X}$	S	P
Neurological Cretin	6	1794.2	461.1	<0.05
Myxedematous Cretin	11	1415.3	446.7	
School Children in IDD Endemia	67	1091.2	608.7	
School Children in Non-endemia Area	52	858.6	442.0	<0.05
	60	893.0	474.6	

Table 6. Comparison of Neuropsychological Changes Among 252 Cretins of Different Types in Dongsheng Endemia, Inner Mongolia, China

	Neurological (184)	Mixed (48)	Myxedematous (20)
Mental Defect			
Idiot	15.7%(184)	9%(48)	0%(20)
Imbecile	36.1%(184)	66%(48)	15%(20)
Moron	47.8%(184)	11%(48)	85%(20)
Hearing Defect			
Completely Deaf	} 95.1%(184)	} 89.6%(48)	0%(20)
Impaired Hearing			50%(20)
Normal	4.8%(184)	10.4%(48)	50%(20)
Speech Defect			
Completely Mute	} 96.2%(184)	} 85.4%(48)	0%(20)
Impaired Speech			30%(20)
Normal	3.8%(184)	14.6%(48)	70%(20)
Neuromuscular Defects			
Abnormal Gait	62.5%(184)	75%(48)	30%(20)
Exaggerated Knee Jerk	37%(123)		10%(20)
Spastic Paralysis	2.7%(184)	0%(48)	0%(20)
Deformity of Extremity	9.9%(11)		0%(20)
Squint	14.6%(184)	6%(48)	0%(20)

By Dr. Liu, W.M.

In some very typical myxedematous cretins in Hetan, the neurological signs are very mild, and difficult to detect by ordinary clinical measurements.

Some myxedematous cretins could hear and understand well on ordinary vocal utterance and even whispering. Some myxedematous cretins could work for us as an interpreter between Mandarin and the minority nationality language. The vestibular function test showed that only 7.7% are abnormal. Pyramidal signs such as exaggerated knee jerk are usually absent; if present, it is very mild. Patients had no definite spasm nor rigidity, no tendency to parkinsonian posture, and they could dance beautifully and harmonically following their folk music. They have very short stature, especially extremities. Both sexual glands and secondary sexual characteristics are underdeveloped. Frank myxedema and other signs of hypothyroidism are common. Clinically they show a picture very much similar to that of childhood or adolescent myxedema patients of non endemic areas. In those endemias where we found such typical myxedematous cretins, the mixed and neurological cretins are still in the majority. We have yet to give an answer as to why iodine deficiency may cause such different types of cretins in the same endemia.

II. The Problem of Subclinical Cretins

Cretins can be diagnosed by ordinary clinical observations, so we could call them the clinical cretins. Besides these, there may be individuals who are harmed by iodine deficiency but hardly recognized with ordinary clinical observations. Twenty years ago, visiting a village primary school in an IDD endemia, I was very much impressed by the

complaint of the school master about the disabilities of his pupils: one
fifth of his pupils could not pass the regular grades; some of them had
repeated their first grade lessons for several years. These pupils were
not naughty at all, on the contrary they were extraordinarily obedient;
some of those pupils had difficulties in certain skills and they were
sluggish in physical exercises and even unable to touch the ball in any
basket ball games. At first I thought this might be an incidental
occurrence, but in the following years, I heard similar complaints from
other endemias. Besides the low intelligence and poor performance, some
teachers also complained about the poor hearing of their pupils.
Sometimes I noticed the children in the endemias were somewhat shorter
than the average height. I used to puzzle: is this really a general
factual phenomenon and what is its nature: Then from the papers of
certain scientists in Latin America[5] as well as in Belgium[6], I learned
that the populations in their endemic areas were also noticed to be not
sharply divided between obvious cretins and completely normal individuals.
There seemed to be some gradation between these two. By the definition of
endemic cretin given by PAHO meeting in Lima 1983, a patient could only be
diagnosed as a cretin if its clinical picture could fulfill the diagnostic
criteria. But the individuals described above could not fulfill the
criteria, yet they are certainly somewhat abnormal. These abnormalities
are definitely caused by iodine deficiency and have the same
characteristics as the endemic cretins but milder and couldn't be detected
by ordinary clinical observations. Thus we could call it the subclinical
cretin or to be brief the subcretin.

Thus we couldn't neglect the presence of the "subcretin"; yet it is
still not so easy to define the term definitely. Like the criteria of
endemic cretinism, the subcretin should be associated with endemic goiter
and severe iodine deficiency. The characteristics of the subcretin might
include:

(1) Subclinical mental retardation
(2) Mild psychomotor defect
(3) Subclinical hearing impairment
(4) Mild physical underdevelopment
(5) Chemical hypothyroidism

It seems necessary to discuss the characteristics and the assessments
of these abnormalities particularly to be included in"subcretin" as
follows:

1.   Mild Intellectual and Psychomotor impairments
We would like to introduce the term "mental retardation" to represent
the condition of "mental deficiency"[7]. In the recent twenty years there
have been striking advances in this field. From the psychiatric point of
view the cretin as well as the subcretin are part of the problem of mental
retardation. Therefore in the study of mental retardation caused by
iodine deficiency, we can utilize up-to-date knowledge of mental
retardation.

It is much more important to utilize objective tests of intellectual
and psychomotor functions for mental retardation in the subcretin than for
severe mental retardation in the cretin. By using these measurements
psychologists can identify and assess mild mental retardation as well as
mild motor impairments of the subcretins.

(1) The Intelligence Quotient Tests (IQ Tests)

The Intelligence Quotient Tests for preschool children, school
children and adults such as Stanford-Binet, Wechsler group tests, etc.

Table 7.   Relations Between Classifications of Mental
Retardation and IDD

I.Q.	Mental Retardation	Stanford-Binet	IDD
0-19	Profound	Idiocy	
20-35	Severe		Cretin
35-49	Moderate	Imbecility	
50-69	Mild	Mental weakness	Subcretin
70-89		Borderline defect or low average	
90-107		Average	

have a very lengthy history and an enormous amount of information has been collected on the performance on different subjects, using different tests. However, most of the standardizations of these tests have been carried out in developed countries among urban population and cannot simply be applied to remote areas of developing countries. Each test should be properly restandardized for a particular country, which however would be very expensive and time consuming.

Recently a series of non-verbal tests has been developed. Dr. Collin measured the intelligence of some school children in an endemic area in China using the Hiskey-Nebraska Test to avoid the language interference. Dr. DeLong and Dr. Lu L. had measured the intelligence of some cretin patients using the Beery test of visuomotor integration which has been considered free of cultural interference.

Because the outcome of the IQ tests for subcretins (IQ 50-69) is easily influenced by some factors, we should pay great attention to the following facts:

1) Any community has about 3% mildly mentally retarded individuals due to other causes.
2) All the intelligence scales are standardized for urban populations. The IQ of rural populations is usually 10% lower than urban populations.
3) Because the intellectual impairment of the subcretin is in the realm of mild mental retardation, social and environmental factors will be especially significant in influencing the results obtained. Hence beside the medical and psychological investigation, social history is very important. We have to keep in mind that most of the cretin families lack social stimulation.
4) Test results might be more comparable for measuring the same group of individuals by the same psychologists at different times but much less comparable for different groups of individuals by different psychologists.

Table 8. Comparison of IQ of Primary School Students in IDD Endemias and in Control Areas in China

| Author | Place | Year | Scale | Iodine Deficiency Endemias | | | Control Areas | | |
				N	IQ ($\bar{X}$)	69-50%	N	IQ ($\bar{X}$)	69-50%
Wang D.	Chengde	1983	S-B3	67	78.1±12.2	16.40	52	85.15±9.08	1.90
Chi J.L.	Anshun	1983	S-B2	93	80.8±9.1	9.00	70	85.4±.9.50	100
Nie R.H.	Changde	1983	S-B2	192	87.7±14.5	6.25	180	91.06±12.08	4.26
Zhang C.Q.	Chifeng	1979	S-B2	1213		4.29	1216		0.33
Wang D.	Hegu	1983	S-B3	28	79.4±8	7.7	24	84.67±9.37	0.00
Shi F.K.	Ningwu	1984	S-B3	30	76.1±8.82	20.0	37	85.08±6.18	0.00
Wang D.	Xinzhou	1985	S-B3	50	77.3±9.1	15.4	70	86.9±11.3	4.3
Wang D.	Tongan	1986	S-B3		78.5	16.4		87.2	5.7
An Y.Q.	Guide	1985	S-B3		78.8	15.6		85.50	0.00
Fu Z.L	Shenyang	1987	S-B3	119	79.6	12.61	87	88.20	3.45
Zhu J.X.	Jingde	1986	W	40	84.4±17.6	20.00	40	96.13±14.84	5.00
Geng P.B.	Jingde	1986	W	44	83.8	18.20	43	100.00	2.30
Sun G.H.	Chaohu	1987	S-B3	600		30.00			
Weng Y.L.	Liuan	1987	S-B2	60	77.9±13	25.00	37	86.6±13.94	2.70

S = Stanford-Binet
W = Wechsler

Hence true subcretinism in an endemia should be much less than the individuals with IQ 50-69. Therefore IQ is only a screening indication but not the final diagnostic standard of the subcretin.

(2) The Psychomotor Tests

For the cretin patients there are a lot of frank motor defects, yet for the subcretin individual frank neurological motor defects are almost undetectable. The subtle motor defects of the subcretin individual are involved and incorporated in each individual performance and those performances at the same time are organized and directed by the individual mind. Therefore it is almost impossible to separate the mind and the motor in the infantile period. Even in childhood or in the adult period, all performances are also carried out under such incorporation of mind and motion. So we call this kind of activity psychomotor activity and the assessment for measuring it is a psychomotor test for children and adults[8]. Relatively stringent tests of motor performance are used for school age children and adults of IDD endemias.

(3) The Development Quotient Test (DQ)

The Developmental Quotient measurement in the infant and/or young child is really a kind of psychomotor measurement. DQ may be influenced by delayed somatic growth, poor nutrition or disease, therefore in case of low DQ those organic conditions should be excluded.

Psychomotor DQ development tests are used for 0-3 year old children in IDD endemias. It is most common to use the DDST to measure DQ in China: Dr. Collin has used the Griffiths test.

(4) School Performance Surveys

Because of the limitations of the intelligence tests most psychologists emphasize that both intellectual functioning and adaptive behavior must be impaired before a person can be considered to be mentally retarded. For the severe mental retardation cases, references from the parents or neighbors might be the best source to get information on adaptive behavior. Yet for the mild mental retardation case and the subcretin, the school performance should give the best information. Age of children entering school, history of dropouts, failure in grading, times repeating the same grade, performance in reading, writing, mathematics, and natural sciences as well as physical culture will all be useful data to estimate the school performance. And the general impression of the responsible teachers may be used as the most comprehensive judgement.

(5) Routine Test Schedule Currently Used in China

Designed by Dr. Wang D.[9] and Dr. Chen Z.P. of the Institute of Endocrinology, Tianjin, China.

Infant & Preschool Children - DDST (Denver Developmental Screening Test)

School Children - Stanford-Binet Intelligence test
(Readapted for Chinese 1982)

- Psychomotor Performance Tests
  1) Simple and Choice Reactions
  2) Tapping Speed
  3) Hand Steadiness

- School Performance

2. Mild Hearing Impairment

Dr. Wang Y.Y. examining the hearing activity by audiometer of "normal" school children in Guizhou, China revealed that average hearing levels were 17.4 and 16.1 dB while the normal control was 9.5 dB. He discovered[10] that the hearing activity of the school children in the endemic areas had improved parallel to the amelioration of chemical hypothyroidism after iodization.

Mean Hearing Level of School Children Before and After
Iodized Salt Prophylaxis in Three Villages

	Average Hearing Level dB (Right Ear)		
	Heba	Shilong	Pingliong
Before Iodine Prophylaxis	17.4±7.1	.	16.1±6.77
After Iodine Prophylaxis			
1 Year	13.9±4.9	12.0±5.2	
2 Year	7.6±3.92*		
3 Year	8.16±3.02*	7.27±4.22*	8.93±3.14*
Normal Controls	7.5±3.83	7.5±3.83	7.5±3.83

* $p < 0.001$

3. Mild Physical Development Defect

Yu Z.H. compared the body height and weight between children in an endemic area and control area and showed a significant difference. We observed the mean height of school children was extremely short in the severe endemic areas of Chifeng and Huangzhong. However, physical development can be influenced by many nutritional factors.

4. A Proposed Criteria of Diagnosis of Subcretinism in China

Epidemiology associated with endemic goiter and iodine deficiency

Manifestation: All in the subclinical realm:
1) Mild intelligence impairment IQ 50-69
2) Mild psychomotor impairment-detectable only by psychomotor tests
3) Mild hearing impairment; hearing activity less than 30 dB .
(4000-8000 HZ)
4) Mild physical underdevelopment
5) Mild hypothyroidism: detectable only by hormonal assays.

These criteria were summed up from our experiences in the neurological cretin endemias. Because the clinical type of the myxedematous cretin is so different from the neurological cretin, we could presume that the subclinical type of the myxedematous cretin should be different to a similar degree.

5. The Significance of the Subclinical Cretin Problem in China

Because of the lack of definite diagnostic criteria and the fact that large scale epidemiological surveys are quite rare, data concerning the relative frequency of endemic cretinism and subcretinism are also varied. We have found in a relatively precise survey in a severe endemia the cretin incidence is 2.56% and that of subclinical individuals 14.8%.

WHO SEARO reported in 1985 that in Nepal the cretin incidence was 4.5% and the other IDD incidence (including cretinoidism and measurably reduced mental/motor function) was 31.4%. In Bhutan the cretin incidence was 6.6% while the other IDD was 48.7%.

The epidemiological data presented here confirm the existence of subclinical abnormal individuals in IDD endemias and show the numbers may be many times more than the cretins. Like an iceberg on the sea, the clinical cases seem to be the above-the-sea-level part but the iceberg under the water is much bigger. They are not so miserable for the family as is the cretin. Yet by their large numbers, they may give a very heavy burden to the community and hinder the socio-economic improvement of China.

REFERENCES

1. Ma T. et al.: The Present Status of Endemic Goiter and Endemic Cretinism in China. Food and Nutrition Bulletin 4(4):13 Oct. 1982.
2. Lu T.Z., Zhang J., Ma T., et al.: Clinical Observation on Endemic Cretinism of Chengde Region, Tianjin Med. & Pharm J. 1965 7(1):1.
3. Wang Y.Y. et al.: Significance of Audiometry & Vestibular Function in IDD Survey (In publishing, personal communication).
4. Zeng K.H., et al.: Neuropsychiatrical Changes of 247 cases of Endemic Cretin in South-eastern part of Guizhou, Chin. J. Neuropsychiatry 1982. 15(3):154.
5. Fierro-Benitez, R., Ramirez, I., Estrella, E. and Stanbury J.B., (1974) The Role of Iodine in intellectual Development in an Area of Endemic Goiter. In J.T. Dunn and G.A. Medeiros-Neto(eds.), Endemic Goiter and Cretinism: Continuing Threats to World Health. Pan American Health Organization Sci. Publ., Washington, D.C., pp. 135-140.
6. Delange, F.M., (1986) Anomalies in physical intellectual development associated with severe endemic goiter. In J.T. Dunn et al. (Eds.) Towards the eradication of endemic goiter, cretinism and iodine deficiency, Pan American Health Organization Sci. Publ., Washington, D.C., pp. 49-67.
7. Joint Commission on International Aspects of Mental Retardation (1985) Mental Retardation: Meeting the Challenge, WHO OFFSET PUBLICATION No. 86 pp. 7-17.
8. Wang D. et al.: Influence of children in endemic and non endemic areas of IDD. Proceeding of scientific research works of Tianjin Medical College. 1984
10. Wang Y.Y. et al.: Improvement in hearing among otherwise normal school children in Iodine-Deficient areas of Guizhou, China, following use of iodized salt. Lancet 1985; 518-528.

IODINE DEFICIENCY, IMPLICATIONS FOR MENTAL

AND PSYCHOMOTOR DEVELOPMENT IN CHILDREN[1]

Nico Bleichrodt[1], Francisco Escobar del Rey[2],
Gabriella Morreale de Escobar[2], Isabel Garcia[2],
and Carmen Rubio[2]

[1]Department of Industrial and Organizational Psychology
and Test Development, Free University of Amsterdam
The Netherlands
[2]Unidad de Endocrinologia Experimental, Instituto de
Investigaciones Biomédicas, C.S.I.C. y Facultad de Medicina
U.A.M., Arzobispo Morcillo 4, Madrid

INTRODUCTION

Endemic goiter and endemic cretinism are an important national
health problem in various countries. Although several ecological
factors may contribute to the development of endemic goiter (1-3),
iodine deficiency is assumed to be the major one (4). This assumption
is supported by the positive effects of prophylactic measures-
especially the administration of iodinated salt and iodized oil - on
the mental and physical development of people from severely iodine-
deficient areas. For its hormone production, the thyroid gland requires
a certain amount of iodine, which is taken in the form of iodinated
salts present in food and drinking water. Severe and prolonged iodine
deficiency frequently leads to an enlargement of the thyroid gland
(goiter), the volume of which may vary strongly. Such an enlargement is
found more often in females than in males (5-9). Goiter is usually
defined in the terms proposed by Perez et al. (10), a definition later
adopted by the World Health Organization in a slightly different
wording: "a thyroid gland whose lateral lobes have a volume greater
than the terminal phalanges of the thumbs of the person being examined
will be considered goitrous". The volume of the enlargement is usually
defined according to a four-graded division, ranging from OB (detectable only by palpation and not visible even when the neck is fully
extended) to III (visible from a long distance).

Excessive growth of the thyroid gland may eventually lead to a
narrowing of the bronchial tubes and increase the chance of carcinoma
of the thyroid gland. In addition, iodine deficiency affects the

AUTHORS' NOTE: We would like to thank Peter Dekker for his
assistance in analyzing the data.

chances of survival and the development of fetuses and babies. Iodine deficiency in a fetus is caused by the mother's iodine deficiency, the consequences of which may be an increased chance of still-born babies, abortions, congenital defects and probably retarded brain development (11-14). Research has shown that the development of the nervous system depends on an adequate intake of iodine (13-15). How serious the consequences of iodine deficiency are depends not only on the severity of the deficiency, but also on the time of its inception. A fetus's thyroid gland first begins to function in the tenth week (16) and the nerve cells begin to multiply at some time between the tenth and eighteenth week (17, 18). It is then that iodine deficiency can seriously disturb these processes. Considering, for instance, the results of experimental research on rats (19), it seems likely that the mechanisms that bring about the clinical symptoms of endemic cretinism can be traced to an inadequate production of both maternal and fetal thyroid hormone as a result of iodine deficiency (20). Various studies indicate that there is a connection between the degree of the iodine deficiency and the presence of goiter and endemic cretinism. This is evidenced also by the fact that endemic cretinism, though irreversible, can be prevented by iodine prophylaxis. Endemic cretinism is found mostly in areas where the iodine intake is less than 20 µg per day, whereas endemic goiter already occurs when the iodine intake is less than 50 µg per day (21, 22).

Endemic goiter and endemic cretinism occur particularly in the mountainous areas of a great many developing countries in Asia, South-America and Africa. However, in Europe too there still are some areas where endemic goiter and endemic cretinism occur (23). It is not known exactly how many people suffer from the symptoms mentioned, but it is certain that great numbers are involved. The estimations of Kelly and Snedden (24), dating from 1960, that 200 million people in all are suffering from goiter, are by now superseded. Ever more areas are discovered and mapped and it is estimated that in Asia alone 400 million people suffer from disorders that are the result of iodine deficiency (12). A considerable number of them - in some areas even more than 10% (25) - must be counted among the cretins.

IODINE DEFICIENCY IN SPAIN

It has been known for many years that there are several areas in Spain affected by endemic goiter. An overall picture of the distribution of goiter and cretinism may be found in a special issue of the Spanish journal Endocrinologia (26). An epidemiological survey was carried out in 1971, covering 226,915 inhabitants of 22 different provinces, of all ages. The overall incidence of goiter was reported to be 6.6% for males and 18.7% for females (13.4% for males and females together). From assessments of the iodine content of drinking water the authors concluded that iodine deficiency was likely to be the most important goitrogenic factor. It should be pointed out that this survey did not include the areas traditionally known for severe endemic goiter, such as Las Hurdes, the mountainous regions of Asturias, Galicia, and Cádiz, or areas in the Pyrenees.

Recent surveys show that endemic goiter is still a serious problem in many parts of Spain. Thus, a survey of 3,872 school children from several provinces of Galicia has shown a goiter frequency of 79%, with 85% of the children excreting less than 25 µg I/liter urine. Another study, involving 5,175 school children from seven of the eight provinces of Andalucia, reported a goiter frequency of 34%, with 40% of the subjects excreting less than 40 μg I/L. Of 1,695 subjects older than two years, selected randomly from 255 rural municipalities in an area in Cataluna (comprising a population of 370,000), including some villages in the Pyrenees, almost 36% had goiter; the mean urinary excretion was found to be 79.4 (S.D. = 44.3) μg I/g creatinine. Of 2,018 randomly selected school children from rural areas in Asturias 49.2% was found to have goiter. Near Madrid, in the province of Guadalajara, 58% of the 89 school children from one village had goiter, with 45.5% excreting less than 40 μg I/L. As many as 86% of the 156 children attending school in three of the most affected villages in Las Hurdes were found to have goiter, urinary I concentration being below 20 μg I/L in 71% of them.

All these reports support earlier conclusions regarding the important role of iodine deficiency in endemic goiter in Spain. They also show that in at least two such areas the severity of the iodine deficiency is of Grade-III intensity and that in these areas the risk for babies to be born cretin is high. The presence of cretins and deafmutes in Las Hurdes was already reported early in this century by Goyanes and Marañon. In 1970, Sánchez-Franco, et al. examined 25 defectives, whose ages ranged from nine to 60. All suffered from severe mental retardation and most required institutionalization. Six of them were deafmutes and another seven had hearing defects to varying degrees. Twenty showed spasticity and the gait typical of neurological cretins. The total population of Las Hurdes is about 12,000, but most of these defectives came from an area where half of this population is living.

Some researchers assume that the mental and (psycho)motor development of non-cretinous children living in areas with Grade-III iodine deficiency are affected too. However, the results of their research are not all that unequivocal and it is frequently pointed out that replication and further thorough research are a necessity (27-31).

A PILOT STUDY IN SPAIN

In 1982, three psychologists, Isabel Garcia, Carmen Rubio and Elisa Alonso, did a small-scale survey on the effects of iodine deficiency on the mental and motor development of children in Spain (32). They examined a total of 147 children aged between six and 14, who came from three of the most affected iodine-deficient villages in Las Hurdes. The tests that were administered covered a broad range of both intellectual and motor skills: Cattell's Culture Fair Intelligence Test, scale 1 and scale 2, the Bender Gestalt Test for Young Children, the performal scale of the Wechsler Intelligence Scale for Children, the Oseretsky Motor Development Scale, and two other motor tests: Threading beads and Making dots. The results achieved on most of these tests indicated mental retardation in the iodine-deficient group as well as disorders of some of the specific motor skills. The results of this pilot study were one of the reasons for setting up a wider ranging survey in an iodine-deficient and a non-iodine deficient area in Spain.

DEFINITION OF THE PROBLEM

The objective of this study is to gain better insight into the effects of severe, chronic iodine deficiency on various aspects of the mental and psychomotor development of that part of a population, that does not have the manifest symptoms of endemic cretinism.

In the survey in question three crucial questions would have to be answered:
1. What is the effect of iodine deficiency on the mental and psychomotor functions of the so-called non-cretinous group?
2. Are the effects, if any, to be found in the whole non-cretinous group or only in part of it?
3. Does the administration of iodine have positive effects on mental and/or psychomotor skills?

SUBJECTS

Selecting the areas for the survey was a task that required time and scrutinous attention. Of course, the iodine-deficient villages and the non-iodine deficient villages had to differ as much as possible with regard to the variable iodine, but as little as possible with regard to all other variables that might affect test performance. Aspects such as socio-economic level, degree of isolation, health care, and quality of education were taken into account. This resulted in the selection of seven villages in three different provinces.

The whole iodine-deficient group (aged 0-12) is from the region of Las Hurdes, in the province of Cáceres. Las Hurdes is close to the Spanish-Portuguese border, on the southern slopes of a mountain chain known as the Sierra de Francia. About half of the 12,000 inhabitants of Las Hurdes lives in villages and hamlets where the incidence of goiter has been and still is quite high. The iodine-deficient villages included in the survey are Cácares, Cabezo, and Aceitunilla, together comprising about 1,500 inhabitants. These villages were not the most severely affected. Immediately after the first survey, the children were given 2 ml Lipiodol orally. About thirty-two months later, a group of children (aged 6-12) thus treated and from the same villages were examined again.

The children making up the control group are from two different provinces. The group of primary-school children (aged 6-12) is from two villages, Miranda del Castañar and Cepeda, in the province of Salamanca. Both villages, with a total population of about 1,650, are located on the northern slopes of the Sierra de Francia, 40 km from Las Hurdes. Unlike the iodine-deficient villages, Miranda del Castañar and Cepeda have no day nurseries and therefore the younger group (aged 0-6) was selected from two other control villages, San Lorenzo del Escorial and Mostoles, in the province of Madrid. These villages, though bigger than the other villages in the survey, have day nurseries whose facilities are similar to those of the day nurseries in Las Hurdes and that are run by the same institution.

The composition of the three test groups is shown in Table 1. The groups have nearly the same average age. Similarity is also found in the level of education, since in Spain education is compulsory up to the age of 14. The control group has the smallest number of repeaters.

Table 1. Number tested, age (years and months), and educational level (grade) of the research groups

| | Fase 1 | | | | | | Fase 2 (32 months later) | | |
| | Control group | | | Iodine-deficient | | | Treated group | | |
	N	age	grade	N	age	grade	N	age	grade
2 - 30 months	48	1.6		26	1.4				
2½ - 5 years	63	4.3		34	4.6				
6 - 12 years	82	9.2	3.8	102	9.2	3.3	103	9.2	3.6

The first phase of the survey involved a total of 355 children whose ages varied between two months and 12 years: 162 from the iodine-deficient area and 193 from the non-iodine deficient area. One of the reasons for including babies and infants in the survey is that educational differences do not occur in this age group. Thirty-two months later the second phase of the survey took place, involving a total of 103 children aged between six and 12, who, immediately after the first phase, had been treated with Lipiodol.

Figure 1 represents the results of the research into a number of physiological characteristics of the children from the control and iodine-deficient areas and of the children treated with Lipiodol. Here, only the group of 6-12-year old school children was examined to establish the incidence of goiter (using the classification of Pérez et al.) as well as the children's weight, height, concentration of iodine in urine samples, and serum T4.

The statistical distributions of the **weights** of the children from the three research groups are slightly irregular, which is probably due to random sampling fluctuations. The average weights for the groups are almost equal, for example 28.1 kg for the iodine-deficient group and 28.3 kg for the control group.

Differences in average **height** between groups are also negligible: 130 cm for the iodine-deficient group and 131 cm for the control group. Also children from these groups are not smaller than children in the population.

Although the control area is not completely free of **goiter**, the difference with the iodine-deficient area is considerable. Of the 6-12-year-olds only 23% in the control area show enlargement of the thyroid gland (only 14 % in the other control area), whereas this is 66% for the iodine-deficient villages. After treatment with Lipiodol the goiter rate decreased from 66 % to 39 %. This decrease is mostly due to grade OB goiter, as the grade I goiter was 36% before treatment and 29 % after treatment. Goiter of grade-II or grade-III volume was not found in any of the test villages. It should be remembered though, that these data concern children of six to 12 years old and that undoubtedly bigger thyroid glands will be found in the older inhabitants.

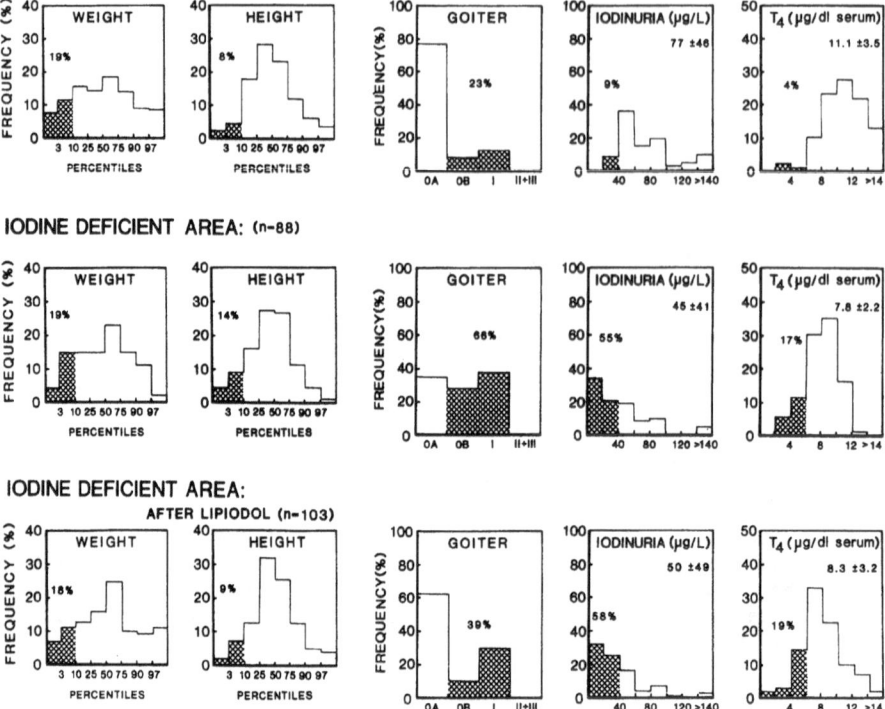

Figure 1. Some physiological characteristics of the group of 6-12-year-
olds from control area 1 (Miranda del Castañar, Cepeda), and
from the iodine-deficient area (Casares, Cabezo, Aceitunilla)

Iodine excretion and serum T4 are markedly lower in the iodine-
deficient group than in the control group. The effects of treatment
with Lipiodol are disappointing; the mean urinary I before and after
treatment is nearly the same and there is no difference in the
frequency distributions. In each group more than half of the children
are excreting less than 40 µg/L. The plasma T4 did not change either,
with 19 % of the treated children and 17 % of the iodine-deficient
children having values lower than 6 µg/dl. There is no difference
between T4 mean values, or between frequency distributions. For a more
detailed description of this study see Escobar del Rey et al. (33).

RESEARCH DESIGN

In order to determine the differences, if any, between the mental
and psychomotor functions of children suffering from chronic iodine
deficiency and those of children who do not and, further, whether the
administration of Lipiodol to the iodine-deficient group has any
positive effect, three groups are compared in this study:
1. A group of children aged between two months and 12 years from areas,
   not (severely) iodine-deficient.

274

2. A group of children aged between two months and 12 years from an area that is severely iodine-deficient.
3. A group of children aged between six and 12 years, who 32 months earlier had been treated with Lipiodol. This group is from the same area as that mentioned under 2.

The results obtained by the three groups on the mental and psycho-motor tests will be compared and differences, if any, analysed.

DESCRIPTION OF THE TESTS

The psychological tests used in the present survey have been described in detail in an earlier article (34). We shall therefore only give a brief summary and then go on to the question whether a reduction or combination of the tests should be considered when the test would appear to have similar psychological meaning.

Nearly all tests used for assessing the children's mental level had, at an earlier stage, been adapted to the Spanish population; that was the case with, for example, the Bayley Scales of Infant Develop-ment, the McCarthy Scales of Children's Abilities, and Cattell's Culture Fair Intelligence Test. Some of the psychomotor tests that were not adapted to Spain were included in the survey anyway because of their expected relevance.
- 0 to 2.5-year-olds: Bailey Scales of Infant Development: mental and motor development scale.
- 2.5 to 6-year-olds: McCarthy Scales of Children's Abilities: verbal, perceptual-performance, quantitative, memory and motor scale.
- 6 to 12-year-olds: Cattell's Culture Fair Intelligence Test, fluency, vocabulary, visual memory, figure comparison, hidden figures, block design, exclusion, and mazes.
  Psychomotor development tests: Oseretsky Motor Development Scale, Bender Gestalt test, pinboard, tapping, threading beads, making dots, simple reaction time, choice reaction time, and muscular strength test.

Information about the tests' significance may be derived from the analysis of the correlations between them. Factor analysis is one method to enhance the accessibility and interpretability of the enormous amount of information contained in the correlation matrix, without loosing too much information. A so-called principal component analysis was performed for both the iodine-deficient and the control group. As the factor structure is not easily gleaned from a matrix with unrotated factors, rotation to simple structure was applied according to Kaiser's varimax criterion (35). Factor analysis could be applied only to the test results of the group of 6-12-year-olds, since this group was administered a great number of different tests and is also sufficiently large to allow application of such a method of analysis.

An important question is whether the test series used has similar psychological meaning for different groups, such as the iodine-deficient and the control group. To find out, separate factor analyses, with rotation to four factors, were applied for each group. Following a varimax rotation and a Procrustes rotation, respectively, using the factor matrix of the control group as the target matrix, the so-called Tucker's phi-coefficients were calculated for each factor (36). These coefficients are a good indication of the factorial invariance of factor matrices. The lower limit for the equality of two factors is usually set at a value of about .85. The coefficients calculated for the four factors are .87, .88, .88, and .80, respectively. Although the phi-coefficient of the fourth factor is slightly lower, it may be concluded that for the most part the factor structures of the two groups run parallel and that, therefore, the test series used in the survey has a meaning for the iodine-deficient group that barely differs from the meaning it has for the control group. Therefore, the further analysis of the test series will rely on just one single factor matrix, based on the combined test group (Table 2), the advantage of which is that the basis is formed by a much larger number of children than would have been the case if separate matrices had been used.

Rotation to four factors appears to offer the best solution for a number of reasons. Most variables have rather high loadings on one of the four factors and appreciably lower ones on the remaining factors. Loadings of more than .45 are underlined. The first factor explaining a large part of the total variance is best described as general intelligence. Both Cattell's general ability test and the more specific tests often included in more comprehensive batteries of intelligence

Table 2. Varimax rotated factor matrix after rotation with Kaiser normalization, for the group of 6 - 12-year-olds (loadings > .45 are underlined)

Test	General intelligence	Manual dexterity	Speed of reaction	Coordination of movements	h²
General intelligence (Cattell)	.80	.05	.10	.17	.68
Fluency	.48	.30	.04	-.12	.34
Vocabulary	.52	.16	.09	.00	.30
Block design	.68	.00	.09	.19	.51
Visual memory	.42	-.10	.14	.32	.31
Figure comparison	.68	.20	.01	.10	.51
Bender Gestalt test	.42	.22	.07	.05	.23
Hidden figures	.50	.41	.20	.04	.45
Mazes	.37	-.01	.20	.36	.31
Pinboard	.24	.62	.14	.10	.47
Threading beads	.22	.64	-.02	.12	.48
Making dots	.09	.53	.13	.10	.32
Tapping	.02	.68	.14	.05	.48
Simple reaction time	.12	.29	.86	.03	.83
Choice reaction time	.21	.34	.75	.04	.72
Visual manual coordination[1]	.06	.26	.01	.39	.22
Dynamic coordination[1]	-.03	-.06	.04	.60	.37
Postural coordination[1]	.04	.26	.06	.39	.23
Speed[1]	.11	.06	.03	.34	.13
Simultaneous movements[1]	.06	.07	-.03	.56	.32
Handdynamometer	.17	-.12	.25	.21	.15
Explained variance	4.62	1.58	1.17	0.99	8.36

1) Oseretsky Motor Development Scale

tests - Fluency, Vocabulary, Block design, Figure comparison, and Hidden figures - have high loadings on this factor. Tests with high loadings on the second factor are Pinboard, Threading beads, Making dots, and Tapping. These tests all call upon the manual dexterity in particular. The third factor is almost exclusively determined by the two tests measuring speed of reaction. The expectation that the five subtests from Oseretsky's motor test would have high loadings on the same factor was not fulfilled. Only two tests (Dynamic coordination and Simultaneous movements) have substantial loadings on this fourth factor, that is best described as coordination of movements.

Considering the above analysis of the psychological meaning of the comprehensive test series used for the groups of 6-12-year-olds, it will be clear that reducing the number of variables will not cause any major loss of information. The general intelligence factor is adequately represented by (the Spanish version of) Cattell's Culture Fair Intelligence Test, with a factor loading of .80. The scores on the tests with high loadings on the three psychomotor factors are combined per factor. As no Spanish age norms are available for these psychomotor tests, the test scores of all the groups studied were corrected for age on the basis of the correlation between age and test scores found in the control group.

As mentioned above, age norms for the Spanish population have been developed for Cattell's test as well as for Bayley's (0-2½-year-olds) and McCarthy's (2½-6-year-olds) tests. Test performance on the two latter test series is represented by a Mental Development Score and a Psychomotor Development Score.

RESULTS

Mental Development

The distributions of the test scores on the Mental Development Test are shown in Figures 2, 3, and 4. These score distributions of the three age groups present similar pictures. Throughout, the test results of children from iodine-deficient areas are lower than those of children from non-iodine deficient areas. This is confirmed again by the differences in average scores between the groups mentioned (Table 3). The differences are significant at the 1% level for all age categories.

The upper parts of the score distributions (Table 3) contain but few children from iodine-deficient areas. Test scores of over 130 do not occur in any group and scores of over 120 do so only sporadically. In the iodine-deficient group, however, low scores are relatively frequent, especially in the two older age categories. If compared with the population, the percentage of children belonging to the categories of borderline cases and the mentally retarded should not exceed 9%, but in the youngest test group it amounts to 11.6%, in the middle group to 20.6%, and in the oldest group to 21.5% (untreated) and 17.5% (treated).

It would seem that the curves of the two youngest age groups are more or less two-topped (Figures 2 and 3), which would mean that the scores of one group of children are extremely low compared to those of the other children. However, since the iodine-deficient groups in

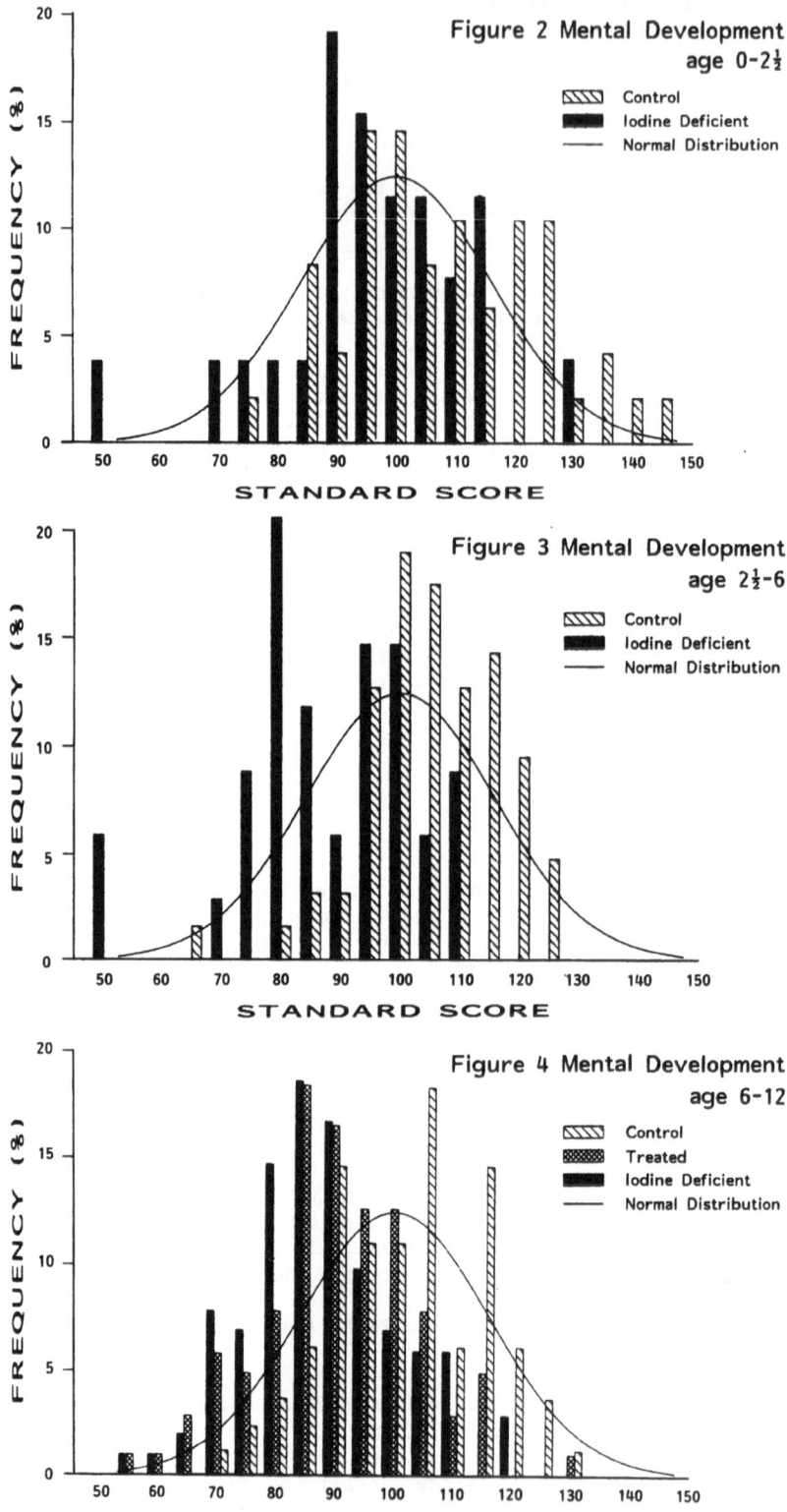

Figure 2-4. Distributions of Mental Development scores

Table 3. Distribution (%), mean and standard deviation of Mental Development scores for the norm group and the research groups (control = co.; iodine-deficient = i.d.; treated = tr.)

Classification	Score category	Norm group	0 - 2½ years		2½ - 6 years		6 - 12 years		
			co.	i.d.	co.	i.d.	co.	i.d.	tr.
Very superior	≥ 130	2.2	8.4	0.0	0.0	0.0	1.2	0.0	0.0
Superior	120-129	6.7	18.7	3.9	12.7	0.0	7.3	0.0	1.0
High average	110-119	16.1	16.7	19.2	23.8	8.8	23.2	3.9	5.8
Average	90-109	50.0	43.9	53.8	54.0	41.2	47.6	37.3	48.5
Low average	80- 89	16.1	10.5	11.6	7.9	29.4	15.9	37.3	27.2
Borderline	70- 79	6.7	2.1	7.7	0.0	14.7	4.9	12.7	10.7
Mentally retarded	≤ 69	2.2	0.0	3.9	1.6	5.9	0.0	8.8	6.8
Mean		100.0	108.1	95.9	105.1	88.1	101.6	88.0	90.2
Standard deviation		15.0	16.0	16.3	11.6	15.0	13.1	12.7	13.3
Number		-	48	26	63	34	82	102	103

particular are rather small, random sampling-fluctuations may have caused all sorts of irregularities in the curves. In fact, only three children had extremely low scores (3 standard deviations below the average): 1 child from the youngest age group and two from the middle group. In the oldest age group with its considerably larger samples the score distributions of the iodine-deficient and treated groups approach the normal distribution.

Performance on intelligence tests of 6-12-year-olds treated with Lipiodol scarcely deviates from the test performance of untreated children; the means are 90.2 and 88.0, respectively. The score distributions (Figure 4) too are very much alike.

Psychomotor development

The distribution of the scores obtained on the psychomotor tests is shown in Figure 5 (0-2½-year-olds), Figure 6 (2½-6-year-olds), and Figures 7-9 (6-12-year-olds). The picture these figures present for the various age categories is not an unequivocal one.

The average test score of babies from the iodine-deficient area almost equals that of babies from the control area (Table 4). The course of the score distributions is slightly irregular, but this may be due to the rather small samples. Differences were found for the group of infants; the control group's average score of 103.7 differs significantly from the mean of 93.6 scored by the iodine-deficient group ($p \leq .01$). Very high scores were not found in the iodine-deficient group, but very low scores were; almost 15% scored below 80.

In the oldest age group no differences were found for coordination of body movements. The average scores are almost equal and the score distributions more or less follow the normal distribution. But the other two psychomotor factors (manual dexterity and speed of reaction) yield significant differences ($p \leq .01$) between the control group on

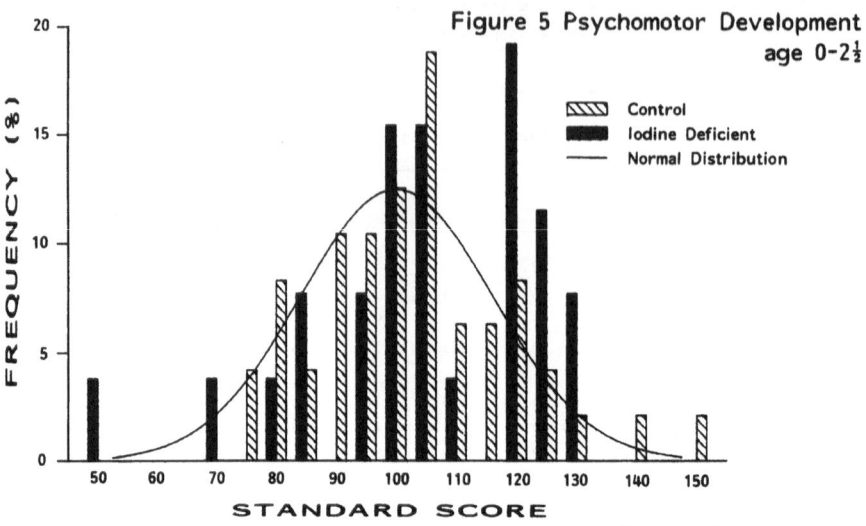

Figure 5 Psychomotor Development age 0-2½

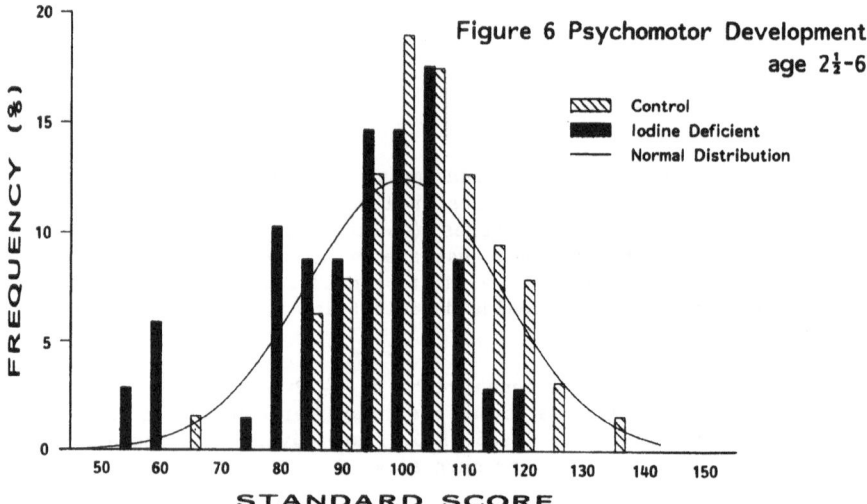

Figure 6 Psychomotor Development age 2½-6

Figure 5-6. Distributions of Psychomotor Development scores

the one hand and the iodine-deficient and treated groups on the other. Children from the latter two groups have lower scores throughout, which means that a relatively large number of children must be counted among the groups of borderline cases and the psychomotorically retarded.

Here too, the two-topped score distribution, notable especially in the youngest groups with small numbers of children, is the result of the scores of only four children. Three of these children also had the lowest scores on the mental development test.

The performance of the group of 6-12-year-olds treated with Lipiodol hardly differs from that of the (untreated) iodine-deficient group. Not all psychomotor tests could be administered to the group of

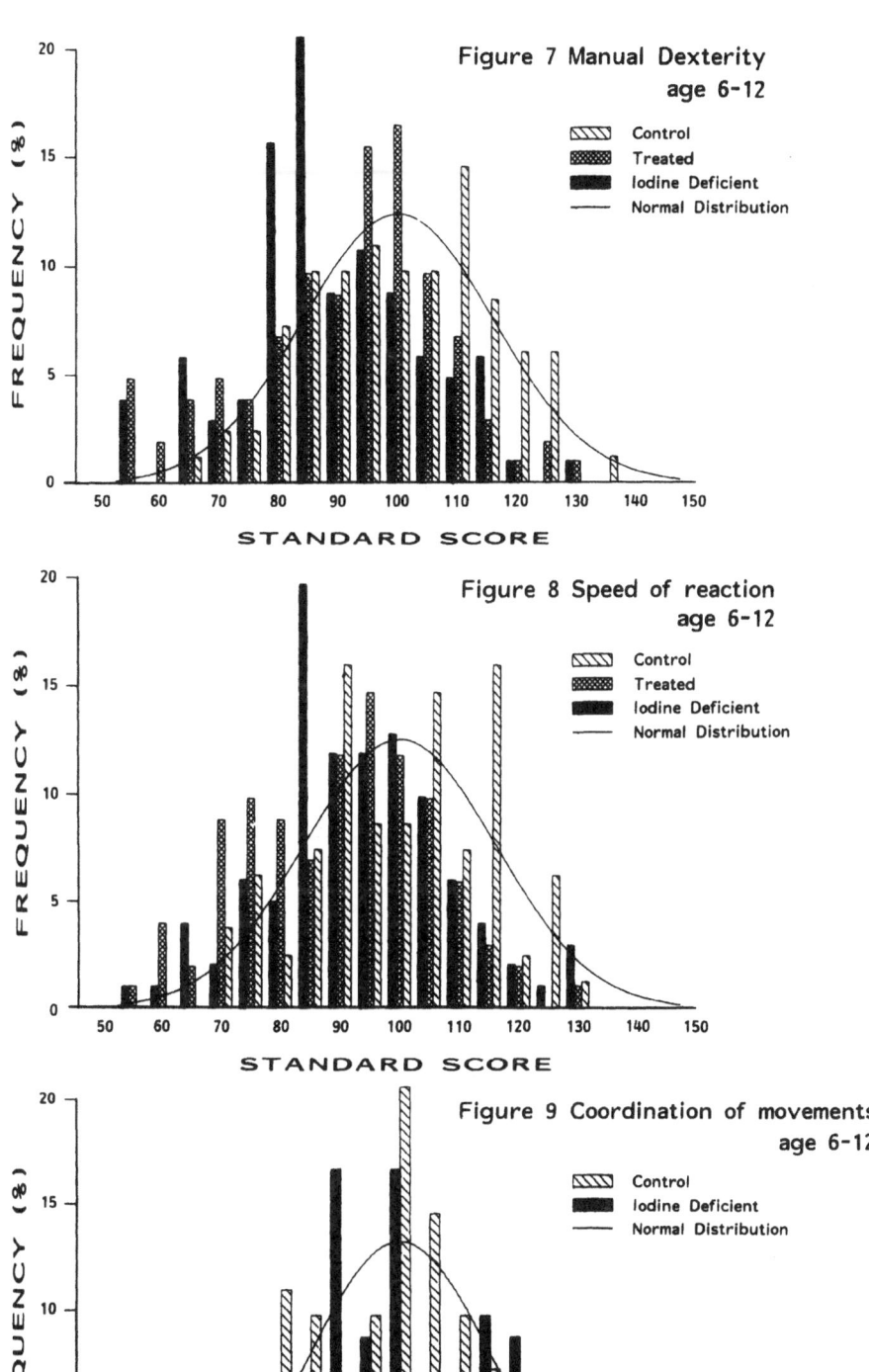

Figure 7-9. Distributions of Psychomotor Development scores

treated children. Scores were obtained for manual dexterity and speed of reaction.

Table 4. Distribution (%), mean and standard deviation of Psychomotor Development scores for the norm group, and the research groups of 0-2½-year-olds, and 2½-6-year-olds

Score category	Norm group	0 - 2½-year-olds		2½ - 6-year-olds	
		control	iodine-deficient	control	iodine-deficient
≥ 130	2.2	8.4	3.9	1.6	0.0
120-129	6.7	10.5	23.0	5.6	1.5
110-119	16.1	16.7	11.5	21.4	10.3
90-109	50.0	45.9	42.3	61.9	54.4
80- 89	16.1	16.7	11.6	7.9	19.1
70- 79	6.7	2.1	0.0	0.0	5.9
≤ 69	2.2	0.0	7.8	1.6	8.8
M	100.0	103.0	104.4	103.7	93.6
S.D.	15.0	15.7	19.8	11.7	15.2
N	-	48	26	63	34

Table 5. Distribution (%), mean and standard deviation of Psychomotor Development scores for the norm group, and the research groups of 6-12-year-olds (control = co.; iodine deficient = i.d.; treated = tr.)

6 - 12-year-olds

Score category	Norm group	Manual dexterity			Speed of reaction			Coordination of movements	
		co.	i.d.	tr.	co.	i.d.	tr.	co.	i.d.
≥ 130	2.2	1.2	0.0	1.0	0.0	1.0	0.0	4.9	2.0
120-129	6.7	8.5	1.0	2.9	9.8	3.9	2.9	4.0	12.7
110-119	16.1	22.0	7.8	7.8	19.5	7.8	6.8	13.4	13.7
90-109	50.0	42.7	34.3	47.6	42.7	44.1	46.6	50.0	40.2
80- 89	16.1	18.3	15.6	20.4	18.3	29.4	16.5	19.5	19.6
70- 79	6.7	6.1	13.7	9.7	7.3	7.8	16.5	6.1	7.8
≤ 69	2.2	1.2	9.8	10.7	2.4	5.9	10.7	1.2	3.9
M	100.0	100.0	88.6	91.3	100.0	93.2	89.7	100.0	98.5
S.D.	15.0	15.0	15.9	17.0	15.0	15.2	15.6	15.0	16.0
N	-	82	102	103	82	102	103	82	102

CONCLUSIONS

The selection of the three groups involved in this survey - the iodine-deficient group, the control group, and the treated group - certainly appears to be sound. The three groups are very similar with regard not only to age and educational level, but also to height and weight. Moreover, the two latter somatic aspects correspond with the population values, which suggests that malnutrition is unlikely to be involved here. For some aspects, however, distinct differences were found that are associated with iodine intake, such as goiter, urinary iodine excretion, and serum T4.

It was very fortunate indeed, that the survey could be carried out with the help of test instruments specifically developed for, or adapted to the Spanish population. Thus, the test performance of the survey's control group deviated only slightly from that of the Spanish norm group. The average scores of the children from the control villages even exceed the norm average of 100. These children probably belong to a privileged group, as are they all from day nurseries. This is also the case with the children from the iodine-deficient villages.

Three questions were formulated in this survey. The first question - does iodine deficiency affect the mental and psychomotor functions of children belonging to the so-called non-cretinous group? - can, with respect to mental functions, be answered in the affirmative. Especially from the age of 2-3 onwards, the effects on the intellectual functions are evident, even if we exclude the three children from the younger age groups scoring extremely low. The average test score of children from the iodine-deficient area lies approximately one standard deviation below that of the control group. Surveys carried out in the past years almost all report comparable results.

Many researchers also found a negative effect of iodine deficiency on psychomotor development. Some report significant differences between the average test scores of children from iodine-deficient areas and those of children from non-iodine deficient areas (a.o. 31), whereas others did not find significant differences for all age groups studied or they did find differences, but not significant ones (28, 30, 37). Connolly (27) reports significant differences for some psychomotor tests, such as Pinboard, Threading beads, and the Lincoln-Oseretsky test, but not for others, such as Tapping, Dotting, and Screw nuts on bolts. The survey carried out in Spain did not yield unequivocal results for all psychomotor tests used nor for all age groups studied. Only from the age of about three years is it possible to perceive any negative effect of iodine deficiency on psychomotor skills, albeit not on all skills. It is especially on the psychomotor factors 'manual dexterity' (Pinboard, Threading beads, Making dots, and Tapping) and 'speed of reaction' (Simple reaction time, Choice reaction time) that, on the average, iodine-deficient children score significantly lower. The differences in the middle age group remain significant when the three lowest scoring children (scores of approximately 60 or less) are excluded. What these six tests determining the two factors have in common is that they are all concerned with the speed and accuracy of finger-, hand- and arm-movements under visual control. As for the overall coordination of body movements, no retardation was found.

It can be established, on the basis of an analysis of the score distributions, that in the age groups where mental and psychomotor retardation occurs as a consequence of iodine deficiency, usually the whole group is involved and not just part of it (question 2). Especially in the groups with larger numbers of children, where random sampling-fluctuations are less important, there appears to be a tendency towards lower scores. Thus, a relatively large number of children fall into the 'borderline' category (scores of 70-80) or the 'mentally retarded' category (scores < 70). Children from these categories will often have problems keeping up with the regular primary school program, with those of the latter group (32), as they grow older, requiring "supervision and guidance when under mild social or economic stress" (38).

Although we tried to include only non-cretinous children, it's possible that (some of) the children belong to the cretin group. It is especially difficult to identify cretins at a very young age. It would be worthwhile, then, to have the children who scored extremely low in our survey undergo another thorough examination. Unfortunately, the group of cretins was not included in the survey, so that it is not clear whether there is a gradual transition in the scores obtained by cretins and non-cretins or whether all cretins have scores of approximately 55 or less. If the latter would be the case, the curve would be two-topped.

The last question posed in the survey concerns the effect of iodine prophylaxis on mental and psychomotor development. Earlier research (a.o. 39, 40, 41) did not yield unequivocal results. In the Spain-survey, the group of treated children was treated with Lipiodol between the age of four and ten, 32 months before the (second) administration of the tests. Test performance was found not to have improved as a result of the iodine treatment. The average scores of the iodine-deficient and the treated groups as well as their score distributions are nearly identical. Apparently, even certain effects of habituation or learning, which might have occurred in the second administration of the tests, did not play any role whatever. It would, at this point, be premature to assume that a treatment with Lipiodol has no positive effects whatsoever on the mental and psychomotor development of children from iodine-deficient areas or that the deficiencies involved are irreversible. Further research is certainly called for. There is still a possibility that the effects may have waned because of the long time interval of 32 months between the administration of the Lipiodol and the psychological tests.

REFERENCES

1. J.B. Stanbury, "Endemic goiter", PAHO Scientific Publication, No. 193 (1969).
2. M.D. Balazs, Influence of metabolic factors on brain development, Br. Med. Bull. 30:217 (1974).
3. F. Delange, C.H. Thilly, and A.M. Ermans, Endemic goitre in Kivu area, Africa, in: "Role of cassava in the etiology of endemic goitre and cretinism", A.M. Ermans, N.M. Mbulamoko, F. Delange, and R. Ahluwalia, eds., International Development Research Centre, Ottawa (1980).
4. A. Querido, Endemic cretinism - a continuous personal educational experience during 10 years, Postgrad. Med. J. 51:591 (1975).

5. J.C. Choufoer, M. van Rhijn, A.A.H. Kassenaar, and A. Querido, Endemic goiter in western New Guinea - I. Iodine metabolism in goitrous and nongoitrous subjects, J. Clin. Endocr. Metab. 23:1203 (1963).

6. R. Fierro-Benitez, W. Penafiel, L.J. de Groot, and I. Ramirez, Endemic goiter and endemic cretinism in the Andean region, N. Engl. J. Med. 6:296 (1969).

7. L.C.G. Lobo, A. Quelce-Salgado, and A. Freire-Maia, Studies on endemic goiter and cretinism in Brazil, in: "Endemic goiter", J.B. Stanbury, ed., PAHO Scientific Publication: No. 193, 194 (1969).

8. P.O.D. Pharoah, The clinical pattern of endemic cretinism in Papua, New Guinea, in: "Human development and the thyroid gland. Relation to endemic cretinism", J.B. Stanbury, and R.L. Kroc, eds., Plenum Press, New York-London (1972).

9. C.H. Thilly, F. Delange, L. Ramioul, R. Lagasse, K. Luvivila, and A.M. Ermans, Strategy of goitre and cretinism control in Central Africa, Int. J. Epidemiol. 6:43 (1977).

10. C. Perez, N.S. Scrimshaw, and J.A. Munoz, Technique of endemic goitre surveys, in: "Endemic Goitre", WHO Monograph Series, 44:369 (1960).

11. P.O.D. Pharoah, S.M. Ellis, R.P. Ekins, and E.S. Williams, Maternal thyroid function, iodine deficiency and fetal development, Clin. Endocrinol. 5:159 (1976).

12. B.S. Hetzel, Iodine deficiency disorders (IDD) and their eradication, The Lancet, November 12:1125 (1983).

13. B.S. Hetzel, and B.J. Porter, Iodine deficiency and the role of thyroid hormones in brain development, in: "Neurobiology of the trace elements, vol. I.", I.E. Dreosti, and R.M. Smith, eds., Humana Press, New Yersey, 83, (1983).

14. P.O.D. Pharoah, K.J. Connolly, R.P. Ekins, and A.G. Harding, Maternal thyroid hormone levels in pregnancy and the subsequent cognitive and motor performance of the children, Clin. Endocrinol. 21:265 (1984).

15. G. Morreale de Escobar, F. Escobar del Rey, and A.R. Ruiz-Marcos, Thyroid hormone and the developing brain, in: "Congenital hypothyroidism", J.H. Dussault, and P. Walker, eds., M. Dekker, New York, 85 (1983).

16. A. Querido, N. Bleichrodt, and R. Djokomoeljanto, Thyroid hormones and human mental development, in: "Maturation of the nervous system, progress in research", M.A. Corner, R.E. Baker, N.E. van de Pol, D.F. Swaab, and H.B.M. Uylings, eds., Elsevier, Amsterdam (1978).

17. J. Dobbing, The later development of the brain and its vulnerability, in: "Scientific foundations of pediatrics", J.A. Davis, and J. Dobbing, eds., Heinemann Medical Books, London, 565 (1974).

18. J. Dobbing, Normal brain development in human, in: "Brain development and thyroid deficiency", A. Querido, and D.F. Swaab, eds., North-Holland, Amsterdam (1975).

19. M.J. Obregon, P. Santisteban, A. Rodriquez-Pena, A. Pascual, P. Carlanega, A. Ruiz-Marcos, L. Lamas, F. Escobar del Rey, and G. Morreale de Escobar, Cerebral hypothyroidism in rats with adult-onset iodine deficiency, Endocrinology 115:1 (1984).

20. A. Querido, "The clinical epidemiology of endemic cretinism", Delhi Conference (in press).

21. A. Querido, The epidemiology of cretinism, in: "Endemic cretinism", B.S. Hetzel, and P.O.D. Pharoah, eds., Proc. Symposium Institute of Human Biology, Monograph Series No. 2, Papua New Guinea (1971).

22. J.B. Stanbury, A.M. Ermans, B.S. Hetzel, E.A. Pretell, and A. Querido, Endemic goiter and cretinism: public health significance and prevention, WHO Cron. 28:220 (1974).

23. J.B. Stanbury, and B.S. Hetzel, eds., "Endemic goiter and endemic cretinism: iodine nutrition in health and disease", Wiley, New York (1980).

24. F.C. Kelly, and W.W. Snedden, Prevalence and geographical distribution of endemic goitre, in: "Endemic Goitre", WHO Monograph Series 44:27 (1960).

25. Ma Tai, Lu Tizhang, Tan Yubin, Chen Bingzhong, and Zhu Xianyi (H.I. Chu), The present status of endemic goitre and endemic cretinism in China, Food and Nutri. Bul. 4,4:13 (1981).

26. F. Escobar del Rey, ed., Bocio Endémico y Deficienca de Yodo en Espana", Numero in memoriam G. Marañon y E. Ortiz de Landázuri, Endocrinologia, Suppl. 2, Ediciones Doyma, Barcelona (1987).

27. K.J. Connolly, P.O.D. Pharoah, and B.S. Hetzel, Fetal iodine deficiency and motor performance during childhood, Lancet 2:1149 (1979).

28. N. Bleichrodt, P.J.D. Drenth, and A. Querido, Effects of iodine deficiency on mental and psychomotor abilities, Amer. J. Phys. Anthrop. 53:55 (1980).

29. P.O.D. Pharoah, K.J. Connolly, B.S. Hetzel, and R.P. Ekins, Maternal thyroid function and motor competence in the child, Develop. Med. Child. Neurol. 23: 76 (1981).

30. C.H. Thilly, Psychomotor development in regions with endemic goiter, in: "Fetal brain disorders - recent approaches to the problem of mental deficiency", B.S. Hetzel, and R.M. Smith, eds., Elsevier/North-Holland Biomedical Press, 265 (1981).

31. R. Fierro-Benitez, R. Cazar, J.B. Stanbury, P. Rodriquez, F. Garces, F. Fierro-Renoy, and E. Estrella, Long-term effect of correction of iodine deficiency on psychomotor and intellectual development. in: Towards the eradication of endemic goiter, cretinism, and iodine deficiency, J.T. Dunn, E.A. Pretell, C.H. Daza, and F.E. Viteri, eds., PAHO, Washington, 182 (1986).

32. I. Garcia, C. Rubio, E. Alonso, C. Turmo, G. Morreale de Escobar, and F. Escobar del Rey, Alteraciones por deficiencia de yodo en Las Hurdes - II. Evaluación del desarrollo psicomotor de escolares, Endocrinologia 34:74 (1987).

33. F. Escobar del Rey, T. Martin, C. Turmo, J. Mallol, M.J. Obregón, and G. Morreale de Escobar, Alteraciones por deficiencia de yodo en Las Hurdes - I. Deficiencia de yodo y efectos del Lipiodol, Endocrinologia 34:61 (1987).

34. N. Bleichrodt, I. Garcia, C. Rubio, G. Morreale de Escobar, and F. Escobar del Rey, Developmental disorders associated with severe iodine deficiency, in: "The prevention and control of iodine deficiency disorders", B.S. Hetzel, J.T. Dunn, and J.B. Stanbury, eds., Elsevier Science Publishers, New York, 1:65 (1987).

35. J.C. Nunnally, "Psychometric theory", McGraw-Hill Book Company, New York (1978).

36. J.M.F. ten Berge, Optimizing factorial invariance, VRB Drukkerijen, Groningen (1977).

37. E.A. Pretell, T. Torres, V. Zenteno, and M. Cornejo, Prophylaxis of endemic goiter with iodized oil in rural Peru, in: "Human development and the thyroid gland. Relation to endemic cretinism", J.B. Stanbury, and R.L. Kroc, eds., Plenum Press, New York-London, 249 (1972).

38. J.M. Sattler, "Assessment of children's intelligence and special abilities", Allyn & Bacon, Boston (1982).

39. P.R. Dodge, H. Palkes, R. Fierro-Benitez, and I. Ramirez, Effect on intelligence of iodine in oil administered to young Andean children. A preliminary report. in: "Endemic Goiter", J.B. Stanbury, ed., PAHO Scientific Publication, 193:378 (1969).

40. R. Fierro-Benitez, I. Ramirez, and J. Suarez, Effect of iodine correction early in fetal life on intelligence quotient. A preliminary report, in: "Human development and the thyroid gland. Relation to endemic cretinism", J.B. Stanbury, and R.L. Kroc, eds., Plenum Press, New York, 239 (1972).

41. A. Bautista, P.A. Barker, J.T. Dunn, M. Sanchez, and D.L. Kaiser, The effects of oral iodized oil on intelligence, thyroid status, and somatic growth in school-age children from an area of endemic goiter, Amer. J. Clin. Nutr. 35:127 (1982).

EARLY CORRECTION OF IODINE DEFICIENCY AND LATE EFFECTS ON PSYCHOMOTOR

CAPABILITIES AND MIGRATION*

Rodrigo Fierro-Benítez, Ramiro Cazar, Hernán Sandoval,
Francisco Fierro-Renoy, Gonzalo Sevilla, Gustavo Fierro-
Carrión, Jaime Andrade, Francisco Guerra, Ney Yanez, Omar
Chamorro, Edwin Rodriguez, Victor M. Pacheco, and John B.
Stanbury

National Polytechnic School and Central University, Quito,
Ecuador

This is a study of the long-term effects of administration of iodized
oil in two Andean rural communities: Tocachi and La Esperanza. This program of prophylaxis was started in March 1966 (1). At that time these two
neighboring villages were similar in terms of isolation, ethnic composition, degree of iodine deficiency, and socioeconomic conditions (protein-
calorie malnutrition, annual income per capita, cultural deprivation,
etc.) They presented similar prevalences of iodine deficiency disorders.
In March 1966 the total population of Tocachi was treated; La Esperanza was
used as a control. Women of childbearing age and children born in Tocachi
were re-injected or injected in 1970, 1974, 1978, 1982 and 1986. All children born after the program was started were examined at the time of birth
and at key stages of development: both physical growth and neuromotor
maturation were observed until they were 5 years old (2-4). This part of
the program was finished in 1973. In that year the intellectual capacity
of those 3 to 7 years of age was studied (5, 6). In 1980 a subprogram of
screening of neonatal hypothyroidism was started (7). In 1981 in those
children born from 1966 to 1973 (8 to 15 years old) the effects of iodized
oil on neural and psychological development, as measured by a battery of
tests and by school performance, were studied (8).

Periodically, from 1966 to 1988, the urinary excretion of iodine was
evaluated both in the treated and in the untreated communities. Since
1969, Ecuadorean law has required that salt for human consumption be
iodized. However until 1978 the urinary excretion of iodine in the untreated community remained below 50 ug per gram of creatinine. Until 1981
more than half of this population presented iodine deficiency. Only in the
last four years have most been using salt with sufficient levels of iodization.

*This work was supported in part by the Public Health Ministry of Ecuador,
the Pan American Health Organization, the U.S.A. National Institutes of
Health, and the Ecuadorean National Council of Universities and Polytechnic
Schools.

BACKGROUND TO THE PRESENT STUDY

First Part: To understand the results obtained until 1973, i.e., children up to 7 years of age, the following classification is used: a) "Early treated children": refers to those whose mothers were treated with iodized oil prior to the second trimester of pregnancy; b) "Late-treated children": refers to those whose mothers were treated during the second or third trimester of pregnancy; and c) "Untreated children": refers to those born in the control village, also from 1966 on.

The results can be summarized as follows: In early and late-treated children: 1) goiter and deaf-mutism were prevented; 2) there was an improvement in gross motor function and maturation of reflex activity, which was significantly different from that in untreated children; 3) both in early-treated, late treated and untreated children neuro-motor development as a whole was retarded in relation to the standards employed (9, 10); 4) no case of what we called "major development retardation" (3) was found in early and late-treated children. Two percent of untreated children had major retardation in all four major areas of neuro-motor development (personal-social, reflex maturation, linguistic, and motor areas). Six out of ten died during the first two years of age. Mental capacity of the four surviving children was severely retarded (IQ less than 40), and they were diagnosed as endemic cretins (one of the four children diagnosed as having endemic cretinism was hypothyroid. His mother and his maternal grandfather and grandmother were also hypothyroid (11); 5) physical growth was similar for the three groups of children; and 6) both in early-treated, late-treated and untreated children general performance in terms of intellectual capacity was poor as judged by scores in the adapted Stanford-Binet Scale (11, 12). In early-treated children: 1) when the distribution of IQ scores was charted the curve showed a clear tendency toward normality; and 2) no cases of children with profound mental deficiency (IQ less than 50) were found. In late-treated children: 1) the curve of IQ scores tended to be skewed in the direction of mental deficiency; 2) eight percent of them presented IQ's of less than 50 (profound mental deficiency) thus, there were cases with obvious mental retardation but without deaf-mutism; and 3) the spectrum of intellectual capacity was wide, covering all mental categories, as in the untreated children.

Second Part: In October 1981, the situation of the children born under the program from October 1966 to October 1973 (when they were 8 to 15 years old) appears in Table 1. Here the term "treated children" refers to those whose mothers were treated before conception or during the first three months of pregnancy, and "untreated children" refers to those born in the control village in the same period. The school performance of 128 non-migrating treated children and 283 non-migrating untreated children was studied, i.e., children who had stayed in school at least a whole year and had started a new year, even if they had not completed the second year. Also the following neuro-psychological tests were applied to those subjects whose school performance was studied: the Terman-Merrill (12) and the Wechsler (13) intelligence tests, modified by Navas (14, 15) for Ecuadorean children; the Goodenough intelligence test (16); the Goddard test (17), which assesses psychomotor development, manual ability, visual motor coordination, and ability to recognize shapes and meanings; the visual-motor test of Bender (18), and the progressive matrices test of Raven (19, 20). Results can be summarized as follows: the percentage of untreated children who were taken out of school for mental incapacity was more than twice that of untreated children (13.3% versus 5.4%). Scholastic achievement was better in treated children when measured in terms of school years reached for age, school dropout rate, failure rate, years repeated, and school marks. There was no difference between the treated and untreated children in the Terman-Merrill test, the Wechsler scale, or the Goodenough test results. The untreated children did not do as well on the Bender test, which evaluates integration of functions on the basis of visual-motor items, and on the Goddard test, which basically measures psychomotor development. Both groups performed poorly on the Raven test, but the treated children did slightly less well. Treated and untreated children seemed impaired in school performance, especially in reading, writing, and mathematics, but more notably the untreated ones. The problems found in these areas were not attributable to dyslexia or dysgraphia, but to a global retardation accompanied by a low level of comprehension of what they read or wrote, low levels of abstraction and generalization, deficient vocabulary, motor instability, and the most impressive limitation: a weak memory.

Table 1. Situation of the treated and untreated children born from October 1966 to October 1973, in October 1981.

	Treated Children		Untreated Children	
	N	%	N	%
Died	20	8.4	53	12.9
Migrating	74	30.9	33	8.0
Non-migrating	145	60.7	325	79.1
Non-migrating who entered school	138	95.2	286	88.0
Non-migrating who did not enter school	7	4.8	39	12.0

Third Part: Regarding the screening of neonatal hypothyroidism, no cases of hypothyroidism were identified in the treated community. The prevalence of neonatal hypothyroidism was 1.68% in the untreated population. Further, 3.5% of the untreated pregnant women were diagnosed as having hypothyroidism. Two out of eight children diagnosed as hypothyroid were born from hypothyroid mothers.

STUDY DESIGN

The present study, begun in October 1981, addressed the following question: what happened to the children who had entered school up to 1981 (see Table 1)?

We knew school performance of most of them and had an evaluation of their psychomotor capability. These were "early-treated children" whose mothers were treated before conception or during the first trimester of pregnancy, in respect to Tocachi village. Almost all of those who were born in the untreated village, at least until 1978, lived under iodine deficiency from conception. In 1981 we had found evident improvement in the treated children, in comparison with the untreated children, in school performance and psychomotor capability.

Of the early-treated children 30.9% had emigrated as compared to only 3.0% of the untreated children (see Table 1). This caught our attention, especially after noticing how the treated village had deteriorated: houses

were in disrepair or abandoned, and there was a predominance of older
people and children.

With these antecedents we tried, starting in October 1987, to locate
and obtain detailed information on the 138 treated children who had entered
school up to 1981. For purposes of comparison we also tried to locate and
obtain reliable information on a similar number of untreated children, also
born between October 1966 to October 1973, who had not emigrated by 1981
and who had entered school. With great effort we could find and obtain
reliable information on 134 of the 138 treated subjects, and--until
February 1988--120 of the untreated subjects (Table 2).

Table 2. Number, mean age and range of the treated and untreated
subjects who were studied in October 1987. Subjects
born from October 1966 to October 1973.

	Treated Subjects N:134		Untreated Subjects N:120	
	Male N:75	Female N:59	Male N:62	Female N:58
Mean Age (years)	17.68	17.59	17.59	17.53
Range	14-21	15-21	15-21	14-21

FINDINGS

Education: Tables 3 and 4 show the educational level attained by the
treated and untreated subjects. It seems apparent that the socioeconomic
reasons and migration were the main reasons for the treated subjects leav-
ing school. In relation to the larger percentage of untreated subjects
that entered Seamstress School (Table 4) it must be pointed out that the 12
subjects (24.5%) are female and that there is a Seamstress School only in
the untreated village. A higher percentage of treated subjects entered a
vocational high school. Among the treated population there was more interest for entering high school and college (the Ecuadorean system includes
six years of elementary school and six years of high school before entering
directly the university to follow professional studies: medicine, engi-
neering, etc.).

Migration: Table 5 shows the percentage of migrating and non-
migrating subjects among the treated and untreated populations. Non-
migrating subjects were defined as those who ate and slept in their original village. Migrating subjects are subdivided into temporary migrants
(those who returned to their village at least once a year) and permanent
migrants, those who never returned. While it is true that among untreated
subjects there is a high ratio of migrants, migration among treated subjects is unquestionably higher, with the aggravating circumstance that
almost a fourth (22.4%) of them never return (Table 5).

As to the places where they migrate to, there was little difference
between treated and untreated subjects (Table 6). However, two treated
subjects went to the coast, two to the Amazonian jungle, and one woman went
to Cuba in order to study. These five cases are interesting because it
demands great effort for a person born in the Andean highlands to move to

the coast, the jungle, or abroad. Most of the treated and untreated subjects migrated to Quito, capital of the country.

Table 3. Status in terms of education of the 134 treated and 120 untreated subjects at beginning of the study (October 1987).

	Treated Subjects %	Untreated Subjects %
Left school because of incapacity (very low performance)	0.74	1.6
Left school for socio-economic reasons and migration	14.9	8.3
Completed elementary school but did not continue studies	46.2	49.2
Completed elementary school and continued studies	38.0	40.8

Table 4. Studies realized by 51 treated and 49 untreated subjects,
who completed elementary school and continued studies
(October 1987).

Studies	Treated Subjects %	Untreated Subjects %
Seamstress School	7.8	24.5
Vocational High School	5.9	2.0
High School	76.4	67.3
University	9.8	6.1

Table 5. Migration in the treated and untreated subjects
(October 1987).

Migration		Treated Subjects N:134 %	Untreated Subjects N:120 %
Non-migrating		20.9	50.8
Migrating	Temporary	56.7	44.2
	Permanent	22.4	5.0
	Total	79.1 (N:106)	49.2 (N:59)

Table 6. Sites of migration of treated migrating and untreated
migrating subjects (October 1987).

Migrated to:	Treated Subjects N:106 %	Untreated Subjects N:59 %
Another site of same county	5.7	6.8
Another county of the same province	16.0	8.5
The capital of the same province: Quito (which is also the capital of the country)	66.6	71.2
Other Andean provinces (both rural and urban sites)	7.5	13.6
Other regions of Ecuador (coast and Amazonian jungle)	3.8	----
Another country (Cuba)	0.9	----

Migration and psychomotor capabilities: As mentioned under
"Background to the Present Study," we studied by means of a battery of
tests the mental capacity and psychomotor function of a high percentage of
those whose school performance was evaluated in 1981. We compared the
scores of those tests in the four groups of subjects: a) treated non-
migrating; b) treated migrating; c) untreated non-migrating; and d) untreated migrating. There were no differences in the scores of the
Goodenough and Raven tests. There were differences between treated and
untreated migrating subjects in the scores of the Terman-Merrill and
Wechsler tests (Table 7, Figure 1).

Table 7. Number, mean IQ scores, SD and "Z" test value of p, of the
treated migrating and untreated migrating subjects, tested
with Terman-Merrill and Wechsler tests modified by Navas.

Group	Number	IQ Mean Value (range)	SD	"Z" Test Value of p
Treated migrating subjects	78	93.50 (64-126)	11.92	
				$p<0.01$
Untreated migrating subjects	57	88.16 (63-111)	11.42	

Comparing the scores on the Goddard Test, used mainly to evaluate
psychomotor development, there were differences between the non-migrating
and migrating treated subjects ($p <0.05$), and between treated and untreated
migrating subjects (Table 8, Figure 2).

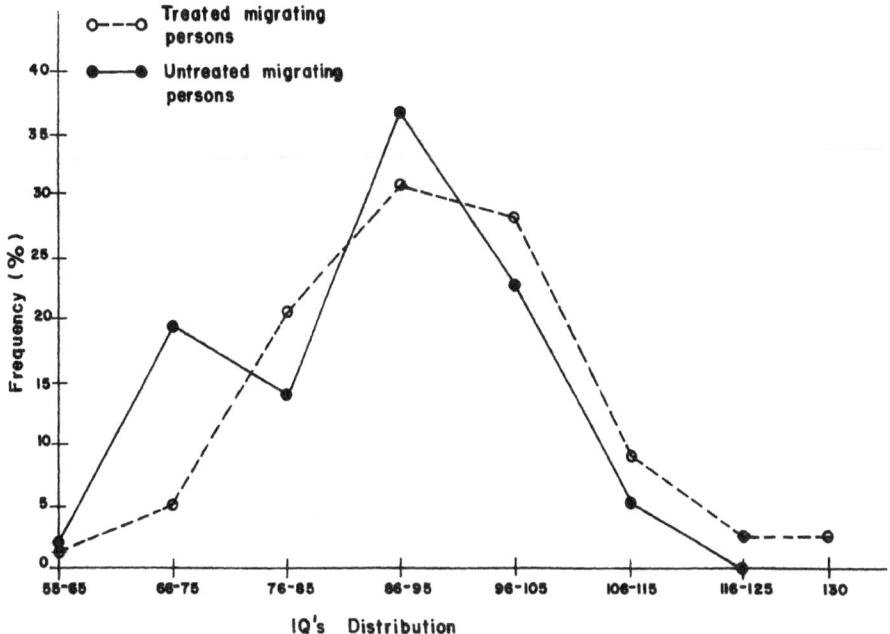

Figure 1. Distribution of IQ scores for treated migrating and untreated migrating persons with the Terman-Merrill and Wechsler Test, as a whole.

Table 8. Number, mean IQ score and range, SD and "Z" test value of p, of the treated migrating and untreated migrating subjects, tested with the Goddard Test.

Group	Number	IQ Mean Value (range)	SD	"Z" Test Value of p
Treated migrating subjects	78	93.38 (60-113)	11.20	
				p<0.01
Untreated migrating subjects	57	84.65 (50-112)	15.15	

With the Bender Gestalt Test, which evaluates integration of functions on the basis of visual-motor items, 43.8% of the untreated and 26.9% of the treated migrating subjects showed lack of maturity (SI) or were abnormal (A) (Table 9). In addition, the configuration of the frequency distribution between the two groups was quite different (Figure 3). Similar differences were found between treated and untreated non-migrating subjects.

These results suggest that the treated migrating subjects were more able in regard to mental ability and psychomotor function.

<u>Migration, occupation and salary</u>: Table 10 shows the occupation of treated non-migrating subjects and untreated non-migrating subjects. There

295

is a higher percentage of artisans among treated non-migrating subjects. The presence of 16.3% of students among untreated non-migrating subjects presents a false picture of reality; the explanation is that only in the untreated village, there is a Seamstress School, where young women can attend after elementary school. As to migrants (Table 11), most of the treated subjects (38.7%) are either servants or underemployed. In an underdeveloped country like Ecuador this means that they do not have any special skills. An underemployed person works part-time, and changes his occupation from time to time.

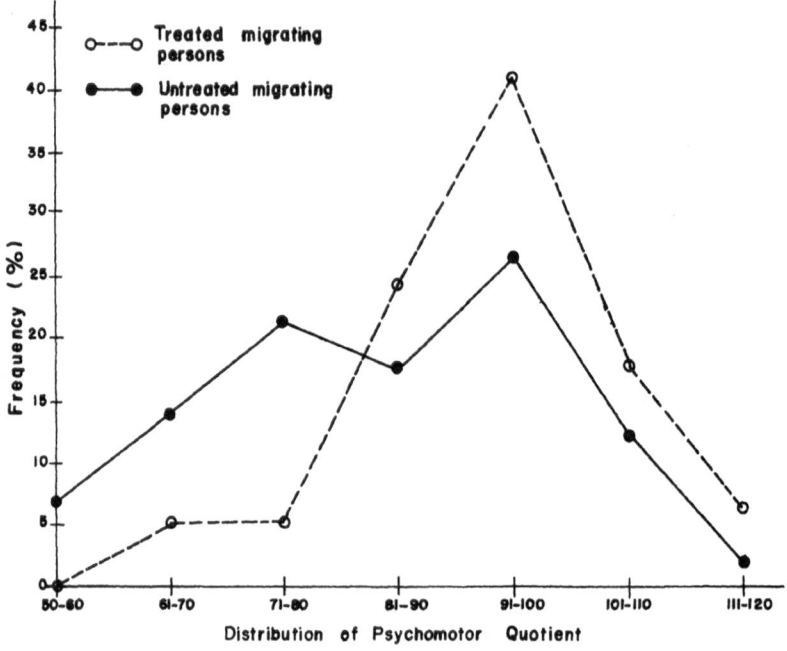

Figure 2. Distribution of IQ scores for treated migrating and untreated migrating persons with the Goddard Test.

Table 9. Results obtained with the Bender Gestalt Test in 78 treated migrating and 57 untreated migrating subjects.

Bender Classification	Treated Migrating Subjects %	Untreated Migrating Subjects %
Normal (N)	43.6	33.4
Normal inferior (NI)	29.5	22.8
Signs of Immaturity (SI)	23.1	29.8
Abnormal (A)	3.8	14.0

Among treated migrants there is the highest ratio of students, both at high school and university. Three of the treated migrating subjects have

296

their own businesses, one a grocery, another a bakery and a third is a
farmer which is exceptional for a 21-year-old person. Among untreated
migrants, one has his own tractor for business, which is also exceptional.

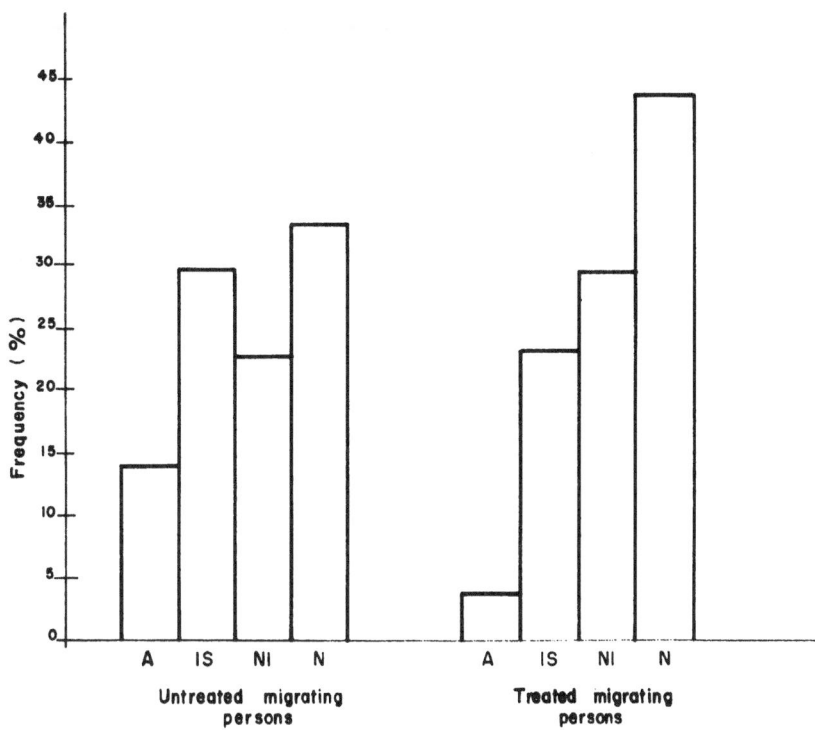

Figure 3. Frequency distribution of the findings observed in untreated
migrating and treated migrating persons through use of the
Bender Gestalt Test: A-abnormal; IS-immaturity signs;
NI-normal inferior; and N-normal.

Table 10. Occupation of 28 treated non-migrating and 61 untreated
non-migrating subjects (October 1987).

Occupation	Treated Non-migrating Subjects %	Untreated Non-migrating Subjects %
Farm laborers and shepherds	42.8	16.3
Other agricultural tasks (floriculture, etc.)	14.2	14.7
Artisans	32.0	19.6
Domestic chores	10.7	18.0
Students	----	16.3*

*All are women who are studying at the Professional Seamstress School of
the untreated village.

Table 11. Occupation of 106 treated migrating and 59 untreated migrating subjects (October 1987).

Occupation	Treated Non-migrating Subjects %	Untreated Non-migrating Subjects %
Farm laborers and shepherds	8.5	3.4
Other agricultural tasks (floriculture, etc.)	0.9	6.8
Artisans	28.3	40.7
Servants	17.9	13.6
Underemployed	20.8	11.9
Private business	2.8	1.7
Army	3.8	5.1
Students	17.0	11.9

Table 12 shows the range of salaries per month in the treated and untreated subjects. Because of limitations it was not possible to determine the per capita income and the average income for each of the studied groups.

Table 12. Range of the salaries per month in treated and untreated subjects.

	Treated Subjects		Untreated Subjects	
	Non-migrating	Migrating	Non-migrating	Migrating
Salary per month (U.S. Dollars)*	26.9-32.3	47.3-52.7	35.5-40.1	58.0-61.6

*At the rate of 270 "sucres" per dollar prevailing when the data were obtained.

An example will show how the range of salaries per month was determined: masons in this area receive from 10,000 to 24,000 sucres a month. These amounts were multiplied by the number of individuals that work in this occupation in each group. We followed the same procedure with the other activities. Then we added minimum and maximum salaries and the results were divided by the number of workers in each group. This was the only way to obtain an estimate that could be used for comparison. We had these facts in front of us: 1) A considerable number of subjects were not engaged in the same activity the whole year. Some months they work, some months they do not; some weeks they work; some weeks they do not. There are even workers who work some days and other days they do not. This defines underemployment, which is especially noticeable in cities. This underemployment has created in Ecuador and other Latin American countries the so-called "under-proletariat." 2) On the other hand, there is also a small number of workers who are covered by Social Security and receive bonuses during the year. For estimating the range of salaries per month we did not include self-employed subjects, given the impossibility of determining their income.

With limitations, the findings of Table 12 can serve as a basis for comparison: treated subjects received lower salaries than the untreated subjects. Migrants received salaries which were 50% higher than those of non-migrating subjects.

The puzzle presented by the lower salaries received by treated subjects may be explained as follows: non-migrants receive a lower salary because the treated village is poorer (1). Migrant treated subjects receive lower salaries because there is a high percentage of subjects that emigrated before they were 12 years old (Table 13). This percentage is four times higher than in the case of untreated migrants (33% versus 8.4%). Furthermore 85.7% of them emigrated by themselves without their families. Emigration to Quito, the capital of the country, the most frequent goal, under these circumstances implies accepting the first job that is offered, at a very low pay, and with no hope of advancement. It can only produce a stagnation of the immigrating child.

Minimum monthly salary is at present U.S. $57.4. Migrants do not even receive this much.

Table 13. Age at which the migration occurred. Percentage in 106 treated and 59 untreated subjects.

Subjects	Before 12 Years of Age	After 12 Years of Age
Treated	33.0%	66.9%
Untreated	8.4	91.5

Prevalence of goiter: The thyroid size was evaluated in 129 treated and in 118 untreated subjects. Thus, five treated and two untreated subjects were not evaluated because we did not find them. In the treated subjects (Table 14) the prevalence of goiter was 5.4%, and in the untreated subjects, 17.7%. Thus, the treated persons are free of endemic goiter. We used the classification of Pérez, et al. (21), adopted with modifications by the PAHO (22).

COMMENTS AND CONCLUSIONS

When we wrote the "Comments and Conclusions" section of our previous paper "Long-term effects of correction of iodine deficiency on psychomotor and intellectual development" (8), we presented the results obtained by the use of iodized oil, and pointed out that while iodine deficiency had been corrected, children continued to suffer the consequences of severe malnutrition and an environment which is a culturally limited, non-stimulating milieu: a complex of social ills.

Now we will address the socio-economic consequences of iodine deficiency correction, 21 years after its beginning. We worked under the assumption that when the correction is early, i.e., before conception or

during the first three months of pregnancy, the following are prevented:
endemic goiter, endemic cretinism and endemic neonatal hypothyroidism, and
that, in general, subjects present better intellectual and psychomotor
capabilities than the untreated subjects. We also took into account that
the economic conditions of the country not only had not improved but had
actually worsened in this decade (23).

Table 14. Prevalence of "visible goiter" in 129 treated and 188
          untreated subjects.

Thyroid Size Grade	Treated %	Untreated %
0a	65.1	56.7
0b	29.4	25.4
I	5.4	14.4
II	---	3.3
Goiter	5.4	17.7

It is true that more treated children have entered school and fewer
have left it because of incapacity. Among the better subjects, among those
who entered school and showed greater mental capacity and better psychomotor capabilities, there has been much greater migration than among untreated subjects. Less gifted subjects remain in the countryside.

Migration from the countryside to the city is an increasing phenomenon
in Latin American countries with their deteriorating economies (23). In
the last twenty years, capital cities of these countries have grown in an
anarchic way due to an uncontrollable immigration by peasants (24, 25,
26). All these cities are surrounded by slums that threaten to strangle
them. In these cities live peasants who prefer to migrate to the city
because conditions in the countryside are too hard for them.

In the cities these peasants without professional skills become the
"under-proletariat" characterized by underemployment (27). The inhabitants
of the slums are peasants who by migrating have worsened their condition
from poverty to destitution. An increasing percentage never return to the
countryside (28).

Socio-economic consequences of peasant migration contribute to the
underdevelopment of the Third World countries, most of which depend on
agricultural production. It is a vicious cycle which is very difficult to
break in societies without a planned economy.

What has happed to the treated population is an example of something
that is now well known: public health is the most sensitive indicator of
the socio-economic conditions of a country. Public health policies and
actions must be part of the context of the general policies of the
country. Otherwise results can be counter-productive. This does not deny,
however, the high priority of health actions conducive to the prevention of
iodine deficiency disorders.

SUMMARY

Children of mothers from iodine-deficient communities in Andean
Ecuador who were treated with intramuscular iodized oil prior to the second
trimester of pregnancy were compared with children of those not so treated
when they were 14 to 21 years of age. Children of treated mothers had
advanced farther in education, had migrated more often to urban centers,
and performed better on tests of neuromotor function, but had a lower level
of income. The complex social dynamics of the program are discussed.

REFERENCES

1. Fierro-Benítez, R., I. Ramírez, E. Estrella, C. Jaramillo, C. Díaz, and
   U. Urresta, Iodized oil in the prevention of endemic goiter and
   associated defects in the Andean region of Ecuador. I. Program
   design, effects on goiter prevalence, thyroid function, and iodine
   excretion, in: "Endemic Goiter," J. B. Stanbury, Ed., PAHO
   Scientific Publication 193, Washington, D.C. (1969), pp. 306-340.

2. Ramírez, I., R. Fierro-Benítez, E. Estrella, C. Jaramillo, C. Díaz, and
   J. Urresta, Iodized oil in the prevention of endemic goiter and
   associated defects in the Andean region of Ecuador. II. Effects
   on neuro-motor development and somatic growth in children before
   two years, in: "Endemic Goiter," J. B. Stanbury, Ed., PAHO
   Scientific Publication 193, Washington, D.C. (1969), pp. 341-359.

3. Fierro-Benítez, R., I. Ramírez, E. Estrella, A. Querido, and J. B.
   Stanbury, The effect of goiter prophylaxis with iodized oil on the
   prevention of endemic cretinism, in: "Further Advances in Thyroid
   Research," K. Fellinger and R. Hofe, Eds., Verlag der Wiener
   Medizinischen Akadamie, Wien, (1979), Vol. 1, pp. 61-77.

4. Ramírez, I., R. Fierro-Benítez, E. Estrella, A. Gómez, C. Jaramillo, C. Her-
   mida, and F. Moncayo, The results of prophylaxis of edemic cretinism with
   iodized oil inrular Andean Ecuador, in: "Human Development and the
   Thyroid Galnd RElation to Edemic Cretinism," J. B. Stanbury and R.L. Koc,
   Eds., Plenum Press, New York (1972), pp. 223-237.

5. Fierro-Benítez, R., I. Ramírez, and J. Reinhart. The role of iodine on
   intellectual deficiency in areas of chronic iodine deficiency and
   protein-calorie malnutrition. Tenth International Congress of
   Nutrition, Kyoto (1975), p. 37.

6. Fierro-Benítez, R., I. Ramírez, E. Estrella, and J. B. Stanbury, The
   role of iodine in intellectual development in an area of endemic
   goiter, in: "Endemic Goiter and Cretinism: Continuing Threats to
   World Health," J. T. Dunn and G. A. Medeiros-Neto, Eds., PAHO
   Scientific Publication 292, Washington, D.C. (1974), pp. 135-142.

7. Fierro-Benítez, R., A. Jijón, J. C. Zevallos, F. Fierro-Renoy, S.
   Guijarro, F. Chiriboga, F. Gándara, M. Martinod, V. M. Pacheco, M.
   Román and J. B. Stanbury, A pilot program for screening congenital
   hypothyroidism in an undeveloped country, in: "Iodine Deficiency
   Disorders and Congenital Hypothyroidism," G. Medeiros-Neto, R. M.
   B. Maciel, and A. Halpem, Eds., Ache, Sao Paulo (1986), pp. 261-266.

8. Fierro-Benítez, R., R. Cazar, J. B. Stanbury, P. Rodriguez, F. Garcés,
   F. Fierro-Renoy, and E. Estrella, Long-term effects of correction
   of iodine deficiency on psychomotor and intellectual development,
   in: "Towards the Eradication of Endemic Goiter, Cretinism, and
   Iodine Deficiency," J. T. Dunn, E. A. Pretell, C. H. Daza, and F.

E. Vitery, Eds., PAHO Scientific Publication 502, Washington, D.C. (1986), pp. 182-200.

9. Gesell, A, "Children from One to Four Years," Paidos. Ed., Buenos Aires (1966).

10. Gesell, A., and C. S. Amatruda, "Diagnostico del Desarrollo Normal y Anormal del niño," Paidos, Ed., Buenos Aires (1966).

11. Stanbury, J. B., R. Fierro-Benítez, E. Estrella, P. S. Milutinovic, M. U. Tellez, and S. Refetoff, Endemic goiter with hypothyroidism in three generations, Journal of Clinical Endocrinological Metabolism, 29:1596-1600 (1969).

12. Terman, L. M., and M. A. Merrill, "Stanford-Binet Intelligence Scale," Houghton-Mifflin Company, Boston (1962).

13. Pichot, P., "Los Tests Mentales," Paidos, Ed., Buenos Aires (1963).

14. Navas, M., "Escala de Aptitud Intelectual de Binet-Terman," Quito (1971).

15. Navas, M., "Prueba de Aptitud Intelectual de Wechsler," Quito (1969).

16. Goodenough, F., "Test de Inteligencia Infantil por Medio del Dibujo de la Figura Humana," Paidos, Ed., Buenos Aires (1957).

17. Goddard, S., "Test de Reintegracion Motriz," Paidos, Ed., Buenos Aires (1975)

18. Bender, L., "Test Gestaltico Visomotor: Usos y Aplicaciones Clinicas," Paidos, Ed., Buenos Aires (1977).

19. Raven, J. C., "Coloured Progressive Matrices," London (1962).

20. Bernstein, J., "Tests de Matrices Progresivas de Raven," Paidos, Ed., Buenos Aires (1976).

21. Pérez, C., N. S. Scrimshaw, and J. A. Munõz, Technic of endemic goiter surveys, in: "Endemic Goiter," WHO Monograph Series 44, Geneve (1960), pp. 369-378.

22. Querido, A., F. Delange, J. T. Dunn, R. Fierro-Benítez, H. K. Ibbertson, D. A. Koutras, and H. Perinetti, Definition of endemic goiter and cretinism, classification of goiter size and severity of endemias, and survey technics, in: "Endemic Goiter and Cretinism: Continuing Threats to World Health," J. T. Dunn and G. A. Medeiros-Neto, Eds., PAHO Scientific Publication 292, Washington, D.C. (1974), pp. 267-272.

23. Consejo Nacional de Desarrollo. Tablas de Indicadores Socio-económicos, Quito, CONADE (1985).

24. Molina, J., "Las Migraciones Internas en el Ecuador," Universitaria, Ed., Quito (1965).

25. Estevez de, M., "Inmigración a Quito," Revista Cultura 20:56-72 (1984).

26. Centro de Estudios de Población y Paternidad Responsable, "Inmigración a Quito y Guayaquil," CEPAR, Quito (1985).

27. Debuyst, F., "La Población en America Latina, Demografia y Evolucion de Empleos," Gabiota, Ed., Buenos Aires (1973), pp. 250-283.

FIELD AND EXPERIMENTAL STUDIES OF IODINE DEFICIENCY IN SPAIN

Francisco Escobar del Rey, María Jesus Obregón and
Gabriella Morreale de Escobar

Unidad de Endocrinología Experimental, Instituto
de Investigaciones Biomédicas, C.S.I.C., Facultad de
Medicina, Arzobispo Morcillo 4, 28029 Madrid (Spain)

BRIEF HISTORY

Complete bibliographic references may be found elsewhere [1]. References to only a few studies are given here.

Goiter is widespread in many areas of Spain. According to Isidor Greenwald [2], references to endemic goiter may be found as early as 1497 in the writings of a Toledan Jew, Isaac Caro, who described it in an area near Madrid (Buitrago). Clearer references regarding different regions of Spain appear in 1885, and between that year and 1921, 14 papers were published, including a few from Marañón. In 1922 [3], Marañón described the very poor sanitary and socio-economic situation he found in the region of Las Hurdes, near the Spanish-Portuguese border, pointing out the high incidence of goiter and cretinism. He collected also information from other goiter areas and presented the results in 1928 at the 1st International Goiter Conference in Berne. His observations promoted the interest of the Spanish medical community, so that other authors described their findings in the Province of Salamanca, in Asturias, and Catalonia, a total of 21 papers being published up to 1934 [1]. Already by 1924 one of Marañón's coworkers, Vidal Jordana [4] had prepared some iodized salt and given it to the children of Las Hurdes with good results, and was trying to convince health authorities to implement iodine prophylaxis, at least in Las Hurdes. Whether the problem was understood by the health authorities, or not, little was done as regards a real iodization program, and any budding efforts stopped with the 1936-1939 Civil War which ravaged the country.

After this period there are hardly any publications by Marañón on the problem; it is likely that he lost hope of being able to convince health authorities about the severity of the problem, and the possibility of solving it by an adequate iodization program. But it is also possible that he was not totally convinced that as simple a measure as iodine supplementation could correct the appalling situation he found, and he directed his main efforts towards the building of roads which would open up the very isolated region, and the introduction of other measures which might improve the socio-economic situation. It is, however, quite striking in 1988 to realize that from his own observations in different goiter endemias in Spain, and probably from reading those described by DeQuervain and Wegelin in Switzerland, and by McCarrison in the Himalayas, but without the benefit of iodine determinations or other laboratory data, Marañón in 1928[5] divided endemias into three grades of severity:

"Endemic deafmutism represents the third degree of cretinous degeneration, which is characterized by this symptom, found together with imbecility, but not always associated with endocrine characteristics of cretinism (such as infantilism, myxedema). These cases leave the impression that in them the most severe alteration is the lesion to the nervous system (including defects of development of cortical centers). With the reservations inherent to any schematic proposal in Medicine, we could therefore summarize as follows:
Goitrous endemia... First degree, goiter: the lesion is mainly thyroidal
Second degree, cretinism: the lesion is preferably pluriglandular
Third degree, deafmutism: the lesion is mainly nervous."

To understand Marañón's meaning properly, it should be noted that he used the word cretinism as he was used to observe it, in sporadic congenital hypothyroidism, and that his reference to a pluriglandular lesion included short stature and dwarfism. It appears that it was conceptually difficult for him to accept that a nutritional deficiency could underly such complex and apparently unrelated manifestations, and he concluded that intermarriage, poverty, etc., were the most important etiological factors. This was most unfortunate, as Marañón was greatly respected, but his efforts in favour of iodine prophylaxis lacked complete conviction.

By 1945 new publications appear on goiter in Spain, most of them from Ortiz de Landázuri and his coworkers, reviewed in 1955[6]. They described the endemic goiter area in the high Sierra Nevada mountains near Granada, La Alpujarra. Marañón was aware of these studies and gave his support to the efforts of Ortiz de Landázuri, who was able to show the beneficial effects of iodine prophylaxis, carried out with support of the local health authorities by his coworkers, including ourselves. Endemic goiter regions were studied by other groups, in other parts of Andalusia (Sevilla, Cadiz, Málaga), Castellón, Avila (near Madrid), Aragón, Asturias, Galicia, Navarra (Pyrenees) and the Canary Islands. The approximately 150 papers published between 1945 and 1987 include an increasing proportion of studies using international criteria for the definition of goiter size, and more modern analytical procedures. Determinations of plasma PBI were already applied by us to the La Alpujarra endemia in 1954[7]. Experimental work in laboratory animals was initiated by Ortiz de Landázuri and his group, including transfer of rats to La Alpujarra: those fed local food developed goiter[8].

In 1971 Vivanco et al.[9] carried out an epidemiological study, measuring goiter size in 226 915 persons of all age groups from 21 provinces of Spain, and determined the iodine content of the drinking waters. This study excluded on purpose those provinces with well-known severe goiter endemias. Overall goiter prevalence was 13.4 % (6.6 % in men; 18.7 % in women), and the goiter rate in different areas was inversely correlated with the water iodine content. Later studies by other authors[1] included the determination of urinary iodine concentrations. These studies showed clearly that iodine deficiency is the major cause of endemic goiter in Spain.

Several studies which have been completed during the last decade have been collected in a special issue of the Spanish Journal Endocrinología[10]. They have mostly been carried out in schoolchildren. As a whole, it may be stated that in Spain the proportion of schoolchildren with goiter in a given area is inversely correlated with the urinary iodine excretion. The proportion of schoolchildren with serum T4 (blood spot test) below 6 µg /dl, and with elevated TSH, was increased as compared to schoolchildren from regions without iodine deficiency. The most severe endemias are those of Las Hurdes, which our group has been studying since 1968, and that of Los Ancares and other areas of Galicia, studied since 1975 by Tojo et al.[11]. The endemias in Las Hurdes and Galicia should be considered as of Grade III severity: urinary iodine excretion is very low, and there are cretins in both regions.

STUDIES IN LAS HURDES

We shall now summarize some of the main findings, which we believe might be pertinent to the problem of mental and psychomotor development of the inhabitants, and later describe experimental findings in rats, which might serve as pointers for future field studies.

Together with different coworkers, one of us (F.E.R.) has been carrying out surveys in Las Hurdes for a twenty year period. Initial studies [12,13] showed a very high goiter rate (90 %) in the schoolchildren from some of the villages, a very low urinary iodine excretion (13.3 $\pm$ 10.8 μg / L for boys and 15.0 $\pm$ 10.5 for girls), and high 24 and 48 hr thyroidal 131I uptake (80 % of dose). It was concluded that iodine deficiency was severe enough to account for the goiter rate. Somatic development was very poor, as 35 % of the schoolchildren were below the 10th percentile for weight and height. Skeletal age (X-ray of the hand), was delayed, in some cases of several years [14] The urinary concentration of creatinine was much lower than expected from the decreased weight. This suggested inadequate protein intake, confirmed by the low urinary concentration of total nitrogen [15].

Three villages where goiter frequency was highest (Fragosa-Martilandrán, El Gasco) were surveyed more extensively. In 1968 the total population was 878, 331 inhabitants being 15 yrs old, or older. Among these adults, 44 defectives (called "inocentes" by the local population) were identified. They were obviously mentally retarded, and presented one of the following abnormalities (criteria of the PAHO conference, 1963): irreversible neuromuscular disorders; abnormalities of hearing and/or speech; delayed somatic development; hypothyroidism. Two younger defectives were also found. Permission was obtained to study 25 of the 44 adult defectives with somewhat more detail [16] Although audiometric testing was not performed, most of them (73 %) had problems of hearing and/ or speech. Hyperreflexia was found in 80%, and spastic gait in 76 %. In the subgroup of defectives which had grown to less than 140 cm the plasma PBI was lower, and TSH higher, than in non-cretin adults living in the same area. In the subgroup taller than 140 cm plasma PBI and TSH were the same as in non-cretin adults. Plasma TSH of the adult population of Las Hurdes (n=65) was elevated as compared to Madrid. Although the plasma PBI of non-cretin adults from Las Hurdes was not significantly different, the range of values was greatly spread, with many below 3.5 μg I/dl, a value seldom found in the adult controls of Madrid.

Thus, the dramatic situation observed by Marañón in 1922 was still present almost 50 years later, despite efforts to improve the socioeconomic situation of the area. Our data showed that iodine deficiency was the most likely cause of the goiter rate, of the low plasma PBI and high plasma TSH, and of the high incidence of cretins, in agreement with data from many other endemias.

Our later surveys involved children attending school in the whole of Las Hurdes, and in nearby villages outside Las Hurdes. However, only three villages had been included in all of the surveys. Results obtained over the years are compared in Table 1. As may be seen, the goiter rate hardly improved during the following 10 year period (1968-1978), despite positive changes in the socioeconomic conditions. This could be accounted for by the iodine intake, which was still very low. As health authorities persisted in

Table 1. Schoolchildren (Fragosa-Martilandrán, Ladrillar, La Huetre)

Survey :	Year	n	goiter (OB+I+II)	µg I/L (± SD)	g creat/L (± SD)	µg T4/ dL (± SD)
	1968	76	74 %	14 + 11	.34 + .21	--
	1970	118	86 %	30 + 15	.41 + .19	--
	1978	156	86 %	16 + 15	.30 + .16	7.1 + 3.3
Lipiodol	1981	158	78 %	27 + 20	.63 + .34	7.9 + 2.8
Iodiz. salt	1984	93	61 %	39 + 40	.67 + .34	7.4 + 2.8
	1986	115	41 %	96 + 82	1.09 + .55	--
Madrid	1978	354	1%	112 + 72	.71 + .27	9.5 + 2.7
Miranda	1981	77	23 %	77 + 46	.63 + .34	11.1 + 3.5
Madrid	1984	180	18 %	102 + 67	.94 + .53	11.0 + 4.9

the attitude that the problem would solve itself with improvements in communication, schooling, etc., even without iodine prophylaxis, we carried out a new survey at the end of 1981 [17]: Again, there was practically no change in goiter rate or urinary iodide. The only improvement was an increase in the urinary creatinine, although it was still quite low compared to control children. Between 1968 and 1981-82 there was also some improvement in weight and height, but development was still quite poor. In the 1981 survey we were able to obtain blood spot T4 and TSH data from numerous schoolchildren, using methods developed for neonatal screening. The low T4, and high TSH (which we had earlier found in the non-cretin adult population) are already a frequent finding in young children. The frequency distribution for T4 was clearly shifted towards lower values than in the reference population: 46 % of the children from the iodine-deficient villages had T4 levels below 6 µg /dl, as compared to 7 % of the controls. In 40 % of the children TSH was above the detection limit for the blood spot assay (7.5 µ U/ml), in contrast to only 4 % of the controls. These findings could be attributed to the persistent iodine deficiency: 71 % excreted less than 20 µg I / L, as compared to 0.3 % of the controls. The urinary

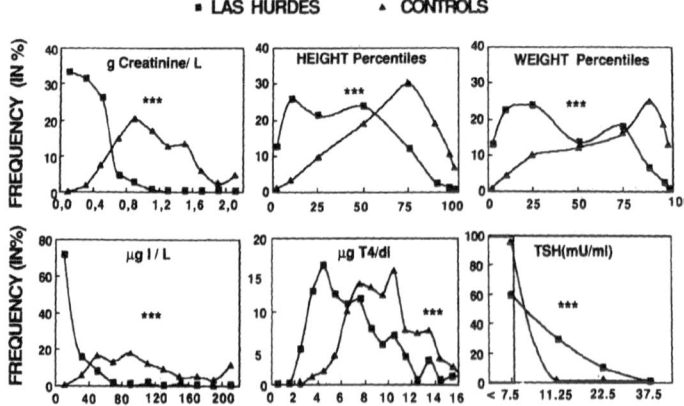

Figure 1. Frequency distributions of urinary iodine and creatinine, plasma T4 and TSH, height and weight percentiles of 156 schoolchildren from Las Hurdes (villages of Table 1), and of 354 from Madrid. 86 % of the children from Las Hurdes had goiter, vs. 1 % for the reference population. All curves for children from Las Hurdes are shifted significantly (***) towards lower values. Drawn from data by Escobar del Rey et al. [17].

creatinine excretion of these children was still very low, and their somatic development poor, with 13 % of them below the 3rd percentile, more than 35 % below the 10th percentile, for both height and weight.

It was clear from such data that unless iodine deficiency were corrected, little rapid change was to be expected from the other socioeconomic measures. In view of the persistent severity of the situation, the children were given 2 ml Lipiodol orally, as a protective measure until iodized salt would become available. Because of administrative changes in the area, we were not allowed to perform a follow-up study until 3 years later [18]. The effects on goiter frequency, urinary iodine, plasma T4 and somatic development were disappointing in the three villages of Table 1, although the overall goiter frequency in a more widespread area had decreased from 72 to 48 %. In 1984 iodized salt finally became available, and a small sample of shcoolchildren surveyed in 1986 clearly indicates that iodine intake was improving, although there were still many very low values. The increase in mean creatinine concentration might be merely apparent, as the 1986 survey included a larger proportion of older children than previous ones, and we have observed that urinary creatinine increases with age[18].

Summarizing these findings it becomes clear that, until consumption of iodized salt started, the schoolchildren were markedly iodine deficient. Despite "compensatory" enlargement of the thyroid in most schoolchildren, many of them had low serum T4, high serum TSH, poor somatic development (height, weight, skeletal maturation). Poor mental and psychomotor development was suggested by an initial study performed by García et al.[19] in 147 schoolchildren (6 to 14-yrs old) from three severely affected villages (Fragosa-Martilandrán, Huetre-Casarrubia, Ladrillar). This study presented the same problem as others, namely that the results were compared to standards given for the general Spanish population, not to results of the same tests given to children from a truly control population, similar in all respects except for the iodine intake. To overcome these problems, Prof. A. Querido suggested we request the invaluable guidance and collaboration of Prof. N. Bleichrodt. The controlled study then performed[20,21] clearly showed that iodine deficiency was indeed associated with poor mental and psychomotor development in the children from Las Hurdes, their test scores being compared to those from children in a nearby village (Miranda del Castañar), where iodine deficiency was less severe. Data presented by Prof. Bleichrodt at this Meeting show that the poor performance of the population tested in Las Hurdes is not due merely to a few isolated cases, which lower the mean values for the test scores. The frequency distribution curves for intelligence tests and some psychomotor tests are shifted towards lower values than those of the control schoolchildren.

QUESTIONS ARISING FROM THESE STUDIES

For years investigators interested in iodine deficiency have been faced with the same problems, some of which we should like to discuss here.

1) Does iodine deficiency per se result in poor mental and psychomotor development of the non-cretin population in goiter endemias similar to Las Hurdes?

2) If so, does poor development result only from congenital (and potentially irreversible) lesions, or does cerebral T3 deficiency, prolonged throughout infancy and into adulthood, contribute to it?

3) How severe does the iodine deficiency have to be to affect mental and psychomotor development?

1) Iodine deficiency per se:

It has been convincingly shown in several endemias[22] that adequate iodization programs eradicate the birth of cretins. Administration of iodized oil to women of childbearing age prevents further birth of neurological cretins provided treatment is instituted before, or very early in pregnancy[23]. Treatment even later in pregnancy prevents the birth of the hypothyroid type of cretins typical of the Zaire endemia[24]. Thus, correction of the iodine deficiency is sufficient to prevent the most serious central nervous system damage.

However, we do not yet have similar evidence that the poor mental and psychomotor performance of the non-cretin population will be normalized by adequate iodization programs. In the region of Las Hurdes we were concerned about the possible role of general malnutrition as assessed from low creatinine and nitrogen excretions. Low creatinine concentrations in the urine are frequently reported in severe goiter endemias, and this point has been discussed recently[25]. In the Ecuadorean Andes, Fierro-Benítez et al.[26] compared school performance and mental and psychomotor development of schoolchildren from Tocachi, born from mothers who had been adequately treated with Ethiodol before and during pregnancy and who were treated every four years after birth, with those of schoolchildren from La Esperanza, left untreated. The study confirmed the clear effect of iodine supplementation on the prevention of cretinism and the more severe forms of mental incapacity. School performance, as measured by grades repeated, psychomotor development (Goddard's test), and maturation (Bender Gestalt test) also improved. But the treated and untreated schoolchildren had similar I.Q. scores, both populations being significantly retarded as compared to standards for Ecuadorean schoolchildren. The authors suggest a possible role of choline, and other nutritional and / or social, deficiencies, which would not be corrected by iodine supplementation.

When the preliminary study was performed by García et al.[19] in Las Hurdes we could not exclude that factors other than iodine deficiency had caused the low scores, as the tests were performed on schoolchildren whose data were comparable to those of Fig 1. Although their poor somatic development might have been a consequence of the persistently low plasma T4 levels and, therefore, of the iodine deficiency, it might also have been related to general malnutrition[17]. The same might be true for their mental and psychomotor development[19]. However, the later controlled study performed

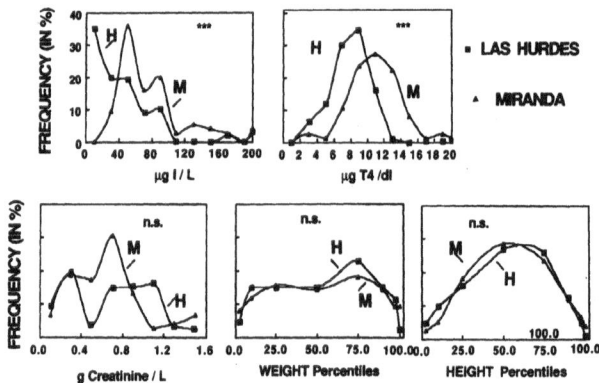

Figure 2; Frequency distributions for urinary iodine and creatinine, plasma T4, weight and height percentiles, corresponding to 6-12 yrs.-old children from Las Hurdes (Casares, Cabezo, Aceitunillo) and from Miranda del Castañar+ Cepeda, tested by Bleichrodt et al.[20,21]. Goiter rates were 66 % and 23 %, respectively. *** indicates P<.001; n.s., P>.05.

by Bleichrodt et al.[20,21] did not suffer from this drawback. Fig. 2 shows the results obtained in the 6 to 12- year old schoolchildren on which the tests were actually performed. Those from Las Hurdes lived in villages which were not as severely affected by iodine deficiency as those of the study of García et al.[19]. As may be seen, they differed from the control schoolchildren of Miranda del Castañar in goiter frequency, urinary iodine excretion and plasma T4, but not in urinary creatinine or somatic development. Thus, we believe this study shows that iodine deficiency alone is enough to affect mental and psychomotor development. This fully agrees with the results obtained by Bleichrodt et al.[27] in Java, where protein nutrition of the iodine-deficient and control villages were the same, as based on 24 hr urinary nitrogen excretion. It is, moreover, quite suggestive that the results from the study in Las Hurdes and in Java are superimposable [21]. The fact that in some endemias iodine deficiency alone is sufficient to result in poor mental and psychomotor development does not exclude that in other endemias the effects of iodine deficiency might be aggravated by some other nutritional deficiency or social deprivation.

The important implication is that correction of the iodine deficiency should ameliorate the mental and psychomotor development of these populations. But will it only benefit the inhabitants born after iodine deficiency has been corrected, or will it also improve the children who were already born under iodine-deficient conditions?

2) Congenital versus post-natal timing of the lesions
The damage to the CNS resulting in the neurological manifestations of the endemic cretin is believed to have started relatively early in pregnancy, at the beginning of the second trimester[28]. It is considered irreversible, even if prompt treatment with thyroxine were initiated soon after birth. But there is little information regarding the timing of the damage which results in poor mental and psychomotor development of the non-cretin part of the population, or about the potential reversibility of the processes leading to it. Poor motor coordination has been correlated with the degree of maternal hypothyroxinemia[29]. This would suggest that the damage was congenital. But it is also possible that the women with the lowest T4 levels were those with the lowest iodine intake, and that their children would also be exposed to the most severe iodine deficiency not only in utero, but during infancy as well. Thus, the poor mental and psychomotor development of children living under conditions of severe iodine deficiency might not be the result of damage occurring only in utero, as appears to be the case for the more severe CNS damage of the cretin, but also of continuing damage caused by cerebral hypothyroidism prolonged into adulthood. Results obtained in a rat experimental model might be pertinent.

ANIMAL EXPERIMENTS

Fetuses from rats on a low iodine diet (LID) have low concentrations of both T4 and T3 in all tissues studied, up to the end of gestation, as compared to fetuses from mothers on iodine-supplemented LID (LID + I)[30]. Cerebral T4 is very low. Despite a marked increase in the activity of typ II 5' Iodothyronine deiodinase (5' D-II) (Obregón et al., unpublished), cerebral T3 is also quite low, as there is simply not enough of the substrate, T4, to maintain normal brain T3 levels. Thus, in utero, the rat fetal brain is severely deficient in T3. Body and brain weights of LID pups near term are smaller.

Once the pups are born[31] several mechanisms appear to protect the brain from T3 deficiency during an important period of maturation, which is post-natal in rats. The number of pups is reduced in LID (8.8 ± 0.6) versus LID + I litters (11.2 ± 0.6). Although the maternal intake of iodine is about 4 % of that of the LID + I mothers, the milk iodine content is only reduced to 22 %, that is, there is a five-fold increase as regards the

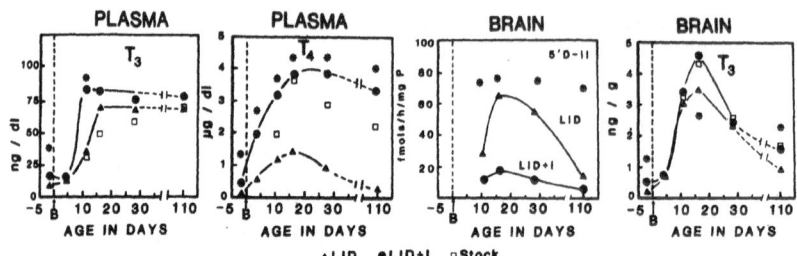

Figure 3. Plasma levels of T3 and T4, and brain 5' D-II activities and T3
concentrations, corresponding to rats born and bred on LID, LID + I, or
the stock diet. Statistically significant differences between the LID
and age-paired LID + I groups are identified by an asterisk. Drawn
from unpublished data by Obregón et al., and from Escobar del Rey et
al.[31], with permission of the publishers.

iodine content which would be expected from the intake. This, combined with
the reduction in the number of lactating pups, results in a relative
increase of iodine availability during the suckling period, as compared to
availability in utero. There is a relative increase in the plasma T4 of the
suckling LID pups, though levels are still reduced as compared to those of
LID + I pups. This relative post-natal increase, coupled with a marked
increase in cerebral 5' D-II, results in normal brain T3 concentrations during most of the suckling period (Fig. 3). This is especially evident at 11
days of age, when brain T3 was normal despite a markedly decreased plasma
T3.

However, once the pups are feeding on the LID themselves, and no longer
protected by the relative increase in iodine intake through maternal milk,
they are unable to maintain this relative increase in plasma T4, and
circulating T4 levels decrease markedly. Thus, despite persistently elevated
cerebral 5' D-II activity, after the suckling period T3 is lower in the
brain of LID rats (110 days of age Fig. 3).

It is known that the gluten-based LID diet that we used is nutritionally
inadequate, even if supplemented with KI[31], probably in more than one
essential aminoacid: The rats born and bred on LID + I grow poorly as
compared to C animals (Table 2). However, the iodine-deficient LID pups grew
even worse. This was already apparent by 16 days of age and becomes
increasingly evident up to 2.5-3 months of age. There was moreover a
striking individual variability of body weights, especially in males. The
coefficient of variation around the mean value was as high as 30 %. This
could not be attributed to differences between litters, as within a litter
body weights could vary from 92 to 260 g. This large coefficient of
variation was not observed either in LID + I or C rats (7.6 and 7.3 %,

Table 2. Body weights (+SD) and pituitary GH contents (+SEM) of male
rats born and bred on LID, LID + I, or stock diet (C)

Age	Body weight (g)			GH (mg $/gland)		
(days)	LID	LID + I	C	LID	LID + I	C
70	148+45 #,$	163+12 $	332+25	---	---	---
110	263+30 #,$	289+20 $	413+28	5.2+0.4 #,$	11.6+0.5 $	9.1+0.5

$ Given in weight equivalents of the NIAMDDK r-GH reference preparation
# Indicates a statistically significant difference between LID and LID
  + I.
$ Indicates a significant difference between LID, or LID + I, and C.

respectively). As the same animals grew older variability decreased and by 3.5-4 months of age the coefficient of variation had decreased to 12 %. A persistent difference in pituitary GH content was also found between LID and LID+ I rats by 28 days of age. Again, this difference was due to the iodine deficiency, as the other nutritional deficiencies resulting in poor growth of the LID + I rats did not affect their pituitary GH content (Table 2).

The T4 and T3 concentrations were measured in several tissues from adult males[32] (Fig. 4). Plasma T4 was markedly decreased in LID rats, plasma T3 being the same. Though not shown, plasma TSH was very high. Not only T4, but also T3 concentrations were lower in all tissue samples from LID rats as compared to both LID + I and C controls, the brain included. Cerebral 5' D-II was higher in LID (64 + 9.2 fmoles I- / hr / mg P) than in LID + I (13.4 ± 1.7) or C (18.6 ± 4.7) rats. These changes could be entirely attributed to the low iodine content of the LID, and were not caused by other nutritional deficiencies of the diet.

Thus, it appears that rats born and fed on LID all their life suffered from a deficiency of both T4 and T3 in all tissues studied, despite normal plasma T3 levels. Their anterior pituitary reacted to this situation as that of a hypothyroid rat, as far as may be assessed from the increased plasma TSH and decreased pituitary GH. Their growth was affected to a greater extent than might be attributed to nutritional deficiencies other than iodine deficiency.

As far as may be assessed from the increase in 5' D-II activity, the brain also reacted as in hypothyroidism. Although biological end-points of thyroid hormone action were not measured, the other T3-deficient tissues might also be hypothyroid. Therefore, we believe that animals on LID are suffering from a generalized cellular thyroid hormone deficiency (and hypothyroidism?) during most of their development and throughout life. This is entirely caused by the iodine deficiency, and not by other nutritional deficiencies, as the T4 and T3 concentrations in tissues from LID + I rats were not lower than those of C animals. Juvenile- and adult-onset cerebral T4 and T3 deficiency affects pyramidal neurons of the cerebral cortex, as shown in rats thyroidectomized as adults by Ruiz-Marcos et al.[33,34]. It is possible that cerebral hypothyroidism, prolonged throughtout life, damages brain development and function, whether or not some damage already occurred in utero.

Other experimental data further support this possibility. Years ago we started adult rats on LID[35,36]. After 10-11 weeks on the LID diet, their data were quite comparable to those just discussed for the adult rats which were born and bred on LID: The plasma T4 was very low, plasma TSH increased, plasma T3 was the same as that of LID + I rats. Pituitary GH content decreased, also suggesting that the anterior pituitary was hypothyroid. The

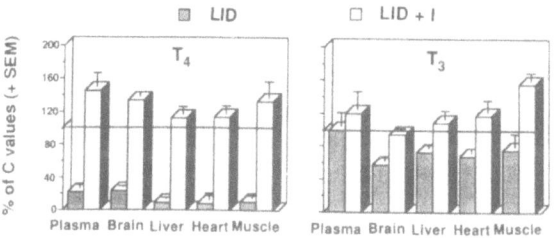

Figure 4. T4 and T3 concentrations in plasma and several tissues from male rats on LID or LID + I. Data are means + SEM, expressed as % of the mean values for age-paired C rats on stock diet (horizontal line). All differences between LID and LID + I rats were statistically significant, plasma T3 excepted. From Escobar del Rey et al.[32].

total and nuclear liver T3 decreased, and so did an hepatic thyroid hormone-dependent response, namely the activity of intramitochondrial α-glycerophosphate dehydrogenase. This result suggested that the liver was not only thyroid hormone-deficient, but hypothyroid as well[35]. Total and nuclear T4 and T3 in the brain also decreased in the LID rats started on the diet as adults[36]. Moreover, the biological response which was measured, namely the number and distribution of spines along the apical shaft of pyramids from the cortex[33,34], indicated a certain degree of cerebral hypothyroidism. Some of these points are illustrated in Fig. 5 [36]. The LID diet itself, if supplemented with KI, did not decrease the number of spines as compared to C rats (not shown), whereas the iodine-deficient LID decreased this number as markedly as thyroidectomy. It appears likely that iodine deficiency throughout life would have at least a similar, if not more severe, adverse effect on this cerebral parameter. Again, it is important to notice that the adverse effect was entirely due to iodine deficiency; other possible nutritional deficiencies of the iodine-supplemented LID did not affect spine number[36].

If any of the changes caused by iodine deficiency in rats (LID vs. LID + I groups) are pertinent to inhabitants of areas with iodine deficiency, it would appear that:

i) The fetal brain is likely to be suffering from T4 and T3 deficiency throughout pregnancy.

ii) During the suckling period, the brain might be protected by a relative increase in iodine intake through maternal milk. It is not known whether this occurs in human babies. However, even if it did, the timing of this protective effect would be different for both species, starting in man after a greater part of brain development has been completed, as compared to rats.

iii) Once the child is exposed to the same iodine deficiency as the rest of the population, it would continue to suffer from cellular thyroid hormone deficiency, brain included. This generalized tissue T3 deficiency could explain poor somatic development, including skeletal maturation, and might also play a role in poor mental and psychomotor performance.

These points are not merely academic, but might be important from the point of view of the benefits which might be expected from adequate implementation of iodine prophylaxis. If the damage resulting in poor mental and psychomotor development is not entirely congenital, but has been aggravated by hypothyroidism during infancy, it is possible that adequate iodization programs may still benefit children who are already born[37], and not only future generations. The disappointing results from schoolchildren

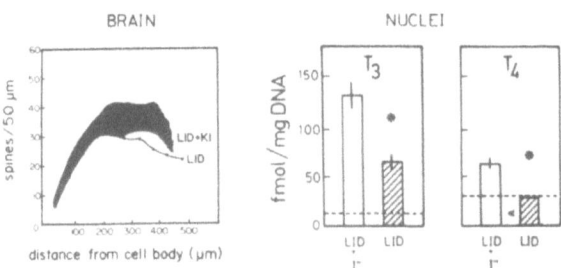

Figure 5. Distribution of spines along the apical shaft of pyramidal neurons from the cerebral cortex of rats started on LID or LID + I as adults. and concentrations of T3 and T4 in cerebral nuclei. Statistical significance of differences is indicated by an asterisk. Drawn from data by Obregón et al.[36] , with permission of the publishers.

of Las Hurdes given Lipiodol, presented at this Meeting by Bleichrodt et al. should be viewed with caution, as the treatment had not corrected the iodine deficiency throughout the study period.

3) Degree of iodine deficiency

At present we can only speculate regarding the degree of iodine deficiency required to affect mental and psychomotor performance of the population. If a situation of low circulating T4, despite normal plasma T3, prolonged throughout development and into adulthood were harmful for somatic and mental development, it is possible that children might be affected even in areas where iodine deficiency is not severe enough to result in the birth of cretins. Thus, in another goiter area of Spain[38], which we classified as of Grade II intensity on the basis of urinary iodine excretion and the absence of reported cases of cretinism, 37 % of the schoolchildren had plasma T4 (blood spot test) levels below 6 µg/ dl, a frequency comparable to that shown in Fig. 2 for the schoolchildren tested by Bleichrodt et al.[20,21] As assessed 27 months after the administration of Lipiodol, the goiter rate had decreased from 56 % to 22 %, the plasma T4 had improved, only a few children having values lower than 6 µg /dl. There was a statistically significant increase of their weight percentiles. This type of endemias are quite frequent, and we believe it would be of great interest to carry out a controlled study to test whether the mental and psychomotor development of such children is affected, and whether their performance would improve with permanent correction of their low plasma T4.

IN SUMMARY

A brief overview of historical aspects of the problem of endemic goiter in Spain indicate that there has been no lack of interest on behalf of the medical and scientific communities. On the other hand, the fact that publications still appear shows that the problem has not been solved, because health authorities so far have shown very little interest in the eradication of iodine deficiency, the major cause of endemic goiter in Spain.

Studies carried out in the region of Las Hurdes, where iodine intake has been very low and cretins were numerous, show that the poor mental and psychomotor development of the general population can be attributed to iodine deficiency.

Experimental studies show that rats on a low iodine intake suffer from generalized tissue T3 deficiency, caused by their iodine deficiency, and not by other nutritional deficiencies of the diet. There are many direct and indirect data which show that these animals suffer from cellular hypothyroidism, the brain included. To observe adverse effects of iodine deficiency on the brain, it is not necessary that the rats were born from mothers on LID. It is sufficient to feed LID chronically, even to adult animals.

If experimental data are pertinent to man, it is possible that the poor mental and psychomotor performance of children from areas with iodine deficiency is not due entirely to damage occurring in utero. If so, children born under conditions of iodine deficiency might still benefit from later institution of adequate iodization programs.

REFERENCES

1. Ferreiro Alaez L, Escobar del Rey F, 100 años de literatura sobre bocio endémico en España, Endocrinología 34:4 (1987)
2. Greenwald I, Notes on the history of goiter in Spain and among the Jews, Bull Hist Med 32:121 (1958)
3. Marañón G, Informe sobre el estado de Las Hurdes, Medicina Ibéra 15:232 (1922)

4. Vidal Jordana J, La profiláxis del bocio endémico por el yodo y su aplicación en Las Hurdes, Arch Esp Endocrinol Nutr 1:69 (1924)
5. Marañón G, El bocio en España y sus condiciones patogénicas, Anales Acad Med Quir Esp 15:1 (1928)
6. Ortiz de Landázuri E, Bocio endémico, Rev Iber Endocrinol 1:315 (1955)
7. Morreale de Castro G, Ortiz de Landázuri E, Mora Lara RJ, Escobar del Rey F, Aranzana A, La iodemia en el bocio endémico, Rev Clin Esp 52:247 (1954)
8. Palenzuela JM, Ortiz de Landázuri E, Mora Lara JR, Estudios sobre el bocio. Características estructurales del bocio endémico experimental, Rev Clin Esp 37:156 (1950)
9. Vivanco F, Palacios Mateos JM, Ramos Duce F, Busturia MA, Muro A, El bocio endémico en Espana, Estudio de la correlación incidencia de bocio y contenido de iodo en las aguas de bebida. Rev Clin Esp 123:426 (1971)
10. Bocio Endémico y Deficiencia de Yodo en España. Número in memoriam G. Marañon y E. Ortiz de Landázuri. Escobar del Rey F (ed). Endocrinologia, Suppl 2. Ediciones Doyma, Barcelona (1987)
11. Tojo R, Fraga JM, Escobar del Rey F, Rodríguez Martínez A, Vázquez E, Esquete C, Estudio del bocio endémico en Galicia. Repercusión sobre el crecimiento y desarrollo. Endocrinología 34:48 (1987)
12. Escobar del Rey F, Ferreiro Aláez L, Sánchez Franco F, Cacicedo L, El bocio endémico en Las Hurdes. I. Incidencia del bocio en la población escolar y pre-escolar, Rev Clin Esp 128:101 (1973)
13. Escobar del Rey F, Sánchez Franco, F, Ferreiro Aláez L, El bocio endémico en Las Hurdes. II. Parámetros que indican la existencia de una deficiencia del iodo, Rev Clin Esp 130:7 (1973)
14. Ferreiro Aláez L, Escobar del Rey, Contribución al estudio de la endemia bociosa en Las Hurdes (Tesis), Rev Univ Madrid 1:382 (1969)
15. Sánchez Franco F, Cacicedo L, Morreale de Escobar G, Escobar del Rey F, Nutrition and iodine versus genetic factors in endemic goiter, J Endocrinol Invest 6:185 (1983)
16. Sánchez Franco F, Ferreiro Aláez L, Cacicedo L, Garcia MD, Morreale de Escobar, Escobar del Rey F, Alteraciones por deficiencia de yodo en Las Hurdes. III. Cretinismo. Endocrinología 34:89 (1987)
17. Escobar del Rey F, Gómez Pan A, Obregón MJ, Mallol J, Arnao-R MD, Aranda A, Morreale de Escobar G, A survey of schoolchildren from a severe endemic goiter area in Spain, Quart J Med 198:233 (1981)
18. Escobar del Rey F, Martin T, Turmo C, Mallol J, Obregón MJ, Morreale de Escobar G, Alteraciones por deficiencia de yodo en Las Hurdes. I. Deficiencia de yodo y efectos del Lipiodol, Endocrinología 34:61 (1987)
19. García I, Rubio C, Alonso E, Turmo C, Morreale de Escobar G, Escobar del Rey F, Alteraciones por deficiencia de yodo en Las Hurdes. II. evaluación del desarrollo psicomotor de escolares. Endocrinología 34:74 (1987)
20. Bleichrodt N, García I, Rubio C, Morreale de Escobar G, Escobar del Rey F, Mental and motor development of children from an iodine deficient area, in: "Iodine Deficiency Disorders and Congenital Hypothyroidism", Medeiros-Neto G, Maciel RMB, Halpern A (Eds), ACHE Press, Sao Paulo, 1:65 (1986)
21. Bleichrodt N, García I, Rubio C, Morreale de Escobar G, Escobar del Rey F, Developmental disorders associated with severe iodine deficiency, in:"The Prevention and Control of Iodine Deficiency Disorders", Hetzel BS, Dunn JT, Stanbury JB (eds), Elsevier Science Publishers, New York, 1:65 (1987)
22. Hetzel BS, An overview of the prevention and control of iodine deficiency disorders, in:"The Prevention and Control of Iodine Deficiency Disorders", Hetzel BS, Dunn JT, Stanbury JB (eds), Elsevier Science Publishers, New York, 1:1 (1987)
23. Pharoah POD, Buttfield IH, Hetzel BS, Neurological damage to the fetus resulting from severe iodine deficiency during pregnancy, Lancet i:308 (1971)

24. Thilly C, Roger G, Lagasse R, Bourdoux P, Ramioul L, Berquist H, Ermans AM, Fetal hypothyroidism and maternal thyroid status in severe endemic goiter, J Clin Endocrinol Metab 47:354 (1978)

25. Thilly C, Bourdoux P, Swennen B, Bebe N, Due D, Ermans AM, Assessment and planning for IDD control programs, in:"The Prevention and Control of Iodine Deficiency Disorders", Hetzel BS, Dunn JT, Stanbury JB (eds), Elsevier Science Publishers, New York, 1:181 (1987)

26. Fierro-Benítez R, Casar R, Stanbury JB, Rodríguez P, Garces F, Fierro-Renoy F, Estrella E, Long-term effects of correction of iodine deficiency on psychomotor and intellectual development, in:"Towards the Eradication of Endemic Goiter, Cretinism, and Iodine Deficiency", Dunn JT, Pretell EA, Hernán Daza C, Viteri FE (eds), PAHO, Washington,DC,182 (1986)

27. Bleichrodt N, Drenth PJD, Querido A, Effects of iodine deficiency on mental and psychomotor abilities, Am J PhysiolAntropol 53:55 (1980)

28. DeLong R, Neurological involvement in iodine deficiency disorders, in: "The Prevention and Control of Iodine Deficiency Disorders", Hetzel BS, Dunn JT, Stanbury JB (eds), Elsevier Science Publishers, New York, 1:49 (1987)

29. Connolly KJ, Pharoah POD, Hetzel HB, Fetal iodine deficiency and motor performance during childhood,Lancet ii:1149 (1979)

30. Escobar del Rey F, Pastor R, Mallol J, Morreale de Escobar G, Effects of maternal iodine deficiency on the L-Thyroxine and 3,5,3' -Triiodo-L-Thyronine contents of rat embryonic tissues before and after onset of fetal thyroid function, Endocrinology 118:1259 (1986)

31. Escobar del Rey F, Mallol J, Pastor P, Morreale de Escobar G, Effects of maternal iodine deficiency on thyroid hormone economy of lactating dams and pups: maintenance of normal cerebral 3,5,3'-triiodothyrone concentrations in pups during major phases of brain development, Endocrinology 121:803 (1987)

32. Escobar del Rey F, Obregón MJ, Ruiz de Oña C, Bernal J, Morreale de Escobar G, Generalized deficiency of 3,5,3'-triiodothyronine (T3) in tissues of rats on a low iodine intake, despite normal circulating T3 levels, XVI Meeting European Thyroid Association, Lausanne, 1987

33. Ruiz Marcos A, Sánchez Toscano F, Escobar del Rey F, Morreale de Escobar G, Reversible morphological alterations of cortical neurons in juvenile and adult-onset hypothyroidism in the rat, Brain Research 185:91 (1980)

34. Ruiz Marcos A, Cartagena P, Escobar del Rey F, Morreale de Escobar G, Rapid effects of adult-onset hypothyroidism on dendritic spines of pyramidal cells of the rat cerebral cortex, Experim Brain Res, in press (1988)

35. Santisteban P, Obregón MJ, Rodríguez Peña A, Lamas L, Escobar del Rey, Morreale de Escobar G, Are iodine-deficient rats euthyroid? Endocrinology 110:1780-1789, 1982

36. Obregón MJ, Santisteban P, Rodríguez Peña A, Pascual A, Cartagena P, Ruiz-Marcos A, Lamas L, Escobar del Rey F, Morreale de Escobar G, Cerebral hypothyroidism in rats with adult-onset iodine deficiency, Endocrinology 115:614 (1984)

37. Dodge PR, Palkes H, Fierro-Benítez R, Ramírez I, Effect on intelligence of iodine in oil administered to young Andean children: a preliminary report, in: "Endemic Goiter", Stanbury JB (ed), PAHO, Washington, DC, 378 (1969)

38. Escobar del Rey F, Mallol J, Turmo C, Jiménez Bustos JM, García García A, Gómez Pan A, Morreale de Escobar G. El bocio endémico en la zona de Guadalajara y su evolución tras la administración de aceite yodado (Lipiodol), Endocrinología 34:53 (1987)

IODINE DEFICIENCY, MATERNAL THYROXINE LEVELS IN PREGNANCY AND

DEVELOPMENTAL DISORDERS IN THE CHILDREN

Kevin J. Connolly

Department of Psychology
University of Sheffield
Sheffield S10 2TN

Peter O.D. Pharoah

Department of Community Health
University of Liverpool
PO Box 147, Liverpool L69 3BX

Endemic cretinism is a complex syndrome involving a triad of severe sensory, motor and cognitive defects. The condition is related to iodine deficiency[1] and it can be prevented by iodine prophylaxis of the mother before or very early in pregnancy [2, 3]. The frequency of goitre in populations at risk for endemic cretinism is also extremely high. The Pan American Health Organisation [4] defined endemic cretinism in terms of the following clinical manifestations; mental deficiency together with either (a) a predominant neurological syndrome consisting of defects of hearing and speech and with characteristic disorders of stance and gait of varying degrees, or (b) predominant hypothyroidism and stunted growth. Follow-up studies of the controlled trial of iodinated oil [2] revealed a relationship between maternal iodine status during pregnancy and the performance of children on a number of motor tasks [5]. This observation is supported by the results of a comparison of children from a severely iodine deficient village with children from a control village in Central Java [6]. The children from the iodine deficient village were generally poorer on a number of measures of motor competence. Similar differences have also been found between children from iodine deficient and non-deficient villages in Spain[7]. In Ecuador a significant difference was found on Stanford-Binet scores between two groups of children one group having been born to mothers given an iodine supplement [8]. The children whose mothers received a supplement did better.

Further follow-up and analysis of some of the children whose mothers were part of the controlled trial in Papua New Guinea uncovered a relationship between maternal thyroxine levels during pregnancy and the motor performance of the children 10 years later[9]. Evidence has also been obtained suggesting a similar relationship between maternal thyroxine levels in pregnancy and aspects of the child's cognitive capacity [10]. Whatever the physiological mechanism linking iodine deficiency to functional consequences neurologically and psychologically, the range of effect is great; from a high proportion of still births[11] and a striking difference in the 15 year cumulative survival rate between treated and control groups[12], through an extremely high incidence of gross neurological damage of the kind found in endemic cretinism[2] to a variety of sub-clinical deficits in performance[5].

There is evidence of sub-clinical defects associated with iodine deficiency on all three of the principal dimensions of cretinism; hearing loss [13], motor competence[5] and cognitive function [10]. That iodine deficiency diseases present a spectrum of developmental consequences which vary from the gross to those detected only by careful and precise quantitative measures is of considerable significance biologically and socially. The less severe manifestations may in fact be of greater social significance because many more individuals are affected and put at developmental risk.

Assessing the intellectual capacity and performance of people in preliterate societies is to say the least difficult. The predominant approach to the measurement of intelligence has been that provided by the psychometric tradition. Here the performance of an individual on such tests as vocabulary, analogies, the manipulation of numbers, the mental rotation of geometric objects and abstract shapes, block design etc. is compared with the performance of an appropriate standardisation sample. This procedure enables individuals to be ranked in relation to each other. However, sub-tests such as those listed above would be inappropriate for preliterate or deprived populations because such groups are likely to have little, or in respect of certain skills, no acquaintance with mental culture of this kind. No tests have been developed specifically for use with the Highland people of Papua New Guinea and few attempts have been made to adapt performance scales (non-verbal tests) for PNG Highlanders. Having regard to the various constraints (language barriers and fundamentally different social and material cultures) Pharoah et al.[10] in a follow-up study which linked maternal thyroxine levels during pregnancy with the abilities of the progeny in middle childhood used the Pacific Design Construction Test (PDCT). This test was developed by Ord [14] and used in Papua New Guinea to select army recruits and young people for technical training. The test is based on the Kohs Block Design Test [15] and has been shown to have some validity as a measure of intelligence in preliterate peoples [16]. While useful for certain selection purposes the test is of limited value diagnostically because it does not relate to any specific cognitive process.

It is now a generally accepted view that an individual's mental ability or intelligence reflects both his knowledge of the world and other basic general information processing skills which do not themselves depend on the content of the information being processed. The form and extent of a person's education and of course the nature of his culture will be reflected in his performance on the familiar Western psychometric tests, such as the Stanford-Binet and Wechsler scales. None of these are culture free and as measures of early biological insult they are probably insensitive. A more recent approach to the study of human intelligence is provided by information processing theories of cognitive function[17]. As the name implies the focus of attention is on the *processes* involved in cognitive activity and the efficiency with which these processes are carried out. The basic unit of analysis in cognitive theories is the information processing component. A component is an elementary information process which operates on mental representations of objects or symbols. The essential difference between the psychometric and the information-processing approaches is that the former sets out to identify the structure by which a set of components is interrelated (the factor structure) whereas the latter seeks to identify the processes involved in the behavioural implementation of the component factors.

The efficiency with which an individual can perform different cognitive processes, which is usually measured by the time taken, has a positive relationship with scores on intelligence tests such as the Wechsler Adult Intelligence Scale and Raven's Advanced Progressive Matrices [18]. The stimuli employed in the reaction time tasks used by Vernon [18] do not require subjects to draw on the kind of information or reasoning skills which are tapped by standard intelligence tests. Although measures of speed of processing have little surface content in common with psychometric measures of intelligence they may be linked by the efficiency with which basic cognitive operations are performed. There have been suggestions that the relationship between reaction time measures of processing speed and intelligence as measured by standard psychometric tests is due largely to the fact that many psychometric tests of ability are timed [19,20]. Vernon et al.[21] have examined the relationship between a battery of reaction time tests and a group intelligence test which gave timed and untimed scores. They found speed of information processing an equally good predictor of performance on timed and untimed intelligence tests. Although timed and untimed intelligence tests impose different information processing demands the speed with which an individual can cope with these demands is equally important in both conditions[22].

Mental operations such as encoding information, scanning information held in short-term memory, retrieval from long-term memory, and speed of decision making are components of many forms of cognitive activity. Although not the only components they are among the most basic. Individual differences in intelligence are likely in part to be a consequence

318

of differences in the efficiency or speed with which individuals are able to perform the basic components of information processing. If we are able to measure the time course of information processing in the nervous system then it is likely that we shall have a sensitive measure of its functional efficiency and one which will reflect sub-optimal conditions of development. The assumption gains support from the observation that individuals with low IQs require more time to perform basic mental operations[23]. The Shorter Oxford English Dictionary defines intelligence as 'quickness of understanding' and 'quickness of mental apprehension'; and in English to speak of a person as being 'slow' has for long been a euphemism for low intelligence.

The measure of cognitive function employed in the investigation reported here was decision time; the time elapsing between the subject being presented with a stimulus and deciding to which class of stimuli it belonged. The difficulty of the decision entailed is related to the number of stimulus-response classes. Put simply, as the number of stimulus alternatives increases the difficulty entailed in deciding on the appropriate response increases and this is reflected in the time to make the decision. A theoretical basis for this was provided by Hick [24] who showed that the reaction time was a function of the number of stimulus-response alternatives; and it is given by the equation $RT = a + b \log N$, where N is the number of stimulus-response alternatives. The slope of the regression line, b, reflects the time penalty per unit of uncertainty measured in bits [25].

The findings reported here arise from a further follow-up of a group of individuals born to mothers for whom information is available on maternal thyroxine levels during pregnancy. In addition to measuring decision time in a variable sorting task (an analogue of a choice reaction time task) measures were also made on skilled arm-hand movements and dextrous manipulations in order to check whether previously observed effects [9, 10] on motor performance were still present.

MATERIALS AND METHODS

Subjects

During 1970 and 1971, in the course of a controlled trial of the effects of iodine supplementation, blood samples were taken from women of childbearing age in several villages of the middle Jimi valley in the Western Highlands Province of Papua New Guinea. During the 1960s and 1970s the prevalence of endemic cretinism was extremely high in this area. Of this set 66 women were pregnant when the blood specimens were taken, these constitute the original sample. All the pregnancies resulted in singleton births. By 1985, 19 of the 66 children had died giving a 15 year cumulative mortality of 288 per 1000[12]. Of the 47 surviving children 44 were traced and tested in 1985. The remaining 3 were known to be alive at that time but because of temporary or permanent migration to another area outside the valley they could not be contacted. When examined the children were aged between 14 and 16.

Biochemical measures

Details of the analytical techniques used to estimate serum total thyroxine (TT_4), total triiodothyronine (TT_3) and thyroid stimulating hormone (TSH), and the means employed to check the effects of climatic conditions on sera have been reported previously[11].

Measures of physical status

Height was measured using a portable Harpenden anthropometer. Mid-triceps skinfold thickness was measured on right and left arms with Harpenden skinfold calipers. The mean of these was used in the analyses.

Measures of motor performance

Two tasks used on previous occasions, the pegboard and bead threading, were used again along with a new measure, the Minnesota Rate of Manipulation Test.

Pegboard. In this task the subject was required to insert as quickly as possible 10 small mushroom shaped plastic pegs into a row of equally spaced holes in a 10 x 10 pegboard, the caps of these pegs had an external diameter of 1 cm. The pegboard was placed on the table in front of the child in the midline position and a box of pegs was positioned 10-15 cm from the pegboard on the side of the subject's dominant hand. The dominant hand was identified by asking the subject to perform a number of simple dextrous tasks and noting which hand was used. The time between the subject grasping the first peg and inserting the tenth into the pegboard was measured using a stopwatch. The task was demonstrated to the subject and the importance of speed stressed by brief instructions in Melanesian Pidgin. Two practice attempts were given, the subject then carried out 5 successive trials with an interval of approximately one minute between trials.

Bead Threading. This task requires the combined use of both hands in threading small (approximately 8 mm diameter) plastic beads onto a lace one end of which was wrapped to give a stiff section approximately 3 cm long, the other end of which was knotted. A box of beads was placed on a table in front of the subject in the midline. The child was asked to thread as many of the beads as possible onto the lace in a period of 30 secs. The task was demonstrated to each subject and then two practice attempts were given. The subject was given the instruction 'go' and the test period was timed from when the first bead was picked up. After 30 secs the instruction 'stop' was given. The start and stop instructions were given in Pidgin. The subject's score was the number of beads threaded in 30 secs. A bead threaded onto the lace but not yet released was included in the score, a bead lifted from the box but not yet threaded was not included in the score. Five trials were given to each subject.

Minnesota Rate of Manipulation Tests (MRM) This test provides a measure of arm-and-hand dexterity and can be used in several ways. In this investigation two of the sub-tests, the placing test and the turning test were employed (details are given in the examiner's manual [26]). The placing test requires the subject to transfer a series of cylindrical blocks (3.5 cm diameter x 2 cm deep) from a row of holes in a board to a row in a similar board placed in front of it. The distance between the corresponding rows, the travel distance, was 25 cm. The time taken to transfer 15 blocks was measured with a stopwatch. The task was repeated 8 times. The turning test requires the subject to lift each block from its hole, turn it upside down and replace it in its original hole. Each trial consisted of turning 15 blocks and was repeated 8 times, the time to perform a series of 15 responses was again measured by stopwatch. Since each block was made from two discs fixed together, one orange and one yellow, the task was readily understood by the subjects; the orange row should be turned to make it yellow and *vice versa*.

Measurement of cognitive performance

A card sorting task was used to measure information processing rate[27]. Once again this was a task which could be explained fairly easily by demonstration and which did not rely on complex verbal instructions. The level of difficulty of the task, the information load, could be systematically varied by changing the number of stimulus-response sets presented to the subject. In essence the task measures choice reaction time.

Stimulus cards were made by sticking 4 cm squares of thin coloured washable plastic material onto blank playing cards. Standard packs of 24 cards were made up for the four experimental conditions employed; A, B, C and D. In condition A a pack of cards was made up of cards of two colours, 12 of each. In condition B a pack was made up of four colours, 6 cards of each; in C six colours, 4 cards each; and in D there were eight colours with 3 cards for each. A pack of blank cards was used to obtain movement times in each of the four conditions. Thus for pack A the choice was between 2 stimuli, and for B, C and D respectively 4, 6 or 8 stimuli.

Each pack of 24 cards was thoroughly shuffled each time before sorting. The subject held the pack face down and turned each card up one at a time. The pack then had to be sorted into its constituent colours by placing cards of specific colours into the appropriate tray identified by a cue card marked with that colour and fixed to the back of each tray. As each card was turned a decision concerning the tray to which it must be assigned was required. Subjects were instructed by demonstration and by brief instructions in Pidgin to sort the pack as quickly as possible. The trays into which the cards were to be sorted (2, 4 6

Figure 1. Child performing the card sorting task

or 8 dependent on the condition) were arranged on a table in a semi-circular array in front of the subject (see Figure 1). Only the appropriate trays for the given condition were arranged on the table.

In order to perform the task a person must be able to distinguish the various colours used and correctly match them. The ability to distinguish and match was therefore tested before the trials began. All the trays were arranged on the table and the examiner demonstrated the match between stimulus cards and the appropriate cue card attached to a tray. The subject was then given a pack of stimulus cards containing 4 of each colour used and asked to assign them to the corresponding tray. There was no time pressure on this initial matching task. The criterion for correct matching was an error free run through the pack. All the subjects were able to make the necessary discriminations and matches.

The task was explained to each subject by demonstration and the importance of speed emphasised by mime and verbally in Pidgin. Several practice trials were given to ensure that they understood what was required of them. The 16 packs (4 at each level of difficulty, A, B, C, D) were presented to the subjects in a Latin square design to avoid order effects. The time required to sort each pack into its constituent colours, from the first card being turned to the last being assigned to a tray, was measured using a stopwatch. In the case of errors the subjects were told to ignore these and press on with all speed. This served to emphasise the need to perform the task as quickly as possible. If more than 4 errors occurred in sorting a pack the trial was stopped and begun again the subject being reminded of the need to match stimulus and cue card. The balance between speed and accuracy, in these terms, was fairly readily achieved by the subjects.

The time required to sort a pack of cards is made up of two components; *decision time*, the time required to identify the appropriate colour and determine where it should be placed, and *movement time*, the time required to make the necessary turning and delivery

Table 1.   Data on maternal thyroid function in pregnancy and measures of the
measured in the card sorting task.   The table is arranged in terms of ascending

| Serial Number | Maternal Thyroid Measures | | | Child's Physical Status | | School Level |
	TT4	TT3	TSH	Height	Skinfold Thickness	
0396	14	412	50	1.39	3.60	1
0245	16	1199	27	1.27	4.0	3
0430	21	862	130	1.25	4.80	2
1058	24	1050	7	1.33	5.60	3
0398	25	1250		1.40	4.80	1
0489	26			1.24	4.20	2
0054	26	1500	6	1.16	5.40	1
0262	31	1430		1.34	5.20	1
0086	32		7			
0149	33	1550	7	1.27	4.60	3
0435	39	2579	10	1.31	6.6	1
0089	43	1300	8	1.28	5.2	1
0243	43			1.24	6.8	1
0745	49			1.40	5.8	3
0362	59	1419	37	1.31	3.8	1
0010	60	975	6	1.24	7.4	1
0129	66	2519	5	1.43	8.2	1
0001	68	2019	4	1.16	6.2	3
0428	77	1450	7	1.44	6.4	1
0761	82	1635	9	1.42	6.0	2
0918	84	1029	2	1.43	9.0	3
0816	94	750	5			2
0980	95	1743	1	1.25	4.0	2
0026	96	1200				
0901	99			1.40	6.2	2
0170	101			1.27	4.6	1
0997	101			1.45	5.6	1
0140	104			1.33	4.2	3
0944	106	1350	1	1.39	4.6	3
0939	107	1453	2	1.36	5.2	2
0113	108			1.46	6.4	1
0765	108	1050	7	1.33	6.0	3
0975	112	1559	1	1.36	5.0	2
0813	114	1575	2	1.50	6.6	2
0810	119	1575	4	1.28	4.8	2
0237	124			1.40	4.2	1
0408	128	1485	5	1.23	6.0	2
0890	128	1350	3	1.35	6.2	3
0034	137			1.30	5.0	1
0368	137	1050	2	1.28	5.8	2
0965	138			1.26	5.0	2
0214	155			1.37	4.2	1
0949	155	2018	1	1.25	7.0	1
1046	163	1400		1.36	5.2	3
1009	165	1475	3	1.27	6.4	1
0136	167			1.36	4.2	3
0377	225	1059	3	1.35	5.6	1

children's physical status, motor performance and information processing rate
maternal thyroxine level. TT4 ng/ml, TT3 pg/ml. TSH µU/ml. See text for details.

Motor Performance				Card Sorting	Serial Number
Beads	Pegs	MRM trans	MRM turn		
8.2	23.0	23.85	23.55	12.99	0396
9.8	19.12	17.17	20.17	9.43	0245
8.4	27.88	25.92	28.25	10.59	0430
9.4	20.48	19.22	23.00	11.79	1058
7.4	24.28	23.55	22.12	18.10	0398
7.6	23.52	20.02	24.70	8.76	0489
9.4	24.08	21.25	20.12	18.23	0054
7.4	26.68	24.57	29.10	28.83	0262
					0086
11.4	21.08	18.90	19.95	10.28	0149
9.2	20.60	19.97	21.45	9.81	0435
11.0	21.68	17.27	19.97	9.55	0089
9.6	23.92	24.37	26.42	20.32	0243
11.0	18.60	16.67	15.62	6.24	0745
9.2	25.8	28.82	28.47	12.36	0362
9.4	21.84	18.27	20.87	13.94	0010
9.8	22.72	19.10	20.05	9.39	0129
10.4	21.0	17.62	21.57	9.32	0001
11.6	20.08	17.67	17.37	6.37	0428
14.6	16.60	15.07	16.82	7.16	0761
12.6	18.56	16.05	18.57	5.32	0918
					0816
14.0	16.60	16.27	16.95	10.85	0980
					0026
13.8	19.44	15.07	16.27	7.03	0901
9.2	23.2	18.95	20.30	11.36	0170
11.8	22.68	16.82	16.95	9.27	0997
				11.36	0140
12.2	17.76	14.55	15.07	7.59	0944
13.0	20.48	16.35	17.50	7.66	0939
12.0	18.24	17.45	18.15	9.61	0113
10.4	22.72	15.75	17.77	4.48	0765
11.4	17.56	16.10	18.47	5.12	0975
12.6	16.0	15.32	16.87	9.03	0813
12.0	16.68	16.32	18.40	8.60	0810
11.0	19.88	17.50	17.65	11.21	0237
11.8	14.92	14.00	13.30	6.80	0408
14.2	15.76	15.00	15.32	4.63	0890
11.0	19.20	14.97	18.62	9.72	0034
10.6	21.68	15.92	22.22	11.26	0368
11.4	19.40	15.42	17.40	6.87	0965
11.2	23.12	18.77	20.77	10.74	0214
11.8	19.76	20.67	21.97	11.01	0948
9.8	21.48	16.47	21.27	6.72	1046
11.0	21.88	17.87	19.77	9.76	1009
12.6	17.40	15.15	15.42	7.62	0136
9.8	26.68	18.37	23.27	12.15	0377

movements of the card. Movement time for 2, 4, 6 and 8 alternatives was estimated by having the subjects systematically distribute a pack of 24 blank cards between 2, 4, 6 and 8 unmarked trays respectively. Four replicates of each condition were made.

RESULTS

In the Jimi valley only six years of primary schooling is available and this at schools in only a few of the villages. Children may therefore have to spend a considerable time each day in walking to and from school. More often however it is necessary for the child to stay in the village where the school is located and return home to his or her family at the weekend. School is not compulsory and fees have to be paid which results in fewer girls than boys being sent to school. Within the population we studied there was considerable variation in school experience but no satisfactory measure of educational achievement was available. In an attempt to make some measure of educational attainment three levels of schooling were distinguished; no school, attendance within grades 1-3 and attendance within grades 4-6. Passage into a higher grade requires satisfactory completion of the lower grade. These three levels, designated 1, 2 and 3 respectively, were included in the model used to analyse the data. No difference was found between the lower and higher school grades and so only two levels (no school/school) were finally included in the model.

The data collected on the 44 children are summarised in Table 1. Scores on the pegboard, bead threading and Minnesota Rate of Manipulation transfer and turn tests are the means of 5, 5, 8 and 8 trials respectively. Mean decision times were obtained from the card sorting data by subtracting movement time from sorting time for the 2, 4, 6 and 8 choice conditions respectively. A linear regression was then calculated between the number of choices and the corresponding decision time for each subject. This provided the data on the efficiency of cognitive function which was related to maternal thyroid hormone status. These data and the corresponding data for the motor performance tasks were examined in relation to maternal thyroid status by multiple regression analysis with level of school as a dependent variable.

One child (serial number 0140) had a cataract in his left eye and a marked strabismus. Because the consequent lack of binocular vision would probably impair dextrous manipulation his data on the manipulative tests were excluded in order to avoid bias. The data on this child's card sorting performance and measures of physical status were included in the analysis. The data on another child were excluded because the mother's TT_4 (serial number 0377) was an outlier being well above the upper limit of normal. A repeat analysis of the maternal blood sample confirmed the outlying value in the face of normal TT_3 and thyroid stimulating hormone (TSH) levels. A third child (serial number 0362) was excluded because of a high maternal TSH level despite normal TT_4 and TT_3 levels (see Pharoah and Connolly, this volume). When serum analyses were carried out priority was always given to TT_4, TT_3 and TSH were then estimated if sufficient serum was available. In 12 cases there was insufficient serum to assay TT_3. These factors account for the variation in the N for each analysis.

A multiple regression analysis in which TT_4 and TT_3 were treated as independent variables, and measures of the children's physical status and cognitive and motor performance as dependent variables, was carried out. Allowance was made for an effect due to attendance at school on the cognitive and motor tests and for a sex effect on measures of physical status. The results of the multiple regression analysis of the various performance measures on maternal TT_4 are given in Table 2. In the analysis of the measures of physical status a sex difference was observed only for skinfold thickness, this was allowed for in the model. For each of the motor task scores and the information processing score the correlation with TT_4 is highly significant (p < .005). Neither height nor skinfold thickness showed any correlation with maternal TT_4.

The comparable analysis for TT_3 is shown in Table 3. Of all the independent variables only one, the score on the pegboard test, correlated significantly with TT_3 and, even in this instance the proportion of variance explained by TT_3 was less than in the corresponding analysis with TT_4.

Table 2. Correlation of maternal TT_4 levels with children's scores on various motor and cognitive tasks, and with measures of physical status

Dependent variable	Other effect included in the model	Regression coefficient (95% confidence interval)	N	R^2 due to TT_4	Full model
Bead threading	School 2 levels	0.0204 0.011 - 0.2099 $p < .0005$	41	28.4	43.7
Pegboard	School 2 levels	-0.0252 -0.041 - -0.0093 $p < .0005$	41	16.9	37.7
MRM transfer	School 2 levels	-0.0365 -0.0503 - -0.0227 $p < .005$	41	32.9	56.3
MRM turn	School 2 levels	-0.0355 -0.552 - -0.0158 $p < .001$	41	23.4	33.0
Decision time	School 2 levels	-0.0279 -0.0462 - -0.0095 $p < .005$	41	14.1	43.4
Height		0.2625 -0.2852 - 0.8103 $p = .339$	42	2.3	2.3
Skinfold thickness	Sex	0.001 -0.0063 - 0.0082 $p = .785$	42	0.2	18.7

Table 3. Correlation of maternal TT_3 levels with children's scores on various motor and cognitive tasks, and with measures of physical status.

Dependent variable	Other effect included in the model	Regression coefficient (95% confidence interval)	N	R^2 due to TT_3	Full model
Bead threading	School 2 levels	0.0011 -0.0003 - 0.0024 p = .116	32	5.9	35.0
Pegboard	School 2 levels	-0.0024 -0.0046 - -0.0002 p = .036	32	9.9	40.6
MRM transfer	School 2 levels	-0.0015 -0.0040 - 0.0010 p = .219	32	3.4	37.2
MRM turn	School 2 levels	-0.0019 -0.0098 - 0.0009 p = .175	32	5.0	25.0
Decision time	School 2 levels	-0.0014 -0.0038 - 0.0010 p = .238	31	3.3	35.8
Height		0.0164 -0.0857 - 0.0522 p = .629	32	0.8	0.8
Skinfold thickness	Sex	0.0006 -0.0004 - 0.0076 p = 2.07	32	4.4	23.1

DISCUSSION

The relationship between the level of maternal thyroxine during pregnancy and the child's performance on a number of skilled motor tasks was first demonstrated on data collected when the children were aged between 7 and 9 [9]. The finding was replicated on 20 of the set when they were aged between 11 and 13 [10] and is still clearly evident in the results reported here when the children were aged between 13 and 15. Three replications within the same set of children over a period of 7 years suggest, to say the least, that the effect is robust. The failure to find any relationship between behavioural performance and height supports, though not conclusively, the view that the effect is not due to hypothyroidism in the children [28]. Hypothyroidism on the part of the children would be expected to show itself in stunted growth. Similarly the absence of any relationship with skinfold thickness suggests that the effect is not due to general malnutrition. Any evidence of gross malnutrition specifically in those children with relatively poor motor and cognitive performance would be expected to show itself in the measures of skinfold thickness.

The correlation between decision time in the card sorting task and maternal TT_4 but not TT_3 level is an important extension of our earlier findings. Decision time in the card sorting tasks which were used is an analogue of 2, 4, 6 and 8 choice reaction time. The greater the number of choices the more difficult is the decision which must be made [24] and this is reflected in the time taken to make it. This measure of mental efficiency provides firm evidence of a significant sub-clinical relationship between the child's cognitive functions and maternal thyroxine status in pregnancy of a kind parallel to the previously reported effects for motor performance.

The use of a sorting task of this kind as a measure of cognitive performance has a number of advantages. The requirements of the task are quite simple and easily understood. The task itself presupposes little specific knowledge beyond that required to match colours, comply with instructions and perform as quickly as possible. Relatively few cognitive processes seem to be involved; the procedure makes no demand on short-term or long-term memory and very little learning is entailed in the performance. The use of sorting time, movement time, and decision time in relation to complexity all give measures on an absolute scale. Finally, although the age spread in our sample was relatively small the procedure avoided having to vary the task content with age as most psychometric tests of intelligence for juveniles require. A further source of variability is thus eliminated. The correlations which have been reported between choice reaction times and IQ as measured by standard psychometric tests [29, 30, 31] imply that there are basic processes in common to both.

At the cellular level three principal stages in brain development, which follow each other more or less consecutively, can be distinguished. The first is concerned with the origin of the nerve cells; their proliferaiton, specification and migration. The second stage is characterised by the establishment of connections; by axonal and dendritic growth and synapse formation. In the third stage connections are modified, the initial inputs are reorganised and selective neuronal death takes place. As the functional and structural connectivity changes the properties of the neuron and its target are themselves modified [32]. The proliferation of glial cells and the process of myelination are also important. How many of these processes are regulated by thyroid hormones has been the subject of debate. Essentially two models have been proposed. One interpretation is that thyroid hormones regulate only one or two of the cellular processes occurring during brain maturation. For example, either an excess or a deficit of thyroid hormone may affect the rate of neurite outgrowth and so desynchronise the pattern of development. This has been called by Nunez [33] the catastrophe model. The alternative position, the pleiotypic model implies that thyroid hormones exert multiple and distinct effects on many of the parameters of brain maturation [33]. The fetal thyroid begins to secrete hormone around the beginning of the fourth month but it is not fully active until about 20 weeks when in mid-gestation there is a sharp rise in serum T_4 [34]. The most active period of neuroblast multiplication in the developing human brain is between 15-20 weeks [35, 36]. This implies that, at least in the early phase, fetal thyroid function does not exert any great effect on neuroblast multiplication. This being so it would seem that either thyroid hormones are not important for the initiation and control of this process or else they are provided by the maternal system and cross the placenta.

A good deal of attention has been devoted to the question of whether thyroid hormones are transported across the placenta. The prevailing consensus has been that the transfer of both T_3 and T_4 is negligible [37, 38]. However, much of the evidence bearing on the point relates to the period of fetal life following the functional development of the fetal thyroid [39]. The possibility remains that prior to this thyroid hormone may cross from mother to fetus. Any hormone deficiency early in pregnancy, that is before the fetal thyroid is fully active, may therefore lead to permanent effects on brain development which are not ameliorated or redressed when the fetal thyroid becomes fully functional. More recent investigations have indicated that in the rat there is substantial transport of maternal T_4 to the fetus early in pregnancy [40, 41]. Bernal and Pekonen[42] report finding T_3 receptors in the brain of the human fetus at mid-gestation, the concentration was very low at 10 weeks but had increased by an order of magnitude by the 16th week. The argument is also made that for the fetus deiodination of maternal T_4 may be a more important source of T_3 than direct uptake from the maternal circulation. Another related factor may be the circulating thyroid binding proteins. It has been suggested that these may play a specific role in the redistribution of hormone delivery through the body in pregnancy. Ekins [40, 43] has pointed out that the characteristics of thyroxine binding globulin (TBG) fit it very well for transporting T_4 rather than T_3 to particular target tissues during pregnancy. The maintenance of an adequate supply of T_4 to the placenta and hence to the fetus is a function for which TBG is ideally suited.

There is both direct experimental evidence from animal work and indirect circumstantial evidence from clinical investigations indicating that thyroid hormones probably play a vital part in several of the processes involved in the maturation of the brain at a number of stages [44, 45, 46]. Mental retardation which is one of the grosser consequences of neonatal hypothyroidism can be substantially ameliorated if not largely prevented by prompt action. Illig et al.[47] have investigated the consequences of neonatal screening programmes for congenital hypothyroidism and report that the earlier treatment which has been generally made possible by the screening has reduced the detrimental effects on psychological function as measured by intelligence tests. However, the effectiveness of replacement therapy in preventing certain psychological sequelae depends on the severity of thyroid deprivation and when it occurred during pregnancy, as well as on prompt postnatal diagnosis and therapy. Wolter et al. [48] report a retrospective investigation of congenitally severely hypothyroid infants in which treatment early in the postnatal period led to IQs in the normal range but where there were nevertheless indications of impairments in cognitive and motor function.

Although our knowledge of the part played by thyroid hormones in the development of the nervous system is limited there are nevertheless good grounds for believing that a pleiotypic model is the more appropriate. They do appear to exert effects at different times in development and therefore probably on different processes. We have previously suggested[10] that thyroxine of maternal origin is necessary for normal neurological maturation of the fetus before the fetal thyroid itself becomes fully functional. If this is correct then it seems plain that thyroxine is not merely a prohormone for triiodothyronine, it has in addition a distinct physiological role specifically concerned with normal fetal development. Although we have no biochemical measures of the children's thyroid status they were all clinically euthyroid. However, there are grounds for believing that in some cases, those where maternal T_4 levels were extremely low early in pregnancy, the maturation of the brain has taken place under sub-optimal conditions. That not all cases of very low maternal T_4 levels early in pregnancy result in a grossly damaged infant suggests that there are modulating factors. Before we can fully appreciate the nature and significance of these we need to better understand not only what action T_4 has at various stages in brain maturation, how the maternal and fetal systems synchronise with each other but also to have some better idea of the epigenetic patterns which are possible. The growth and development of the brain is to say the least a complex business and the effects of some early insults may be mitigated by other later occurrences or compounded by them. Alongside experimental work with animal preparations which seeks to link molecular events to the behavioural integrity of the animal we also have further need of careful psychological investigations of children with deficits in iodine nutrition. An experimental rather than psychometric approach to this would seem to be the most promising.

ACKNOWLEDGEMENTS

We should like to record our gratitude to the Director of Public Health of Papua New Guinea for his permission to carry out the study, to Dr Michael Alpers and his staff at the Papua New Guinea Institute of Medical Research, Goroka, and to Mrs Alison Heywood of the Madang base of the Institute for their invaluable help in various ways. We are also grateful to Lance Woodward and the late Mrs Margaret Woodward for their unfailing and generous help while we were in the Jimi valley. The hormone estimates were made by the Department of Nuclear Medicine, Middlesex Hospital Medical School, London. The research was made possible by a grant from the National Fund for Research into Crippling Diseases, U.K. which is gratefully acknowledged.

REFERENCES

1   J.B. Stanbury, B.S. Hetzel,"Endemic Goiter and Endemic Cretinism", Wiley, New York (1980).
2   P.O.D. Pharoah, I.H. Buttfield, B.S. Hetzel, Neurological damage to the fetus resulting from severe iodine deficiency during pregnancy, *Lancet*, i:308-310 (1971).
3   P.O.D. Pharoah, I.H. Buttfield, B.S. Hetzel, Effect of iodine prophylaxis on the incidence of endemic cretinism, *Adv Exp Med Biol*, 30:201-221 (1972).
4   Pan American Health Organisation,"Report of the Technical Group on Endemic Goiter", Pan American Health Organisation, Washington (1974).
5   K.J. Connolly, P.O.D. Pharoah, B.S. Hetzel, Fetal iodine deficiency and motor performance during childhood, *Lancet*, ii:1149-1151 (1979).
6   H. Bleichrodt, P.J.D. Drenth, A. Querido, Effects of iodine deficiency on mental and psychomotor abilities. *Am J Phys Anthropol*, 53:55-67 (1980).
7   N. Bleichrodt, I. Garcia, C. Rubio, G. Morreale de Escobar, F. Escobar del Rey, Developmental disorders associated with severe iodine deficiency, *in* "The prevention and control of iodine deficiency disorders", B.S. Hetzel, J.T. Dunn, J.B. Stanbury, eds., Elsevier, Amsterdam (1987).
8   R. Fierro-Benitez, I. Ramirez, E. Estrella, J.B. Stanbury, The role of iodine in intellectual development in an area of endemic goiter, *in*: "Endemic goiter and cretinism: Continuing threats to World Health", J.T. Dunn, G.A. Medeiros-Neto, eds., Pan American Health Organisation, Scientific Publication No 292: 135-142, Washington (1974).
9   P.O.D. Pharoah, K.J. Connolly, B.S. Hetzel, R.P. Ekins, Maternal thyroid function and motor competence in the child, *Dev Med Child Neurol*, 23:76-82 (1981).
10 P.O.D. Pharoah, K.J. Connolly, R.P. Ekins, A.G. Harding, Maternal thyroid hormone levels in pregnancy and the subsequent cognitive and motor performance of the children, *Clin Endocrinol*, 21:265-270 (1984).
11 P.O.D. Pharoah, S.M. Ellis, R.P. Ekins, E.S. Williams, Maternal thyroid function, iodine deficiency and fetal development, *Clin Endocrinol*, 5:159-166 (1976).
12 P.O.D. Pharoah, K.J. Connolly, A controlled trial of iodinated oil for the prevention of endemic cretinism : A long-term follow-up. *Int J Epidemiol*, 16:68-73 (1987).
13 M.B. Goslings, *in*: "Brain Development and Thyroid Deficiency", A. Querido and D.F. Swaab, eds., North-Holland Pub. Co., Amsterdam (General discussion p44) (1975).
14 I.G. Ord,"Manual for Pacific Design Construction Test", Australian Council for Educational Research, Hawthorn, Victoria, Australia (1968).
15 S.C. Kohs, "Intelligence Measurement: a Psychological and Statistical Study Based Upon the Block-Design Tests", Macmillan, New York (1923).
16 I.G. Ord, "Mental Tests for Preliterates", Ginn, London (1971).

17   R.J. Sternberg, "Handbook of Human Intelligence", Cambridge University Press, Cambridge (1982).

18   P.A. Vernon, Speed of information processing and general intelligence, *Intelligence*, 7:53-70 (1983).

19   S. Schwartz, T.M. Griffin, J. Brown, Power and speed components of individual differences in letter matching, *Intelligence*, 7:369-378 (1983).

20   R.J. Sternberg, Toward a triarchic theory of human intelligence, *Brain Behav Sci*, 7:269-315 (1984).

21   P.A. Vernon, S. Nador, L. Kantor, Reaction times and speed of processing : their relationship to timed and untimed measures of intelligence, *Intelligence*, 9:357-374 (1985).

22   P.A. Vernon, L. Kantor, Reaction time correlations with intelligence test scores obtained under timed or untimed conditions, *Intelligence*, 10:315-330 (1986).

23   T. Nettelbeck, N. Brewer, Studies of mild mental retardation and timed performance, *in:* "International Review of Research in Mental Retardation" (vol 10), N. R. Ellis, ed., Academic Press, New York (1981).

24   W.E. Hick, On the rate of gain of information, *Q J Exp Psychol*, 4:11-26 (1952).

25   F. Attneave,"Applications of Information Theory to Psychology", Holt, Rinehart & Winston, New York (1959).

26   Minnesota Rate of Manipulation Test, American Guidance Service, Circle Pines, Minnesota (1969).

27   K.J. Connolly, Response speed, temporal sequencing and information processing in children, *in:* "Mechanisms of Motor Skill Development", K.J. Connolly, ed., Academic Press, London (1970).

28   A. Querido, 1975, *in:* "Brain Development and Thyroid Deficiency", A. Querido and D.F. Swaab, eds., North Holland Pub. Co., Amsterdam (General discussion p46).

29   J.S. Carlson, C.M. Jensen, Reaction time, movement time, and intelligence : A replication and extension. *Intelligence*, 6:265-274 (1982).

30   J.S. Carlson, M.C. Jensen, K.F. Widaman, Reaction time, intelligence, and attention. *Intelligence*, 7:329-344 (1983).

31   G.A. Smith, G. Stanley, Clocking g : Relating intelligence and measures of timed performance, *Intelligence*, 7:353-368 (1983).

32   W.G. Hopkins and M.C. Brown,"Development of Nerve Cells and their Connections", Cambridge University Press, Cambridge (1984).

33   J. Nunez, Effects of thyroid hormones during brain differentiation, *Mol Cell Endocrinol*, 37:125-132 (1984).

34   D.A. Fisher, A.H. Klein, Thyroid development and disorders of thyroid function in the newborn. *New Eng J Med*, 304:702-712 (1981).

35   J. Dobbing, J. Sands, Timing of neuroblast multiplication in developing human brain, *Nature*, 226:639-640 (1970).

36   J. Dobbing, J. Sands, Quantitative growth and development of human brain, *Arch Dis Child*, 48:757-767 (1973).

37   J.H. Dussault, P. Coulombe, Minimal placental transfer of L-thyroxine (T_4) in the rat. *Pediatr Res*, 14:228-231 (1980).

38   D.A. Fisher, J.K. Dussault, J. Sack, I.J. Chopra, Ontogenesis of hypothalamic-pituitary-thyroid function in man, sheep and rat. *Recent Prog Horm Res*, 33:59-116 (1977).

39   E. Roti, A. Grudi, L.E. Braverman, The placental transport, synthesis and metabolism of hormones and drugs which affect fetal thyroid function, *Endocr Rev*, 4:131-149, Elsevier, Amsterdam (1983).

40   R.P. Ekins, A.K. Sinha, R.J. Woods, Maternal thyroid hormones and development of the foetal brain, *in:* "Iodine Nutrition Thyroxine and Brain Development", N. Kochupillai, M. Karmarker, V Ramalingaswami, eds., Tata McGraw-Hill, New Delhi (1986).

41   F. Escobar del Rey, R. Pastor, J. Mallol, G. Morrealle de Escobar, Effects of maternal iodine deficiency on the content of T_4 and T_3 in rat embryos before and after onset of fetal thyroid function, *Endocrinology*, 118:1259-1265 (1986).

42   J. Bernal, F. Pekonen, Ontogenesis of the nuclear 3,5,3'-triiodothyronine receptor in the human fetal brain. *Endocrinology*, 114:677-679 (1984).

43 R. Ekins, Role of serum thyroxine-binding proteins and maternal thyroid hormones in fetal development, *Lancet*, 1:1129-1132 (1985).

44 B.S. Hetzel, R.M. Smith,"Fetal Brain Disorders - Recent Approaches to the Problem of Mental Deficiency", Elsevier/North Holland, Amsterdam (1981).

45 J.H. Dussault, J. Ruel, Thyroid hormones and brain development, *Ann Rev Physiol*, 49:321-334 (1987).

46 G. Morreale de Escobar, M.J. Obregon, F. Escobar del Rey, Fetal and maternal thyroid hormones, *Horm Res*, 26:12-27 (1987).

47 R. Illig, R.H. Largo, Q. Qin, T. Torresani, P. Rochiccioli, A. Larsson, Mental development in congenital hypothyroidism after neonatal screening, *Arch Dis Child*, 62:1050-1055 (1987).

48 R. Wolter, P. Noel, M. Croen, Ch. Ernould, F. Verstraeten, J. Simons, S. Mertens, N. Vanbroeck, M. Vanderschveren-Lodewekz, Neuropsychological study in treated thyroid dysgenesis, *Acta Paediatr Scand Supp*, 277:41-46 (1979).

MATERNAL THYROID HORMONES AND FETAL BRAIN DEVELOPMENT

Peter O. D. Pharoah

Kevin J. Connolly

Department of Community Health
University of Liverpool
P O Box 147, Liverpool L69 3BX

Department of Psychology
University of Sheffield
Sheffield S10 2TN

Originally the classification of endemic cretinism into myxoedematous and nervous varieties was by McCarrison as a consequence of his work in the Chitral and Gilgit valleys in the Himalayas.[1] His description of the clinical features of the myxoedematous type was incomplete merely stating that, 'it corresponds to the form of the affliction met with in Europe and is described in any textbook of medicine'. His clinical description of nervous cretinism is more precise and is as accurate today as it was 80 years ago. The impairment is described as pertaining more especially to the central nervous system. He states that deaf-mutism is as a rule complete, mentality is much disordered and there is a congenital diplegia with increased knee-jerks and a spastic rigidity more severely affecting the lower limbs with a characteristic gait. A coarse nystagmus and internal strabismus were noted in some cases.

The relative proportions of the two syndromes varies in different countries, the hypothyroid or myxoedematous form predominates in many African countries while the nervous type predominates in Papua New Guinea and South America.[2,3] Controversy exists as to whether the two conditions are distinct syndromes or whether they represent polar extremes of a continuous spectrum of abnormality.[4] The current definition of endemic cretinism was formulated at a Pan American Health Organisation meeting in 1974 and encompasses the range of abnormalities that are found. The definition states that the clinical manifestations comprise mental deficiency together with either: (a) a predominant neurological syndrome consisting of defects of hearing and speech and with characteristic disorders of stance and gait of varying degrees, or (b) predominant hypothyroidism and stunted growth.[5] A general classification of cretinism is shown in Figure 1.

The syndrome of endemic cretinism as it is found in Papua New Guinea consists almost exclusively of the nervous type. The predominant clinical features include mental retardation, abnormalities of hearing and speech including severe deaf-mutism, a spastic diplegia which gives rise to a characteristic sitting posture, stance, and gait and strabismus. Clinical hypothyroidism is notable by its absence. This cluster of clinical abnormalities is explicit in numerous descriptive reports of endemic cretinism from both West Irian (formerly Dutch New Guinea) and Papua New Guinea.[6-10]

A little over two decades ago there was doubt concerning the role of iodine in the pathogenesis of the syndrome. Controversy arose from the fact that the prevalence of cretinism had declined prior to any measures of iodine prophylaxis. The decline had been noted in Europe eg. Italy,[11] Sardinia,[12] England,[13] in the United Provinces of India[14] and in Argentina.[15] However, Wespi had shown a correlation between the decline of deaf-mutism and the extent of salt iodination which was introduced independently in several Swiss cantons.[16] This correlation was only time related and other factors could equally have been responsible. Wespi's analysis was subsequently disputed on the ground that no cretins were born about the time the iodine prophylaxis was introduced, indeed, all the cretins were born at least 10 years previously.[17]

In order to resolve the controversy over the role of iodine in the pathogenesis of the nervous type of cretinism observed in Papua New Guinea, a double blind controlled trial was carried out using a single intramuscular injection of iodinated oil. This had previously been shown to be an effective prophylactic for endemic goitre.[18] Furthermore, it had a long duration of action, four years after a single 4.0ml dose, the treated group had significantly higher serum protein bound and urinary iodine excretion values and lower 24 hour radio-iodine uptake than an untreated control group.[19,20,21]

The controlled trial to assess the effectiveness of iodinated oil in preventing endemic cretinism was carried out in the Jimi valley in the

TABLE 1

BIRTH PREVALENCE OF ENDEMIC CRETINISM IN TEST AND CONTROL GROUPS

	Injected before conception		Injected after conception	
	Oil	Saline	Oil	Saline
Number of births	592	487	95	90
Number of cretins	1	26	5	5
Birth-prevalence rate of cretinism per 1000	1.7	53.4	52.6	55.6

highlands of Papua New Guinea, an area with a high prevalence of
cretinism (the 'nervous' variety) and goitre. On the occasion of a
census in 1966, 16,500 people were enrolled in the trial and alternate
families were injected with either iodinated oil or physiological
saline. Each member of the family received 4ml if aged 12 years or more
and 2ml if under 12 years of age.

Subsequent follow up of the test and control cohorts of infants born
into the trial and the criteria used for the diagnosis of cretinism have
been previously described.[22,23] A summary of the results of the trial
is shown in Table 1. Iodine supplementation appears to be clearly
ineffective if given after conception; the birth prevalence rates in
both the test and control groups injected post-conception are very
similar. On the other hand there is a highly significant difference in
test and control groups injected before conception. In addition to
those children in the trial, there were a further 62 children born to
mothers from the same villages who received neither iodinated oil nor
placebo saline because they were unavailable at the time of the
census. There were 3 cretins among the 62 children, a birth prevalence
rate of 48.4 per 1000 which is very similar to the rate in the control
group and in both test and control groups injected post-conception.

The trial was terminated in 1972 when intramuscular iodinated oil
was given to all women of child bearing age and adolescent girls.

Subsequently the cohort of children born within the trial were
followed up. Since 1972, for logistic reasons, the follow up has been
limited to children in 5 out of the original 13 villages in the trial.
The area covered is rugged mountainous country, there are no roads and
the only means of reaching the villages is on foot. In order to spend
more time in examining and assessing each child it was necessary to
reduce the area covered.

Visits were made to the 5 villages in 1974, 1976, 1978 and 1982,
particular record being made of abnormalities of speech and hearing and
signs of upper motor neurone lesions in those children who were
developmentally delayed. A definite cretin was so designated if there
were abnormalities of hearing and/or speech together with evidence of an
upper motor neurone lesion. Possible cretins included those who had
hearing and/or speech abnormalities but in whom there was no clear

clinical evidence of an upper motor neurone lesion, although in the
majority of cases motor milestones were also delayed. On each patrol
note was made of all the children who had died. In the majority of
cases it was not possible to obtain the precise date of death and, for
the purposes of constructing survival curves, the mid-point between the
date when the child was last seen alive and the date when death was
noted was taken to be the date of death. The people of the Jimi Valley
do not have a written language and the passage of time does not have the
importance attached as in our own culture. Measures of time are few and
rudimentary, consequently there is no accurate record of when a death
occurred.

In 1978 and 1982 measures of motor and cognitive function were
incorporated into the assessment schedule. These included bead-
threading and pegboard tests as measures of bimanual and unimanual
dexterity and the Pacific Design Construction Test (PDCT) developed by
Ord and used in Papua New Guinea to select army recruits and young
people for technical training.[24] The test procedures have been
previously described.[25, 26]

The results of the long term follow up of the two cohorts of
children born into the controlled trial have been reported
elsewhere.[27] Although on only a sub-set of the original trial group, it
confirmed the previous observations ie. that iodinated oil was
effective in preventing endemic cretinism provided it was given prior to
conception. Not only was there a highly significant excess of definite
and possible cretins in the control compared with the treated group, but
there was also an excess of children in the control group who performed
poorly on the cognitive and motor tasks. The designation as 'poor
performers' was made on an arbitary criterion, namely a score in the
lowest 10th centile. All the definite cretins and all but one of the
possible cretins performed poorly as judged by this criterion. The
results are summarised in Figure 2.

In addition to iodine supplementation effectively preventing endemic
cretinism, there was also a significant effect on mortality. The
cummulative survival curves of children born into the iodinated (test)
and saline (control) groups are shown in Figure 3. There is a highly
significant difference in 15 year cumulative survival rates in favour of
the test group (p = 0.002, Lee Desu statistic). Unfortunately,

TABLE 2
MATERNAL HORMONAL PROFILES

Serial Number	TT_4 Normal range 60–140 ng/ml	TT_3 Normal range 850–1750 pg/ml	FT_4 Normal range 20–40 pg/ml	FT_3 Normal range 4–8 pg/ml	TSH Normal ⩽5 µU/ml	Gestational age at sampling	Clinical findings
604	5	240	6.5	3.3	320	24 weeks	Stillbirth [infant born born after 1 year birth interval also a stillbirth]
638	8	362	3	2.5	500	36 weeks	Stillbirth/ early neonatal death
516	10	750	6.5	8.5	60	4 weeks	Early neonatal death
568 }	6 12	– 1500	– 3.5	– 8	– 61	4 weeks } 32 weeks }	Stillbirth/ early neo- natal death
245	16	1199	3	9.5	27	38 weeks	Jerky movement of hands, otherwise NAD
430	21	862	2	3.8	130	32 weeks	Partially deaf, slow speech, otherwise NAD
1058	24	1050	6.5	9.9	7	0 weeks	Normal
398 }	25 24	1250 1250	7.5 3	13.6 7.1	– 47	18 weeks } 38 weeks }	Mute, partially deaf Abnormal stance Plantars extensor. Cretin
489	26	–	–	–	–	38 weeks	Normal
054	26	1500	13	8.6	6	4 weeks	Slight delay in walking Otherwise NAD
262	31	1430	9	11.0	–	36 weeks	Normal
396 }	31 14	– 412	– –	– –	– 50	4 weeks } 28 weeks }	Normal
086	32	–	15.5	5.5	7	39 weeks	Normal
149	33	1550	8.5	9.1	7	16 weeks	Normal

Serial Number	TT$_4$ Normal range 60–140 ng/ml	TT$_3$ Normal range 850–1750 pg/ml	FT$_4$ Normal range 20–40 pg/ml	FT$_3$ Normal range 4–8 pg/ml	TSH Normal <5 μU/ml	Gestational age at sampling	Clinical findings
435	39	2579	6	9.5	10	32 weeks	Normal
089	43	1300	18	9.5	8	39 weeks	Normal
243	43	–	–	–	–	38 weeks	Normal
541	47	1575	15	10.3	9	4 weeks	Stillbirth (Iodine Supplement at 32 weeks)
745	49	–	–	–	–	34 weeks	Normal
503	51	–	–	–	–	36 weeks	Severe MD, bilateral congenital cataracts. Died age 8
1055	52	825	27	3.8	1	30+ weeks	No clinical details. Infant death
637	53	1644	–	–	9	30+ weeks	No clinical details. Infant death
362	59	1419	13	14	37	33 weeks	Mute but not deaf large perforations both drums No motor delay
	Repeat TSH = 41, Repeat FT$_4$ and FT$_3$ using new techniques: FT$_4$ 3.7 (range 7–17), FT$_3$ 5.2 (range 3.5–6.5)						
010	60	975	26	8.3	6	36 weeks	Normal
129	66	2519	14	14.4	5	28 weeks	Normal
001	68	2019	–	–	4	36 weeks	Normal
080	69	1600	26	8.8	5	28 weeks	Normal, Died aged 4
428	77	1450	24	9.4	7	18 weeks	Normal
576	77	–	–	–	–	30+ weeks	Infant death
519	78	875	41	7.1	4	12 weeks	Infant death
528	80	1100	36	6.9	5	36 weeks	Stillbirth/neonatal death
761	82	1635	28	10.9	9	34 weeks	Normal

(continued)

TABLE 2 cont.

Serial Number	TT$_4$ Normal range 60-140 ng/ml	TT$_3$ Normal range 850-1750 pg/ml	FT$_4$ Normal range 20-40 pg/ml	FT$_3$ Normal range 4-8 pg/ml	TSH Normal <5 μU/ml	Gestational age at sampling	Clinical findings
535	83	-	-	-	-	30+ weeks	Infant death
918	84	1029	41	7.4	2	39 weeks	Normal
1056	89	1275	23	10.1	1	30+ weeks	No clinical details, died in childhood
816	94	750	38	7.9	5		Normal
980	95	1743	24	9.3	1	28 weeks	Normal
026	96	1200	26	6.4	-	32 weeks	Normal
058	96	2225	76	15.3	3	28 weeks	No clinical details. Died in childhood.
901	99	-	-	-	-	32 weeks	Normal
170	101	-	-	-	-	38 weeks	Normal
997	101	-	-	-	-	32 weeks	Normal
140	104	-	-	-	-	36 weeks	Congenital cataract severe strabismus. Hearing and motor development normal
944	106	1350	40	10	1	39 weeks	Normal
948	106	-	-	-	-	34 weeks	Infant death
939	107	1453	40	9.4	2	12 weeks	Normal
113	108	-	-	--	-	36 weeks	Normal
745	108	1050	31	7.3	7	12 weeks	Normal
935	112	1075	40	8.5	3	24 weeks	No clinical data. Died in infancy/ childhood
975	112	1559	35	8.9	1	24 weeks	Normal
813	114	1575	34	8.9	2	39 weeks	Normal
810	119	1575	27	9.8	4	20 weeks	Normal

erial umber	TT$_4$ Normal range 60–140 ng/ml	TT$_3$ Normal range 850–1750 pg/ml	FT$_4$ Normal range 20–40 pg/ml	FT$_3$ Normal range 4–8 pg/nl	TSH Normal $\lessgtr 5$ μU/ml	Gestational age at sampling	Clinical findings
855	120	1725	38	12.3	3	32 weeks	No clinical data. Died in infancy/ childhood
237	124	–	–	–	–	30 weeks	Normal
890	128	1350	31	7.3	3	20 weeks	Normal
408	128	1485	27	10.4	5	38 weeks	Normal
034	137	–	–	–	–	34 weeks	Deaf mute 70 Db loss both ears Otherwise normal
368	137	1050	38	7.6	2	28 weeks	Normal
965	138	–	–	–	–	18 weeks	Normal clinically Died aged 15 years
214	155	–	–	–	–	39 weeks	Normal
949	155	2018	250	32.1	1	12 weeks	Normal
631	159 } 121 }	1500 **800**	54 51	14.2 6.9	3 –	18 weeks } 36 weeks }	Infant death
1046	163	1400	37	9.2	–	22 weeks	Normal
1009	165	1475	37	10.4	3	22 weeks	Normal
136	167	–	–	–	–	38 weeks	Normal
377	225	1059	–	–	3	38 weeks	Normal

* Abnormal values in heavy type.

information on the cause of death in individual cases could not be obtained.

While the controlled trial provided clear and significant evidence that supplementary iodine was effective in preventing endemic cretinism, other field studies in the Jimi valley provided additional information that dietary iodine deficiency was implicated in the pathogenesis of the syndrome. It appeared that endemic cretinism was of recent onset in the valley. Following first contact with Europeans in 1953 there was a sharp change in the prevalence of cretinism from approximately 0.1% pre-1953 rising through the late 1950's and early 1960's and reaching a peak of 15% of the children born in 1965 - Figure 4. This increase in prevalence was found to represent a true increase in incidence which could be attributed to the substitution of a locally produced salt with a salt introduced by the Europeans. The salt produced by traditional means prior to European contact was obtained by evaporation and crystalisation from saline pools in a volcanic area. Analysis of a sample of the salt revealed a high iodine content of about 1 part iodine to 5000 parts of salt. This is at least twice the concentration of iodine found in commercially iodinated salt.

The controlled trial indicated that iodine supplementation given after conception was not effective in preventing the endemic cretinism. The birth prevalence rates of cretinism among women who were already pregnant when entered into the trial were very similar. Previous reports had also indicated that the impairment arose during intra-uterine development. McCullagh termed the syndrome 'goitre associated congenital defect'.[8] Eggenberger and Messerli state that the deaf-mutism has its origin in the 4th month of fetal life, although evidence supporting this statement is not offered.[29] Costa et al also comment that endemic cretinism seems to develop during intra-uterine life and to be related to the parental environment.[11]

Functionally iodine is important because of its role in the synthesis of the thyroid hormones and it has been postulated that hypothyroidism, either maternal or fetal,[30,31] or even elemental iodine per se, prior to the development of functional competence of the fetal thyroid,[23] is important in the pathogenesis of endemic cretinism. In order to examine the role of maternal thyroid hormones in fetal development, blood samples were taken, in 1970 and 1971, from 106 women

342

of whom 66 were pregnant. Subsequently, on a series of visits made in conjunction with the follow up of the controlled trial, record was made of the outcome of the pregnancies. Stillbirths were included with infant deaths as it was not always possible to distinguish between them from the available history. The surviving children were a subset of those in the controlled trial and they were subject to the same clinical examination and series of test procedures as the remainder of children in the controlled trial.

The hormonal profiles of the blood samples were determined as follows:

Serum thyroid stimulating hormone (TSH) was measured by a modification of the double antibody technique.[32,33] The normal range is < 5 $\mu U/ml$.

Serum total thyroxine (TT_4) was measured by a saturation analysis technique[34] with a normal range during pregnancy of 60-140 ng/ml.

Serum total triiodothyronine (TT_3) was measured by radioimmunoassay[35] with a normal range during pregnancy of 850-1750 pg/ml.

Serum free thyroxine (FT_4) and free triiodothyronine (FT_3) were directly measured by radiommunoassay in serum dialysates[36] with normal ranges of 20-40 pg/ml and 4-8 pg/ml respectively.

The serum TT_4 levels of those women who had received the supplementary intramuscular injection of iodinated oil in 1966 ranged from 94 to 225 ng/ml, ie. all except 4 were within the normal range. Of these 3 were mildly elevated with values of 163, 165 and 169 ng/ml. The fourth was more grossly elevated at 225 ng/ml which was confirmed on repeat assay but the TT_3 was within the normal range. Among those women who had not received supplementary iodine the range of TT_4 was 5-124ng/ml; in 50% of these the serum TT_4 was below the lower limit of normal ie < 60ng/ml and in 9 instances it was 25ng/ml or less.

The serum TT_3 value was determined for 49 women, in 6 of them it was below the lower limit of normal (< 850ng/ml). None of the 6 had received supplementary iodine.

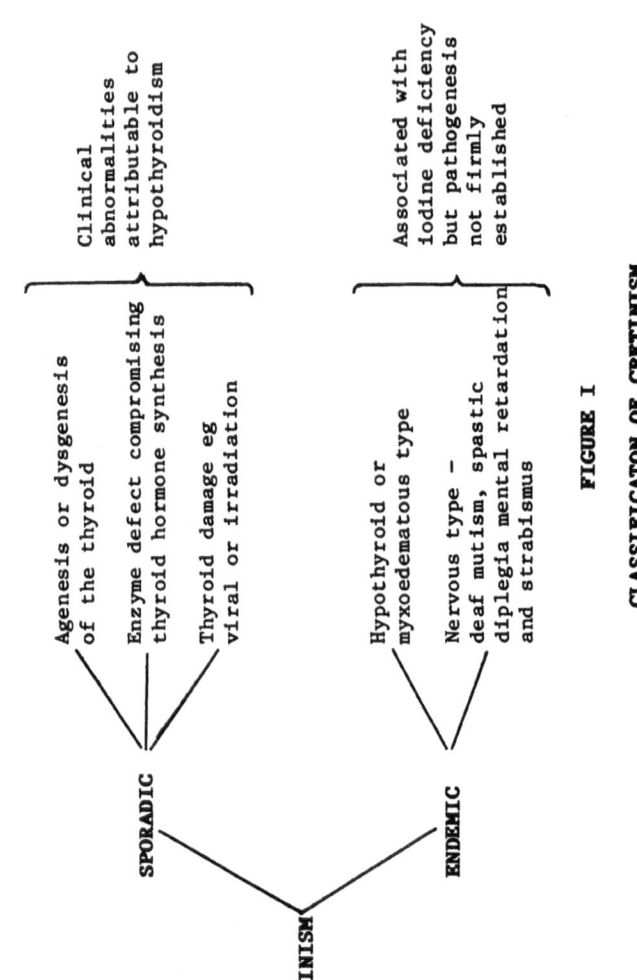

FIGURE I

CLASSIFICATON OF CRETINISM

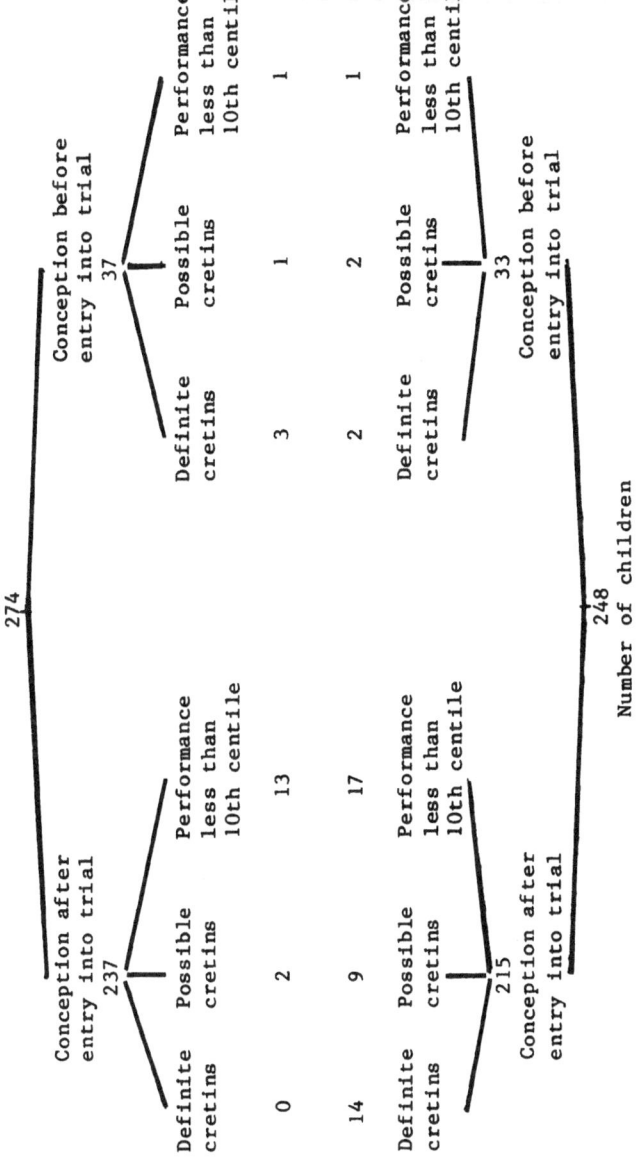

FIGURE 2

MEASURES OF OUTCOME IN TEST AND CONTROL GROUPS

345

The hormonal profiles of the 66 women during pregnancy are shown in
Table 2. It is apparent that the main effect of iodine deficiency on
hormonal levels is a reduction in the serum TT_4 and a concommitant
increase in TSH. The TT_3 levels are maintained in the face of a
considerably lowered TT_4 and it is only when the latter is grossly
reduced that the TT_3 also falls below the normal range.

Table 2 also shows the clinical outcome of each pregnancy. The data
have been arranged in ascending order of maternal TT_4. The gestational
age at which the blood sample was taken was estimated by subtracting
from 40 the number of weeks that elapsed from the sample being taken and
birth of the child. This estimation is based on the assumption that
every pregnancy proceeded to term at 40 weeks. A further source of
imprecision lay in the recording of the date of child's date of birth,
the error of which could be up to 4 weeks. The majority of blood
samples were withdrawn from women in whom the pregnancy was clinically
obvious on inspection ie at a gestational age of about 32 weeks or
more. In an endeavour to obtain samples at an earlier gestational age,
advantage was taken of the observation that the local tendency was to
have a birth interval of 3-4 years. Women in the child bearing age
range who had not delivered a child in the previous 3-4 years were
sampled in the expectation that some would be at an early stage of
pregnancy.

The first 3 cases (serial nos 604, 638, 515) in the table had
hormonal profiles entirely consistent with severe hypothyroidism
although these women did not exhibit the clinical features of
hypothyroidism. In all three cases the pregnancy outcome was a
stillbirth/early neonatal death. In the succeeding 3 cases (serial nos
568, 245, 430) the maternal total and free thyroxine values were low but
triiodothyronine values were within the normal range. In one of these
the outcome was a stillbirth/early neonatal death and one child has a
hearing/speech defect but no other abnormality and, by our definition,
cannot be labelled as an endemic cretin.

Another case (serial no 398) is of particular interest in that the
child exhibits the clinical features of cretinism. The maternal
thyroxine and TSH values were abnormal and consistent with
hypothyroidism but the triiodothyronine values were within the normal

range. The maternal hormonal profile is of further interest because measurements were made at two gestational ages, namely 18 and 38 weeks. On both occasions both TT_4 and FT_4 were reduced but TT_3 and FT_3 were within the normal range.

One child (serial no 362) who was mute but not deaf had large perforations of both eardrums and on one occasion pus was observed in the external auditory meatus. Initial analysis of the maternal serum relating to this child revealed a reduced FT_4 and considerably elevated TSH although the TT_4 was almost within the normal range. Repeat analysis of the serum confirmed a reduced FT_4, elevated TSH and normal FT_3 and TT_3. Unfortunately there was insufficient serum remaining to validate the TT_4 value.

The data presented in Table 2 suggest that biochemical hypothyroidism as evidenced by a reduction in both T_4 and T_3 and an elevation in TSH is relatively uncommon but may be manifest with very severe iodine deficiency. Under these circumstances, survival of the infant is severely prejudiced. In lesser degrees of iodine deficiency TT_4 and FT_4 are reduced, TSH elevated but TT_3 and FT_3 are maintained within the normal range. In such circumstances the development of the fetal nervous system is threatened and the infant is predisposed to present with the neurological features of cretinism.

Drawing conclusions from this data set is problematical primarily because the assessment of the maternal hormonal profile was made at a single point (or in a minority of cases at two points) in pregnancy. Usually this was rather late in gestation and therefore may not be representative of hormonal function at an earlier time when the fetus appears to be particularly at risk of neurological damage, perhaps during the phase of neuroblast proliferation. However, it could be argued that if the serum thyroid hormone levels were low late in pregnancy, they must also have been low at an earlier stage, but the converse is not necessarily so. The data presented here support the thesis that maternal thyroid hormones have an essential role in the maturation of the fetal nervous system, moreover, the view that hormones are not transported across the placenta from mother to fetus needs to be reexamined.

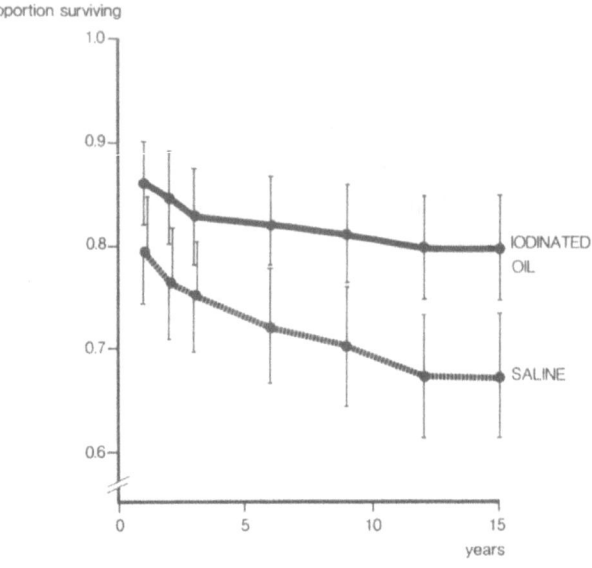

FIGURE 3

Reproduced by kind permission of the Editor,
International Journal of Epidemiology.

TRENDS IN THE PREVALENCE OF ENDEMIC CRETINISM

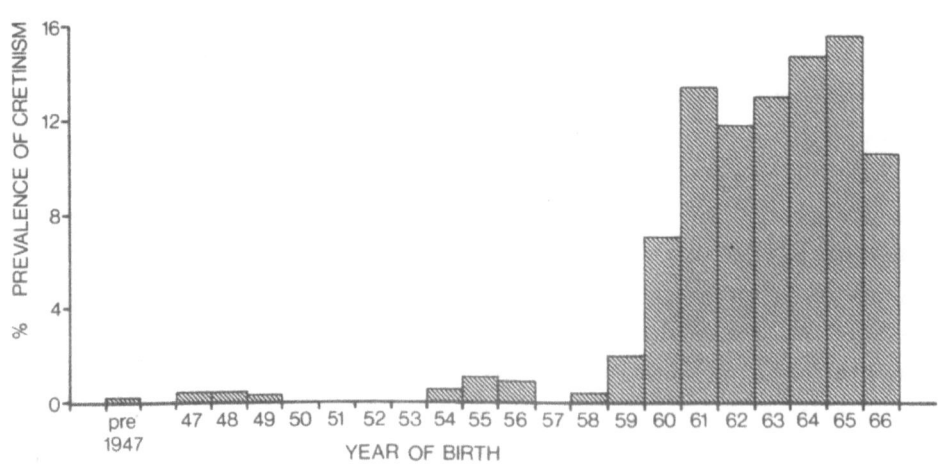

FIGURE 4

Reproduced by kind permission of the Editor, The Lancet.

One further facet that is of interest is whether the neurological endemic cretin is hypothyroid and whether its thyroid is capable of normal functioning. To this end blood samples were taken from cretins and clinically normal female adults. Serum TT_4 was measured by competitive protein binding analysis[37] giving a normal range of 2.5-7.0 μg/100 ml. (These analyses were carried out in a different laboratory using a different technique than reported earlier in this section hence the difference in the normal range and the units given). The results are illustrated in figure 5. In those cretins not given supplementary iodine there is evidence of a reduction in TT_4 with 50% below the lower limit of normal. However, clinically normal adults from the same region show a similar proportion with reduced TT_4. Moreover, in both clinically normal adults and cretins, the administration of supplementary iodine leads to correction of the TT_4 so that it comes to lie within the normal range. It is clear that any reduction of TT_4 is a feature of iodine deficiency whether it is in neurologically impaired cretins or the normal population and that the thyroid in the cretin can respond to iodine repletion by a normal synthesis of thyroxine.

SUMMARY

1. In a double blind controlled trial using intramuscular iodinated oil in an iodine deficient population, the infant and childhood cummulative mortality over the first 15 years of life was significantly greater in the control than in the test group.

2. The neurological variety of cretinism can be prevented if the iodine deficiency is alleviated prior to conception. The impairment of fetal neurogenesis associated with iodine deficiency appears to take place in the first half of pregnancy.

3. Using quantitative measures of motor and cognitive function, subclinical differences in performance are demonstrable between children born into the test and control groups and these show the test group to be significantly superior.

4. Iodine deficiency results predominantly in a reduction of serum thyroxine but serum triiodothyronine remains within the normal range. Only when iodine deficiency is extremely severe are the serum triiodothyronine levels compromised.

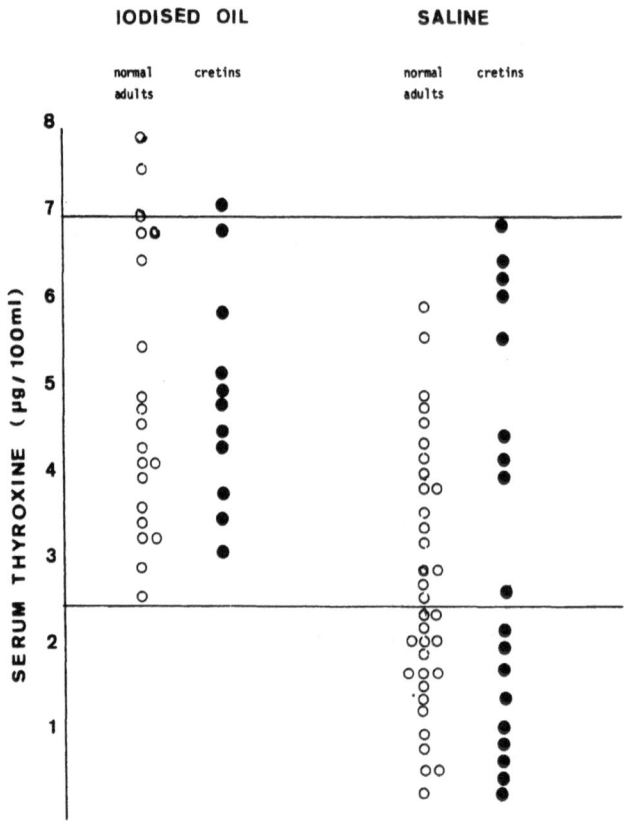

SERUM THYROXINE LEVELS IN CRETINS & NORMAL ADULTS

FIGURE 5

5. There is a spectrum of impairment ranging from death through
    clinical endemic cretinism and subclinical deficits of motor and
    cognitive performance, to normality. This spectrum of effects
    appears to be associated with maternal thyroxine during pregnancy.
    No association with maternal triiodothyronine was found.

6. In all endemic cretins correction of the hormonal deficiency
    requires only the administration of iodine. This indicates that
    the cretin's thyroid is functionally competent.

Acknowledgements

    We should like to record our gratitude to the Director of Public
Health of Papua New Guinea for his permission to carry out the study, to
Dr Michael Alpers and his staff at the Papua New Guinea Institute of
Medical Research, Goroka, and to Mrs Alison Heywood of the Madang base
of the Institute for their invaluable help in various ways. We are also
grateful to Lance Woodward, the late Mrs Margaret Woodward and the staff
at the Anglican Mission, Koinambe for their unfailing and generous help
while we were in the Jimi valley. The hormone estimates were by
courtesy of Professor Roger Ekins, Department of Molecular,
Endocrinology, Middlesex Hospital Medical School, London. The research
was made possible by grants from the Commonwealth Scientific and
Industrial Research Organisation, Adelaide, the Medical Research
Council, London, the Wellcome Trust and the National Fund for Research
into Crippling Diseases, UK which are gratefully acknowledged.

REFERENCES

1. R. McCarrison, Observations on endemic cretinism in the Chitral
        and Gilgit valleys. Lancet 2: 1275 (1908).

2. P.O.D. Pharoah, Geographical variation in the clinical
        manifestations of endemic cretinism. Trop Geog Med 28: 259
        (1976).

3. P.O.D. Pharoah, F. Delange, R. Fierro-Benitez and J.B. Stanbury,
        Endemic cretinism, in: J.B. Stanbury, B.S. Hetzel eds Endemic
        Goiter and Endemic Cretinism. Wiley, New York (1980).

4.   A. Querido, I.H. Buttfield, J.B. Stanbury, et al., Hypothyroidism
      in endemic cretinism, in:, B.S. Hetzel, P.O.D. Pharoah, eds
      Endemic Cretinism. Institute of Human Biology Papua New
      Guinea Monograph Series, 2: 127 (1971).

5.   Pan American Health Organisation. Report of the technical group
      on endemic goitre. Pan American Health Organisation
      Washington: (1974).

6.   J. C. Choufoer, M. Van Rhijn and A. Querido, Endemic goiter in
      Western New Guinea. II Incidence and Pathogenesis of Endemic
      Cretinism. J Clin Endocr, 25: 285 (1965).

7.   M. Van Rhijn, Een endemie van struma en cretinisme in het
      centrale bergland van West Nieuw-Guinea. Avanti-Zaltbommel.
      Utrecht (1969).

8.   S. F. McCullagh, The Huon Peninsula endemic IV. Endemic goitre
      and congenital defect. Med J Aust 1: 884 (1963).

9.   P.O.D. Pharoah, Endemic cretinism in New Guinea. Papua New
      Guinea Med J 14: 115 (1971).

10.   R. Hornabrook, Endemic cretinism in: R.W. Hornabrook, ed, Topics
      in Tropical Neurology. Davis, Philadephia (1975).

11.   A. Costa, F. Cottino, M. Mortara and U. Vogliazzo, Endemic
      cretinism in Piedmont. Panminerva Med 6: 250 (1964).

12.   Sardinian Royal Commission: Rapport de la commission creee par s
      m le roi de Sardaigne pour etudier le cretinisme. Imprimerie
      Royale Turin (1848).

13.   C.H. Fagge, Sporadic cretinism occurring in England. Med - Chir
      Trans 55: 155 (1871).

14.   H. Stott, B.B. Bhatia, R.S. Lal and K.C. Rai, The distribution and
      cause of endemic goitre in the United Provinces. Indian J Med
      Res 18: 1059 (1930).

15.   I. Greenwald, The history of goitre in the Inca Empire: Peru,
      Chile and the Argentine Republic. Tex Rep Biol Med 15: 874
      (1957).

16.   H.J. Wespi, Abnahme der taubstummheit in der Schweiz als volge
      der kropfprophylaxe mit jodiertem kochsalz. Schweiz med Wschr
      28: 625 (1945).

17.   P. Koenig and P. Veraguth, Studies of thyroid function in endemic
      cretins, Pitt-Rivers, ed., Advances in Thyroid Research.
      Pergammon Press, London (1961).

18.   S.F. McCullagh, Goitre control project. Papua New Guinea Med J
      3: 43 (1959).

19.   K.H. Clarke, S.F. McCullagh and D. Winikoff, The use of an
      intramuscular depot of iodised oil as a long-lasting source of
      iodine. Med J Aust 1: 89 (1960).

20.   I.H. Buttfield and B.S. Hetzel, Endemic goiter in Eastern New
      Guinea. Bull Wld Hlth Org 36: 243 (1967).

21. I.H. Buttfield and B.S. Hetzel, Endemic cretinism in Eastern New Guinea its relation to goitre and iodine deficiency. In Endemic Cretinism. B.S. Hetzel, P.O.D. Pharoah, eds., Institute of Human Biology, Papua New Guinea Monograph Series No, 2: 55 (1971).

22. P.O.D. Pharoah, I.H. Buttfield and B.S. Hetzel, Neurological Damage to the fetus resulting from severe iodine deficiency during pregnancy. Lancet 1: 308 (1971).

23. P.O.D. Pharoah, I.H. Buttfield and B.S. Hetzel, The effect of iodine prophylaxis on the incidence of endemic cretinsm. Adv. Exp Med Biol 30: 201 (1972).

24. I.G. Ord, Mental tests for pre-literates. Ginn, London (1971).

25. K.J. Connolly, P.O.D. Pharoah and B.S. Hetzel, Fetal iodine deficiency and motor performance during childhood. Lancet 2: 1149 (1979).

26. P.O.D.Pharoah, K.J. Connolly, R.P. Ekins and A.G. Harding, Maternal thyroid levels in pregnancy and the subsequent cognitive and motor performance of the children. Clin Endocrinol 21: 265 (1984).

27. P.O.D. Pharoah and K.J. Connolly, A controlled trial of iodinated oil for the prevention of endemic cretinism: a long-term follow-up. Int J Epid 16: 68 (1987).

28. P.O.D. Pharoah and R.W. Hornabrook, Endemic cretinism of recent onset in New Guinea. Lancet 2: 1038 (1974).

29. H. Eggenberger and F.M. Messerli, Theory and results of prophylaxis of endemic goiter in Switzerland. Trans 3rd Int Goiter Conference 64 (1938).

30. M.P. Koenig, M.P., Die konigenitale hypothyreose und der endemische kretinismus. Springer-Verlag. Berlin-Heidelberg; New York (1968).

31. J.B. Stanbury, Research on endemic goiter in Latin America. WHO chronicle 24: 537 (1970).

32. W.D. Odell, J.F. Wilber, and W.E. Paul, Radioimmunoassay of human thyrotropin in serum. Metabolism 14: 465 (1965).

33. R.D. Utiger, Radioimmunoassay of human plasma thyrotropin. J Clin Invest 44: 1277 (1965).

34. R.P. Ekins, E.S. Williams, and S Ellis, The sensitive and precise measurement of serum thyroxine by saturation analysis. Clin Biochem 2: 253 (1969).

35. R.P. Brown, S.M. Ekins, W.S. Ellis, and W.S. Reith, The radioimmunoassay of triiodothyronine, in: K. Fellinger and R. Hofer, eds., Further Advances in Thyroid Research. Springer-Verlag, Vienna (1971).

36. S.M. Ellis, and R.P. Ekins, The direct measurement by
     radioimmunoassay of the free thyroid hormone concentration in
     serum. 9th Acta Endocrinologica Congress, (1974).

37. B.P. Murhpy, and C. Jachan, The determination of thyroxine by
     competitive protein binding analysis employing an ion-exchange
     resin and radiothyroxine. J Lab Clin Med 66: 161 (1965).

IMMUNOCYTOCHEMICAL MAPPING OF NUCLEAR T3 RECEPTORS USING A MONOCLONAL

ANTIBODY IN THE DEVELOPING AND ADULT RAT BRAIN

Min Luo and Jean H. Dussault

Unite d'Ontogenese et Genetique Moleculaires
CHUL, Ste-Foy, P.Q.
Canada

By using a monoclonal antibody raised against rat nuclear T3 receptors
(NTR), the immunocytochemical localization of NTR has been achieved in the
developing and adult rat brain. In 16-day-old embryo, only weakly NTR-
immunostaining neurons were detected in the globus pallidus (GP), amygdala
(AA) and hypothalamus areas. In the 18-day-old embryo, the NTR-
immunostaining neurons were increased in both density and intensity. The
NTR-staining neurons were mainly distributed in the striatum, GP, AA,
thalamus (TH) and hypothalamus areas. Thereafter, the number of NTR-
staining neurons was progressively increased in cerebral cortex and in the
regions previously described. After birth, the highest density of NTR-
staining cells was found in the 6-day-old rat. The high density of the
medium to strongly NTR-staining neurons was seen in cerebral cortex (lamina
II and III, particularly in cingulate and interhemispheris cortex), nucleus
habenula medialis (Hbm), amygdal area and some nuclei of the hypothalamus
(nucleus (N) paraventricularis, N. dorsomedialis, N. ventromedialis, N.
supraopticus and N. preopticus). Moderate density was found in the
hippocampus dentate gyrus, most parts of the thalamus, some parts of
cerebral cortex (piriform and suprarhinal cortex) while only a few of the
weakly to median staining cells were observed in the cerebellum (only in
the Purkinje's cells) and the brain stem. In 14-day-olds, the density and
intensity of NTR-immunoreactivity were slightly decreased except in the
hippocampus and cerebellar cortex. In the adult brain, the distribution
was similar to that in the 14-day-old, but the number of NTR-staining cells
in the cerebellar cortex (particularly in granular layer) was evidently
increased. These studies provide a map of the NTR in the developing, adult
brain and indicate that the NTR is selectively concentrated in certain
regions in both developing and adult brains.

INFLUENCE OF TRIIODOTHYRONINE (L-T3) ON THE DEVELOPMENT OF FETAL BRAIN

CHOLINERGIC NEURONS CULTURED IN A CHEMICALLY DEFINED MEDIUM.

Puyrimat, J., Garza, R. and Dussault, J.H.

Ontogenese et Genetique Moleculaires
Centre Hospitalier de l'Universite Laval
Ste-Foy, Quebec, Canada

In cerebral hemisphere cultures initiated from 15 day old rat embryos, the number of acetylcholinesterase (AChE) positive cells increased from 6.8 +/- 1.6 on day 3 to 112 +/- 16 on day 15. With time in culture, AChE-cells increased in size and in neurite length. The addition of L-T3 (3.10^{-8} M) had no effect on the number of AChE-neurons. Morphometric analysis were performed in 8 day old cultures. L-T3 significantly increased the size of AChE-neurons since the number of AChE-neurons with a surface less than 150 microns^2 decreased from 46% to 20% whereas the number of large AChE-neurons increased from 8% to 31% in T3 treated cultures. The number of AChE-neurons which have at least one neurite with a length above 150 microns increased from 17% in control cultures to 79% in T3-treated cultures (Chi-2 test $p > 0.0001$). These morphological effects of L-T3 are associated with several biochemical effects. T3 increased ChAT (choline acetyltransferase) and AChE activities in both a dose and time dependent manner. Treatment of cultures with L-T3 at different times in culture demonstrated the presence of a critical period which occurs between day 15 and day 20 since the stimulatory effect of L-T3 on ChAT activity is lost after 15 days in vitro. Studies of the time necessary for L-T3 to increase both ChAT and AChE activities demonstrated that at least 2 days and 15 days were required for L-T3 to significantly stimulate both enzyme activities respectively.

To rule out the possibility that some neuronal effects of T3 are indirectly mediated through glial cells, we studied the effect of conditioned medium prepared from glial cell cultures of different ages initiated from newborn rats and grown in the presence or in the absence of L-T3 on ChAT and AChE activities. No significant effect was found with either CM on ChAT or AChE activities despite an increase of AChE-neurite lengths which were observed with both conditioned mediums. Thus L-T3 appears to have a direct effect on cholinergic neurons first by affecting the morphology of the cells and subsequently by an action on neurotransmitter metabolizing enzymes.

THYROID HORMONE CONTENTS AND 5 & 5' MONODEIODINASE ACTIVITIES IN DEVELOPING HUMAN CEREBRAL CORTEX

M. G. Karmarkar, D. Prabakaran, M. M. Godbole, and
M. M. S. Ahuja

Thyroid hormones play an important role during brain development and their deficiency leads to derangement in the brain. We have analysed twenty eight human cerebral cortices obtained during medico legal abortions between the 11th and 25th week of gestation from iodine sufficient areas. The cortices were analysed for T4, T3, and rT3 contents and 5 and 5' monodeiodinase activities. The thyroid hormones as well as enzyme activities were detectable at the 11th week of gestation. T4 and T3 concurrently increased till the 18th week while rT3 increased till the 2nd week. After this period the thyroid hormone contents decreased till the 25th week. 5 and 5' monodeiodinase activities increased up to the 22nd week and then significantly decreased till the 25th week.

Twenty two cerebral cortices from areas known to be iodine deficient were subjected to similar analysis. The results indicated that 5 and 5' monodeiodinase activities and T3 content show different pattern as compared to iodine sufficient area. It appears that in iodine deficient cortices T3 levels are maintained at adequate levels for neural development during the 14th to the 20th week of gestation.

ONTOGENESIS OF NUCLEAR T3 RECEPTOR IN HUMAN FETAL BRAIN

H. L. Su, P. Ling, R. K. Yang and H. C. Chao

Institute of Endocrinology
Tianjin Medical College
Tianjin, PRC

T3 receptor, endogenous T3 and related parameters in 23 human fetal cerebra of 12-36th pregnant weeks and 12 cerebella of 18-36th pregnant weeks were analyzed after the methodological investigation has been performed successfully and the following results were obtained.

(1) Average affinity constants of nuclear T3 receptor to T3 in cerebra as well as cerebella all ranged $(2.54 +/- 0.63) \times 10^{10}M^{-1}$ and didn't vary with different brain regions and developmental stages.

(2) Maximum binding capacity (MBC) of nuclear T3 receptors in both cerebra and cerebella increased with the increase of pregnant age and a linear correlation was shown between them.

      for cerebra, MBC=177.065+22.930xPA, r=0.953, p<0.01
      for cerebella, MBC=15.9934+5.6355xPA, r=0.987, p<0.05
      (PA: pregnant age in weeks)

Cerebral MBC increased from 104.64 of 12th week to 648.87 fmol/mg DNA of 36th week, or by 6.2 times, whereas in cerebella, it increased only by 2.4 times from 18th to 36th weeks of pregnancy.

(3) Endogenous T3 contents in fetal cerebra and cerebella all increased with development within 20th-36th weeks and showed a significant correlation with MBC of nuclear T3 receptor. This, therefore, corresponded closely to the dependence of fetal brain development on thyroid hormones in pregnancy.

      for cerebra, MBC=7.2611+1.6791xEnT3, r=0.905, p<0.01

(4) Based on that any two parameters among T3 receptor MBC, endogenous T3 and pregnant age correlated significantly, the following multivirate regression equation was developed,

      MBC=179.728+15.577xPA+0.708xEnT3, r=0.976, p<0.01

(5) The occupancy of T3 receptor in fetal brain was found to be 27.51 +/- 2.4% (m +/- SE) and unchanged with development.

In conclusion, we have obtained a whole set of data about the relation between the fetal brain development and T3 and its nuclear receptor. It has also been shown clearly that the dependence of tissue development on thyroid hormone was mediated chiefly through the regulation of nuclear T3 and its receptor levels rather than altering its receptor occupancy.

QUANTITATIVE HISTOLOGY STUDY ON BRAIN NERVOUS CELLS OF NEUROLOGICAL ENDEMIC

CRETINS

Yan Yuqin, Guan Chunxiang, Leng Lili and Li Jianqun

Chinese Research Center for Endemic Disease Control
Harbin Medical College
Harbin, China

The quantitative histology study was made on pyramidal and stellate cells in cortical motor area and auditory area of 6 neurological endemic cretins (20--49 years old) and 4 normal adults using Rapid Golgi Silver-impregnated technique. The results showed: 1) the body size of pyramidal and stellate cells in these two cerebral cortex areas of cretins were smaller than that of the normal; 2) the number in branch of apical dendrite and primary dendrite of pyramidal cells, especially in peripheral dendrite of stellate cells was decreased markedly in comparison with the normal; 3) the result above was relative to the severity of the cretinism, i.e. the more severe cretinism is, the more obvious the changes above are, particularly in number of peripheral dendrite of stellate cells. We also found the branch of apical dendrite of purkinje cells in cerebellum of cretins was short and sparse.

These findings indicated that there was severe retardation in growth and differentiation of brain nervous cells of cretins, therefore this paper will give the morphological evidence for explanation of clinical manifestations of neurological endemic cretinism.

ENDEMIC CRETINISM IN QINGHAI PROVINCE

G.F. Maberly[1], S.C. Boyages[1], J-P. Halpern[2], J.G. Morris[2], J.K. Collins[3], C.J. Eastman[1], D. Yu[4], C. You[4], C. Jin[5] and Z-L. Wang[5]

Endocrine[1] & Neurology[2] Units, Department of Medicine, Westmead Hospital, Australia. Department of Behavioural Sciences [3], Macquarie University, Australia. Qinghai Provincial People's Hospital [4] and Institute of Endemic Diseases[5].

Endemic cretinism has traditionally been divided into neurological and myxoedematous types. Myxoedematous cretinism occurs less commonly and is best described from Zaire. The pathophysiological mechanisms that results in severe hypothyroidism are unresolved. We report here the findings from a predominately myxoedematous endemia (Qinghai Province, PRC), but where examples of neurological and "mixed" cretinism are evident. The aim was to define the salient characteristics of this endemia and to correlate these findings with current thyroid function. Clinical, psychometric, radiological and electrophysiological assessments were undertaken. 69 endemic cretins were selected by local health care workers and classified according to PAHO criteria (15 neurological, 29 "mixed", & 25 myxoedematous). Major results are tabulated below.

There was considerable overlap between groups. Deafness and intellectual impairment (mean IQ 28.9 [16.7]) was evident in all clinical group. All

Table 1

Variable	Neurological (n=15)	Mixed (n=29)	Myxoedematous (n=25)	F	p<
Age (yrs)	22.3[9.9]	24.4[11.1]	31.6[12.0]	4.2	.02
TSH (mIU/1)	3.9[2.3]	28.1[76.5]	123.8[112.2]	12.9	.0001
TT4 (nmol/1)	122.5[26.3]	105.8[32.7]	53.9[34.7]	26.8	.0001
FT4 (pmol/1)	20.4[4.2]	19.5[6.1]	8.4[6.4]	30.3	.0001
Height (cm)	145.9[8.5]	131.0[16.0]	131.4[18.7]	4.9	.01
Sitting Height (cm)	79.5[5.7]	73.9[8.7]	74.5[10.6]	2.0	NS
Bone Delay (yrs)	2.2[2.0]	4.0[6.3]	5.3[7.6]	1.2	NS
Pubic Tanner	3.1[1.3]	2.3[1.4]	2.1[10.9]	3.5	.05

Results are shown as mean + [SD]. *F-value derived from one way analysis of variance (variable versus traditional label).

GRADATION OF NEUROLOGIC DEFICIT AND DEVELOPMENTAL DELAY IN AN IODINE

DEFICIENT POPULATION

E.M. Dulbert*, K. Widjaja**, R. Djokomoeljanto[+], L. Belmont*,
and B.S. Hetzel[++]

*Gertrude H. Sergievsky Center, Columbia University,
 New York City, U.S.A.
**Department of Pediatric Neurology, Dr. Kariadi Hospital
 Semarang, Indonesia
[+]Department of Medicine, diponegoro Univrsity, Semarang, Indonesia
[++]ICCIDD, Department of Human Nutrition
 CSIRO, Adelaide, Australia

This study was conducted in a historically severely iodine deficient
village in central Java, Indonesia, five years after much of its population
had received iodized oil injections. 898 children age 1-16 years living in
five of eight hamlets (pop. 2496) of the village were included in the
study. Children were examined for neurologic endemic cretinism. Also, a
survey determining the age the children began walking was carried out
independently of the examination.

This report pertains to: (1) an iodine deficient (ID) cohort of children
(n=519) who were born prior to the introduction of any iodine prophylaxis;
(2) a cohort of children (n=124) of variable iodine status (IV) who were
born of mothers who had <u>not</u> been injected with iodized oil but had possibly
used iodated salt by the time of the index child's conception, and (3) an
iodine sufficient (IS) cohort of children (n=143) who were born of mothers
who had been injected with iodized oil prior to the index child's
conception.

Evidence for gradation of neurologic defect and developmental delay was
obtained as follows:

(1) Examination of 40 neurologic endemic cretins born prior to the
    introduction of iodized oil (i.e. in the ID cohort) showed a gradation
    of severity in each of the three major types of impairments of
    neurologic endemic cretinism: mental retardation, characteristic
    hearing loss and/or speech defect, and neuromuscular defect. Analysis
    of these data also showed a wide range and uniform distribution of
    overall degree of severity of impairment as measured by a composite
    index of severity.

(2) In cretins, increasing severity of impairment as measured by
    examination, was associated with increasing degree of retardation of
    motor development, as measured by age at walking.

(3) In the non-cretinous portion of the population of the same treatment
    cohort (i.e. the ID cohort) a less severe degree of retardation of
    walking age was also observed.

cretins displayed degrees of neurological abnormality irrespective of current thyroid function or traditional classification. Severe hypothyroidism correlated with clinical findings of myxedema, sexual and skeletal immaturity and stunting of growth. However, all cretins were growth retarded and in those with fused epiphyses, growth arrest lines were seen, indicating past hypothyroidism. Thyroid atrophy in severely hypothyroid cretins was confirmed by thyroid ultrasound, the results of which will be presented.

In conclusion, classical myxoedematous cretinism is found in the Qinghai endemia. Nevertheless, all cretins displayed degrees of neurological abnormality. In this endemia the traditional classification was not useful and the clinical syndrome of cretinism was modified by the severity and duration of ongoing hypothyroidism. The factors that influence postnatal thyroid function are unidentified, but the finding of thyroid atrophy implies an autoimmune process.

ELECTROPHYSIOLOGICAL STUDIES IN ENDEMIC CRETINISM IN QINGHAI PROVINCE, PEOPLE'S REPUBLIC OF CHINA

Jean-Pierre Halpern, Steven C. Boyages, Con Yiannikas, John G.L. Morris, Lee C. Lim, Yan-You Wang* and Jian-Li Lui#

Neurology & Endocrine Units, Department of Medicine, Westmead Hospital, Australia. *Department of Otolaryngology, Tianjin Medical College, Tianjin, People's Republic of China and #Institute of Endemic Diseases, Qinghai Province, People's Republic of China.

There have been few attempts to perform systematic electrophysiological studies in endemic cretinism[1]. Somatosensory evoked potentials and brainstem auditory evoked responses (BAERs) have been reported in congenital hypothyroidism[2] and acquired hypothyroidism[3,4]. We report the results of electrophysiological studies on 46 cretins from a predominantly myxoedematous endemia in China, where nevertheless, all forms of cretinism are represented.

Approximately 50% of the sample were currently hypothyroid. Studies of median and tibial somatosensory evoked potentials showed minimal disturbances. Peripheral conduction studies unexpectedly showed only minor abnormalities, despite the fact that some individuals have been severely myxoedematous for decades. 56% of those patients tested had evidence of hearing deficit. Analysis of BAERs pointed to a peripheral cause of deafness.

REFERENCES

1. Ramirez I., Cruz M., Varea J. Endemic Cretinism in the Andean Region: New Methodological Approaches. In, Cassava Toxicity and Thyroid Research in Public Health Issues. 1983. 73.
2. Laureau E., Vanasse M., Herbert R., Letarte J et al. Somatosensory Evoked Potentials and Auditory Brain-Stem Responses in Congenital Hypothyroidism. 1. A Longitudinal Study Before and After Treatment in Six Infants Detected in the Neonatal Period. Electroen. Clin. Neurophys. 1986. 64. 501-510.
3. Nishitani H & Kool A.K. Cerebral Evoked Responses in Hypothyroidism. Electroen. Clin Neurophys. 1968. 24. 554-560.
4. Ladenson P.W., Stakes J.W., & Ridgway E.C. Reversible Alteration of the Visual Evoked Potentials in Hypothyroidism. Amer J Med. 1984. 77. 1010-1014.

(4) In children whose mothers had received iodized oil prior to the index child's conception (i.e. the IS cohort), there was complete prevention of cretinism, and there was also a reduction in their age at walking as compared to children whose mothers were iodine deficient throughout the index child's gestation (i.e. the ID cohort).

PITUITARY TUMOURS IN ENDEMIC CRETINISM (CT AND DYNAMIC STUDIES OF PITUITARY

FUNCTION)

S.C. Boyages[1], G.F. Maberly[1], J-P. Halpern[2], C.J. Eastman[1],
D. Yu[3], C. You[3], C. Jin[3], Z-L. Wang[4]
Endocrine[1] & Neurology[2] Units, Department of Medicine,
Westmead Hospital, Australia. Qinghai Provincial People's
Hospital[3] and the Institute of Endemic Diseases[4].

Enlargement of the sella turcica in primary hypothyroidism has been
recognised since the last century[1]. Pituitary adenomas secondary to
hypothyroidism are well described in animal studies but occur rarely in
humans. The factors that determine this latter outcome have not been
resolved. Furthermore, the effects of prolonged thyroid hormone deficiency
on hypothalamic-pituitary secretion are poorly understood. This study was
undertaken in Qinghai Province, China, a predominantly myxoedematous
endemia, but where neurological and "mixed" cretinism is also found. 42
endemic cretins were selected, 9 neurological, 14 "mixed" and 19
myxoedematous according to McCarrison's criteria. Serum TSH, LH, FSH,
prolactin and GH were measured. Radiology included lateral skull views and
coronal CT views [n=42] of the pituitary fossa.

Pituitary tumours, ranging in size from 8 to 11 mm, occurred more
frequently in those with higher levels of circulating TSH and did not
correlate with duration of disease i.e. age. Dynamic pituitary testing
with TRH and LHRH in 4 patients with enlarged pituitaries (mean TSH:
117mIU/1) gave a flat TSH response but significant responses to prolactin,
GH, LH and FSH. A partial empty sella was detected in older hypothyroid
cretins (mean age: 41.3 [7.4]). Basal ganglia calcification was an
additional abnormality peculiar to myxoedematous cretins [n=6; p < .05].
In conclusion, severe chronic hypothyroidism may result in pituitary

Table 1

	Neurological (n=9)	Mean [SD] Mixed (n=14)	Myxoedematous (n=19)
SXR:			
enlarged sella	1 (11%)	1 (14%)	11 (58%)
CT: tumour	1 (11%)	0	5 (26%)
CT: empty sella	0	0	8 (42%)
TSH (mIU/1)	4.7[2.5]	44 [93]	129 [105]
TT4 (nmol/1)	112 [22]	103 [36]	47 [30]
FT4 (pmol/1)	19.4[3.1]	18 [5.5]	7.5 [6.0]
PRL (mIU/1)	135 [41]	288 [531]	578 [489]
AGE (yrs)	23.3[11.2]	31.0[9.9]	33.2 [11.7]

HYPOPHYSEOTROPIC HORMONES ACTION AT PITUITARY LEVEL IN LONG-STANDING

UNTREATED ENDEMIC CRETINS

Meyer Knobel and Geraldo Medeiros-Neto

Thyroid Laboratory
Hospital das Clinicas
05403 Sao Paulo, Brazil

It is firmly established that the two types of endemic cretinism (EC) represent polar opposites of a wide spectral range of clinical abnormality. In endemic areas of Brazil the neurologic type of cretinism predominates but mixed forms or, more rarely, hypothyroid individuals may be seen. We have challenged the hypothalamic-pituitary function of 27 long-standing untreated EC with a combined test using TRH, GnRH, GHRH (1-40) and an alpha-adrenergic agonist (Clonidine) measuring TSH, PRL, alpha and beta subunits, LH and FSH, hGH and plasma Somatomedin-C levels. The mixed group of EC comprised (I) 8 hypothyroid goitrous patients (II), 14 euthyroid subjects with mental deficiency, deaf-mutism and neurological signs, with short proportional stature and (III) 4 deaf-mute individuals with goiter, mental deficiency, euthyroidism and normal stature. Enlarged sella turcica was present in 7/8 hypothyroid patients. All patients in (I) had abnormally elevated basal serum TSH and PRL levels with an exaggerated response to TRH. Accordingly, beta-subunit levels (but not alpha subunit) were also significantly elevated. hGH response to GHRH (1-40) or oral clonidine was significantly lower than normal but restored to a normal response after a brief (3 days) treatment with L-T3. Gonadotrophins response to LH-RH was normal in all 3 groups. In (II) the mean basal serum TSH level (2.3 +/- 0.4 uU/ml) and peak response to TRH (20.0 +/- 4.7 uU/ml) was normal but 6/14 had an exaggerated TSH response to TRH (low thyroid reserve). In (III) 3/5 had also an abnormally elevated peak response to TRH. In both (II) and (III) alpha and beta subunits, PRL and the GH response to GHRH or oral clonidine were normal. In all patients (I) and (II) with short stature a low mean +/- SD plasma Somatomedin-C concentration was found. We conclude that the hypothalamic-pituitary relationship in EC is altered by hypothyroidism or low thyroid function reserve in affecting mainly TSH/PRL release to TRH and the hGH provocative tests. These changes are reversible by a brief treatment with L-T3.

adenoma formation. The severity of hypothyroidism appears to be the major determinant of this outcome. Perturbations of growth, puberty and sexual function seen in endemic cretins, are partly explained by the effects of hypothyroidism on other adenohypophyseal hormones.

REFERENCES

1. Niepce, B (1851) Enlarged sella due to cretinism. Trait du Goitre et du Cretinisme. pp25-45. Balliere, Paris.

THE EFFECTS OF IODIZED OIL AND IODIZED SALT PROPHYLAXIS ON

COGNITIVE FUNCTIONING IN IODINE DEFICIENT ENDEMIAS IN CHINA

J.K. Collins[1], S.C. Boyages[2], G.F. Maberly[2], C.J. Eastman[2],
Liu Derun[3], Qian Qidong[4], Zhang Peiying[4] and Qu Chegyi[4]

Department of Behavioural Sciences[1], Macquarie University,
Australia. Endocrine Unit[2], Department of Medicine, Westmead
Hospital, Australia. Shanxi Medical College[3], Taiyuan,
Shanxi Province, Anti-Endemic Institute[4] and First Teaching
Hospital, Obstetrics and Paediatrics Hospital, China.

Psychological testing was carried out on 499 children and adults (0-45 years) from iodine deficient and replete villages in China, together with urban controls. The results showed a decrement in the cognitive functioning of persons living in the iodine deficient village beyond the expected rural suppression effect usually found in isolated communities.

Urban controls (Taiyuan) tested at the higher end of the normal range. The intellectual deficits correlated with age showing the cumulative effects of isolation over long periods in both villages. Persons from the iodine

TAble 1

COGNITIVE FUNCTION RESULTS

Mean IQ stores [SD]

Region	0-7 (yrs)	7-14 (yrs)	>25 (yrs)
Baiyuhao			
No prophylaxis (n=50)			65.1[19.6]*
Iodized salt (n=141)		72.4[18.5]#	
Iodized oil (n=65)	85.1[18.5]&		
No iodized oil (n=43)	87.4[15.8]&		
Huanglo			
Rural controls	9.39[19.6] (n=51)	84.4[16.9] (n=51)	75.6[16.8] (n=49)
Taiyuan			
Urban controls (n=49)		108.5[16.3]	

Baihuyao versus Huanglo: * = p<0.005; & = p>0.05.
Baihuyao versus Huanglo and Taiyuan: # = p<0.001.

deficient village (Baihuyao) tested lower than those from the iodine replete village (Huanglo) at all ages. The effects of an era of iodized salt prophylaxis and iodized oil injections were examined. Results question the effectiveness of these interventions as remedial measures for the phlegmatic cognitive functioning found in persons in iodine deficient endemias.

TRIAL APPLICATION OF COMPUTER EEG TOPOGRAPHY (EQUIPOTENTIAL DISTRIBUTION

MAPS OF FREQUENCY BAND) OF ENDEMIC CRETINISM IN ZHENLIN COUNTY OF GUIZHOU

PROVINCE, P.R.C.

Lu Liang, et al.

Neurology Department, Guizhou Geriatrics Institute
Guiyang Medical College
Guiyang, Guizhou, P.R.C.

43 examinees in Zhenlin County of Guizhou Province were analyzed by
computer EEG topography (CET). Among them, 33 examinees were cretins
(group A) and 10 examinees were "normal control" from the same endemic area
(group B). Abnormality rate of CET in group A was 81.9% (27 cases), that
in group B was 60% (6 cases). In 12 examinees of group A, there were a
localized high potential beta-2 frequency band and low potential alpha-2
frequency band in the median central or median centro-parietal region.
These abnormalities of CET were discussed in association with brain CT
scan, Hiskey-Nebraska learning aptitude test, Beery and Buktenica visual
motor integration development test.

STUDIES ON IODINE DEFICIENCY AND RELATED DISORDERS IN TUSCANY

A. Pinchera, G.F. Fenzi, L.F. Guisti, F. Aghini-Lombardi,
F. Santini, C. Marcocci, S. Bargagna*, G. Falciglia*,
M. Monteleone*, and P. Pfanner*

Cattedra di Endocrinologia and
*Cattedra di Neuropsichiatria Infantile
University of Pisa, Italy

In this study are illustrated the findings of an epidemiological survey on iodine deficiency and related disorders in Tuscany. To this purpose we collected 3562 schoolchildren (age 6-14 years) and 2258 adults from different villages, statistically representative of the various montane districts. The urinary iodine excretion (U.I.) ranged from 35 to 49 ug/g creatinine (average 40 ug/g creatinine) and the prevalence of goiter was 81% in the adult population and 56% in the schoolchildren. The physical and sexual development of the schoolchildren from these areas was completely superimposable to that of the urban control areas, where U.I. excretion ranged from 64.4 to 91 ug/g creatinine and the prevalence of goiter was 7% in the adults and 5% in the children, respectively. In a particular montane area of Central Apennines several cases of endemic cretinism and neurological defects were found, but only in subjects more than 45 years old. In order to evaluate whether in this moderate iodine deficient area (mean U.I. - 39 ug/g creatinine) also the schoolchildren population had an impairment of neuropsychological development, we have investigated 50 children of the 8 and 10 years age groups. The psychometric tests employed included PM47, Raven test collectively administered, Toulose-Pieron barrage test for 3 individuals, WISC R scales individually administered. All children studied showed no cognitive impairment; some individuals had significantly lower scores for verbal and coding memory with respect to the urban matched control population. However, these preliminary data need to be confirmed using as control an iodine sufficient population more similar in socio-economic status.

In conclusion: i) the whole population of Tuscany, including the urban areas, is iodine deficient, ii) several montane areas with moderate iodine deficiency and high prevalence of goiter were found; iii) very preliminary data indicate that some impairment of neuropsychological development is present, although this finding needs to be more appropriately controlled.

THYROID HORMONE DELIVERY TO TARGET TISSUES: THE ROLE OF HORMONE BINDING

PROTEINS IN SERUM

Roger Ekins, Philip Edwards and Arun Sinha

Department of Molecular Endocrinology
Midlesex Hospital Medical School
University of London
London, England

Considerable controversy currently centers on the free hormone hypothesis of thyroid and steroid hormone delivery. This postulates that hormonal effects are governed by free hormone concentrations as measured in vitro. Meanwhile the physiological role of the specific hormone binding proteins present in serum (such as thyroxine binding globulin [TBG]) remains unknown.

TBG, like many other analogous steroid binding proteins, rises markedly in pregnancy, suggesting that its physiological function (if any) relates to mammalian reproduction. Binding proteins are commonly perceived as restricting hormone efflux from intracapillary blood; TBG in particular has been claimed to prevent thyroid hormone (TH) transport from mother to fetus, its pregnancy-induced rise being commonly viewed as fulfilling this purpose (1). Recently, however, Pardridge et al (2,3) have suggested that TH are specifically released from serum proteins in certain target tissues, causing an in vivo elevation in circulating free hormone levels, and implying that protein-bound TH are selectively directed to certain tissues (e.g., the liver) in pregnancy.

It is readily demonstrable that the entire foundation of this new hypothesis is an oversimple theoretical analysis of the kinetics of TH transport in the body, leading to a misinterpretation of experimental data relating to TH in target tissues (4). Nevertheless, a fuller analysis of such data supports the proposition that, under certain conditions, dissociation of bound hormone from serum proteins exerts rate-limiting effects on hormone delivery to certain target organs. Under such conditions, increase of serum binding protein levels enhances (and does not, as commonly believed, impede) target organ uptake. Furthermore, it can be readily demonstrated that, in these circumstances, TBG specifically increases T4 delivery relative to that of T3.

These physicochemical concepts relating to the role of serum binding proteins from one of the cornerstones of our own working hypothesis, i.e. that the TH binding proteins serve to ensure adequacy of a T4 supply to the feto-placental unit throughout pregnancy, and that T4 itself is transported to, and is required by, the fetus prior to the development of the fetal thyroid gland (5). The existence of a mechanism to ensure the availability of high concentrations of maternal T4 to the feto-placental unit would be expected to confer particular advantage in conditions of iodine deficiency, thereby possible accounting for the relatively recent evolutionary appearance of TBG.

REFERENCES

1. Osorio, C. and N.B. Myrant. The passage of thyroid hormones from mother to foetus and its relation to foetal development. Brit. Med. Bul.; 16; 159-164, 1980.

2. Pardridge, W. M., B. N. Premachandra, and G. Fierer. Transport of thyroxine bound to human prealbumin into rate liver. Am. J. Physiol.: 248 (Gastrointest. Liver Physiol 11); G545-G550, 1985.

3. Pardridge, W. M. Plasma protein-mediated transport of steroid and thyroid hormones. (Editorial Review). Am. J. Physiol.: 252 (Endocrinol. Metab. 15); E157-E164, 1987.

4. Ekins, R. P. and P. R. Edwards. Plasma protein-mediated transport of steroid and thyroid hormones: a critique. Ann. N. Y. Acad. Sci.: in press.

5. Ekins, R. P. Hypothesis: The roles of serum thyroxine binding proteins and maternal thyroid hormones in foetal development. Lancet: (i); 1129-1132, 1985

THE CORRELATION OF MATERNAL HYPOTHYROXINEMIA IN EARLY PREGNANCY IN

RATS AND IRREVERSIBLE BIOCHEMICAL LESIONS IN THE BRAINS OF THE PROGENY

A.K. Sinha, R.P. Ekins, M.J.F. Hubank, M. Ballabio, M. Pickard,
Z. Al Mazidi, D. Gullo*, M.C. Ruiz de Elvira, and M. Khaled

Department of Molecular Endocrinology
Middlesex Hospital Medical School
University of London
London, England
*Department of Endocrinology
University of Catania
Catania, Italy.

Maternal hypothyroxinemia in endemic iodine-deficient regions correlates with
an increased incidence of irreversible neurological disorders in the
offspring. Damage to the CNS is evidenced by subliminal deficit in
psychomotor competence, diplegia, strabismus, deaf-mutism and severe mental
handicap. However, the specific biochemical lesions underlying these
disorders remain to be identified. Our studies have therefore centered on
the long-term irreversible biochemical effects in the brains of the adult
progeny of hypothyroxinemic euthyroid rat dams.

Partially thyroidectomized rate dams were maintained on normal laboratory
diet with food and water _ad libitum_ in a 12 hour light/dark cycle at 37^C
constant temperature, mated with normal males and the progeny allowed to grow
to 7 months of age. The progeny were then killed and brain regions separated
on ice and stored frozen before use. Tissues were thawed, homogenized in
0.32 M sucrose and various biochemical parameters assessed using conventional
techniques. Results were compared with matched control animals.

Protein content in all brain regions was reduced in experimental animals by
between 20 - 22%, but such reduction was significant only in cortex and
midbrain ($P < 0.02$). DNA concentrations were significantly reduced ($P <
0.05$) in several regions (by an average of 30%) as was inorganic phosphate in
cortex and cerebellum ($P < 0.01$ and 0.05 respectively) B-D galactosidase and
acetyl choline esterase were also severely reduced in several regions in
experimental animals ($P < 0.05$ and 0.005 respectively). No changes in SDH
and Mg++ ATPase activities were observed, but marked reductions in NA+K+
ATPase, LDH and aryl sulphatase activities in several brain regions ($P <
0.05$) were recorded.

These results reveal significant irreversible biochemical and enzymatic
disfunction in the brain of the progeny associated with early maternal
hypothyroxinemia.

Index